ENVIRONMENTAL IMPACT ASSESSMENT IN THE ARCTIC

This book is dedicated to my Southern family –
my mother and father Pirkko and Matti
and my sister Tiina

Environmental Impact Assessment in the Arctic

A study of international legal norms

TIMO KOIVUROVA
Northern Institute for Environmental and Minority Law
Arctic Centre, University of Lapland, Finland

Routledge
Taylor & Francis Group
LONDON AND NEW YORK

First published 2002 by Ashgate Publishing

Reissued 2018 by Routledge
2 Park Square, Milton Park, Abingdon, Oxon OX14 4RN
711 Third Avenue, New York, NY 10017, USA

Routledge is an imprint of the Taylor & Francis Group, an informa business

A Library of Congress record exists under LC control number: 2002103984

ISBN 13: 978-1-138-73000-7 (hbk)
ISBN 13: 978-1-138-72999-5 (pbk)
ISBN 13: 978-1-315-18837-9 (ebk)

Contents

Acknowledgements

Writing this doctoral thesis has been a solitary job for me. I have not enjoyed the privilege of having someone at hand doing a thesis on a topic very close to my own and thus enabling the day-to-day exchange of views. I am not sure whether this has benefited or circumscribed my work, but it definitely has made it a maturing experience.

Although the thesis has been a solitary exercise in principle, surprisingly many people have contributed to the form it has ultimately taken. First, I can confirm that my supervisor, Kari Hakapää, is the great personality and scholar everyone thinks he is. He was always there for me, supporting me on every level of human experience. Especially important was Kari's realistic approach to the writing of a thesis, which dragged me back to planet Earth when my feet were firmly planted in the sky. I have also greatly benefited from the profound thinking on international law of my examiner, Professor Erik Franckx. He has provided me with invaluable comments which have had a distinct impact on the structure of the thesis. Likewise, Doctor Tuomas Kuokkanen, also an examiner of the thesis, has made many extremely useful suggestions and comments, which I have taken into account in revising the text.

Lauri Hannikainen was the man responsible for my specialization in international law. As many other students of law can tell, Lauri's integrity as a scholar, wonderful personality and unlimited time for his students have allured many of us to become international lawyers. Although Lauri is nowadays my boss, I still feel more like a student of his than a subordinate. Perhaps the biggest contribution to the ideas presented in this thesis has been made by Jan Klabbers of the University of Helsinki. Initially, I read some articles Jan had written, and then familiarized myself with his thesis, which clearly is one of the best monographs written in international law in recent years. I was also fortunate enough to acquaint myself with Jan himself, our correspondence mainly taking place via e-mail. Although busy as ever, he always found time to reply to my queries. When I was working as an assistant in public and private international law in the Faculty of Law of the University of Lapland, I got to know Professor Maurice Andem. Without the strong encouragement Maurice gave me in many aspects of research on law, as well as life in general, it is possible that this work would have been delayed quite a bit.

Martti Koskenniemi was also an important source of inspiration for the thesis, if not so much for the ideas herein as a person who has made international law an engaging subject of study in general. While I have mostly encountered Martti through his works (of art), I have also had several telephone conversations with him, after which I found myself standing in the sky. Although I stated at the beginning that there was no doctoral candidate with whom to share my thoughts, this is only partially true. My dear friend in personal life, Rauno Korhonen, has been doing his thesis on the law of legal informatics, and we have had countless discussions, from which I have most certainly benefited. The late Frank Horn was actually the one responsible for the topic of my thesis. He was always ready to assist me as thoroughly as only he could. Lastly, but not least, I have had the privilege of working with Richard Foley for several years, first when writing articles and now in completing this thesis. As you read this text, it has already been transformed by the excellence of his proofreading skills. With Rich also start working on his thesis, many interesting conversations materialized; in fact, some ideas of this thesis were greatly deepened by the difficult questions he posed.

Countless other persons have assisted me in many ways, of whom I would like to mention several here. Monica Tennberg from the Department of International Relations at the University of Lapland has deepened my understanding of the theory of international relations as well as of Arctic international affairs. Eyassu Gayim, Paula Kankaanpää, Outi Korhonen, Tarja Långström and Merja Pentikäinen have provided some useful insights on the thesis. Doctoral candidates Maarit Klemetti, Karri Puustinen, Jussi Syrjänen and Pentti Tepsa have been important not only as friends but as persons with whom the discussions have shifted easily from legal theory to philosophical-theological topics in general. Arto Vitikka From the Arctic Centre was irreplaceable in technical assistance.

I was fortunate enough to receive a four-year scholarship in the first doctoral program in law in Finland, hosted by the Helsinki University Institute of International Economic Law. Doctor Olli Mäenpää, the director of the Institute at the time, provided important support especially by organizing the 'infrastructure' for writing a doctoral thesis. My thanks also go to my other employers, the Faculty of Law of the University of Lapland and my present one, the Northern Institute for Environmental and Minority Law of the Arctic Centre, for allowing me to spend enough time working on the thesis. I was fortunate to have had access to the splendid research facilities offered by the Library of the University of Lapland as well as the Law Library of Columbia University in New York.

I have been blessed with a wonderful family. My family and relatives as well as Anniina's have always been there to support me. Anniina, the love of my life, and our little son Joonas have been the lifeline without which all of this would be meaningless.

Rovaniemi, September 2001 Timo Koivurova

Abbreviations

AAC	Arctic Athabascan Council
AC	Arctic Council
ACAP	Arctic Council Action Plan to Eliminate Pollution
ACIA	Arctic Climate Impact Assessment
AEPS	Arctic Environment Protection Strategy
AIC	The Aleut International Association
AMAP	Arctic Monitoring and Assessment Program
CAFF	Conservation of Arctic Flora and Fauna
EA	Environmental Assessment
EC	European Community (European Communities)
ECSC	European Coal and Steel Community
EEC	European Economic Community
EEZ	Exclusive Economic Zone
EFTA	European Free Trade Association
EIA	Environmental Impact Assessment
EPPR	Emergency Prevention, Praperedness and Response
EURATOM	European Atomic Energy Community
EU	European Union
IAEA	International Atomic Energy Agency
IASC	The International Arctic Science Committee
ICJ	International Court of Justice
ICC	The Inuit Circumpolar Conference
ILA	International Law Association

ILC	International Law Commission
ILM	International Legal Materials
ILO	International Labour Organization
INSROP	International Northern Sea Route Programme
IPS	Indigenous Peoples Secretariat
ISA	International Sea-Bed Authority
ITLOS	International Tribunal for the Law of the Sea
LOSC	United Nations Convention on the Law of the Sea
OSCE	Organization on Security and Cooperation in Europe
OECD	Organization for Economic Cooperation and Development
PAME	Protection of the Arctic Marine Environment
PCIJ	Permanent Court of International Justice
RAIPON	The Russian Association of Indigenous Peoples of the North
SAAO	Senior Arctic Affairs Officials
SAO	Senior Arctic Officials
TEU	Treaty on European Union
UNGA	United Nations General Assembly
UN ECE	United Nations Economic Commission for Europe
UNEP	United Nations Environment Program
UNTS	United Nations Treaty Series
WWF	World Wildlife Fund

I INTRODUCTION

Let us assume a situation – entirely hypothetical – in which exploratory drillings in the Iceland Sea show a huge oil deposit beneath the sea-bed. The deposit is located on the Icelandic side of the sea-bed, close to Denmark (Greenland) and Norway (Jan Mayen Land). Since the Iceland Sea and the neighboring Greenland Sea contain abundant fish reserves, both Norway and Denmark become worried about the potential environmental impacts, especially the effects on fish populations, of drilling for oil in the area. The Danish and Norwegian authorities approach Iceland arguing that drilling for oil in these waters would be very risky because of the presence of ice and because oil disperses very slowly in such frigid waters; they make a plea to Iceland asking that it makes a comprehensive assessment of potential impacts on the marine environment located on the Greenland side. The plans for drilling cause alarm in Iceland too. The fishing industry as well as some environmental associations strongly oppose the project because of its potentially grave environmental implications and demand that a comprehensive environmental assessment be carried out.

This imaginary situation captures some of the key issues examined in this thesis. Because of the special characteristics of the Arctic environment, especially the presence of ice and permafrost, industrial activities exploiting the region's vast natural resources pose grave risks. As pollution knows no borders, and as the maritime borders of states have grown closer through expanding coastal state jurisdictions, environmental risks are often transboundary in nature, touching upon the interests of many states. Moreover, the ecosystems of the Arctic are considered to be more vulnerable to man-made pollution than those in more temperate climatic zones. An important environmental management tool for such circumstances is environmental impact assessment (EIA), a procedure which aims at ensuring that a scientific study is made of the potential impacts on the environment of a proposed activity with the involvement of all interested parties before a licence is granted for the activity. Increasingly, states have concluded international treaties to the effect that this EIA procedure can be extended to foreign impacts and foreign actors as well.

This thesis examines the norms of international law that apply to the planning stage of large-scale activities such as hydrocarbon exploitation, mineral extraction and forestry. These stationary activities, i.e., activities that will remain at a single location for a period of time,[1] pose grave risks to the Arctic environment since the development of technology has made it profitable to exploit the natural resources even in such harsh regions. Accordingly, the main aim of the thesis is to examine systematically the norms[2] of international law that apply to one of the most threatening situations for the Arctic environment: plans for environmentally harmful stationary activities. Since the Arctic environment is threatened by a multitude of pollution problems, it will be useful to clarify why it is that this situation merits particular attention. And, as the present focus is restricted to the planning stage of stationary activities, it will be necessary to explain why not all the norms that apply to those activities are studied. The second principal aim of this thesis is to provide an

---

* A note on footnotes is necessary here. In order to decrease the size of footnotes, only the year, name and abbreviated name of treaties are used; i.e., no information is included on the source where the document can be obtained. The reader is thus instructed to seek further information in the index of treaties and informal arrangements provided at the end of the work. A second concern is the use of Internet sources, which, as is well known, lack the permanence of printed sources. The problem has been dealt with in two ways. First, the day on which the source has been found at the address referred to is placed after the address (all of which are found in the year 2001). Second, where the information referred to has been provided by an institution, this institution is indicated along with the source. If the address changes, the reader can at least try to find the institution mentioned and locate the new address of the source.

[1] There are some activities which are hard to qualify either as mobile or stationary. A good example is the recent plan for a floating nuclear power plant in the Eastern Siberian port of Tevek. According to the plan, the nuclear power plant is to be taken to Tevek and towed to Atomflot in Murmansk every four years in order to remove and replace the uranium fuel. The voyage would transit the partly ice-covered northern coast of Siberia and also cross the Barents Sea. See WWF Arctic Bulletin, No. 4 (2000), p. 20.

[2] The concept of 'norm' is used in this study as the most general concept, the sub-concepts of which are 'rules' and 'principles'. Rules are more specific in content and scope than principles.

analytical perspective on how to evaluate the status of the Arctic cooperation process from the viewpoint of international law, for a considerable variety of views has been presented on this matter.

1. Approach of the Thesis

Any study of norms in international law must start from a given background since international law consists of several fields which are often viewed as independent of one another. The norms that regulate the situation to be studied here, the planning stage of environmentally harmful stationary activities, can be categorized, in the main, as part of international environmental law; however, there are other relevant norms which are traditionally considered part of international human rights law[3] and international minority rights law.[4]

The law pertaining to environmentally harmful activities seems to be in a state of flux. The Brundtland Report of 1987 paved the way for the 1992 Rio Conference on Environment and Development, with both providing a new integrative perspective on international law relating to 'environment, economics and civil and political rights'[5], which had previously been seen as separate concerns. This change was reflected in the legal literature as scholars of international law proposed new concepts to capture the new integrative approach. In an article published in 1994, Philippe Sands argued that these formerly separate branches of international law could be seen as one, i.e., an international law of sustainable development.[6] Tuomas Kuokkanen, in his doctoral

---

[3] See *infra* chapter IV (3.3.2.).
[4] See *infra* chapter IV (3.3.3.).
[5] Sands 1994, p. 379.
[6] See Sands 1994, pp. 303-381. He argues (p. 379): 'Certainly, the concept of 'sustainable development' is now established in international law, even if its meaning and effect are uncertain. It is a legal term which refers to processes, principles and objectives, as well as to a large body of international agreements on environment, economics and civil and political rights. 'Sustainable development' allows a broad range of interpretations, with the objective of integration at various levels: North and South; national and international; economic, environmental and social'. Hohmann

dissertation, saw this change as a development from the modern era of international environmental law and the law of natural resources to the postmodern era of environmental integration and sustainable development.[7] In light of these changes, it is essential that the approach adopted in the present study can examine the norms unencumbered by the distinction international/national[8] or the distinctions imposed by the doctrinal traditions of international law.

The main reason why the study focuses on situations specific to the Arctic is the Arctic cooperation process. This international cooperation process essentially began in 1991 in Rovaniemi, Finland, when the eight Arctic states (the five Nordic states, the United States, the Russian Federation and Canada) signed a Declaration on, and adopted a Strategy for, the protection of the Arctic environment. International environmental protection was to be come the focus of Arctic-wide cooperation. The cooperation process culminated in the establishment of the Arctic Council in 1996, at which time the mandate of Arctic cooperation was extended to include the goal of sustainable development.[9]

(1994, pp. 343-344) sees it 'as a change of paradigm in international environmental law' characterized by precautionary legal duties. See also Fitzmaurice 1994, pp. 181-226, especially pp. 224-226.

[7] Kuokkanen 2000. Kuokkanen divides his thesis into three parts: traditional era, modern era and postmodern era. He points out (especially pp. 286-290) that although these eras have been presented in a certain chronological order, the approaches to international environmental management they encapsulate are all available at present as well. Kuokkanen suggests that the tools of each era can be useful if implemented in the right context. He argues (p. 287), for instance, that the 'classification of certain methods and techniques as traditional does not mean that they would be inapplicable anymore. Although the context is different, the legal value of classical doctrines is still basically the same'.

[8] For a general treatment of this issue in international environmental law, see Ebbesson 1996.

[9] As will be shown *infra* in chapter III (1.1.), the AEPS cooperation already had a task force on sustainable development. Yet, it was the Arctic Council which provided impetus to the field of sustainable development by explicitly declaring sustainable development to be its other field of competence. Tennberg has argued that the establishment of the Arctic Council shifted Arctic cooperation from international environmental protection to sustainable development. See Tennberg 1998, pp. 212-213.

Irrespective of how one characterizes the legal status of this cooperation process, it has already spawned monographs and articles studying international law that is applicable in the Arctic.[10] Through this cooperation process, the Arctic has emerged as a distinct region in international affairs in general, and international law in particular. This development is captured by Oran Young[11], a scholar of international relations who has been observing Arctic affairs from the outset:

> Where, then, does the comparative advantage of the Arctic Council lie? It may come as a surprise to some to realize that the council's most important role is probably generative in nature. Through its very existence, the council has become a symbol of the emergence of the Arctic as a distinct region in international society.[12]

Why is it that the planning stage of stationary activities is considered particularly significant in posing threats to the Arctic environment? There are many pollution and other environmental problems facing the Arctic, many of which are global or regional in character and may pose grave threats to the environment.[13] However, while the Arctic states can act jointly in other international fora and try to contribute to the control of such environmental problems, they can only do so in a limited fashion.[14]

---

[10] See, e.g., Franckx 1993; Rothwell 1995 I, 1995 II and 1996; Brubaker 1993; Timtchenko 1996; Koivurova 1999 and 2000.

[11] Oran Young was among the first to envision a new cooperative future for the Arctic, which he describes in a book co-authored with Gail Osherenko 'The Age of the Arctic', 1989.

[12] Young 2000 (chapter 4, recommendation 2, paragraph 2).

[13] The Arctic Council has one project now underway to study the effects of these global pollution problems on the Arctic environment. The mission statement of this project declares: 'The goal of the Arctic Climate Impact Assessment (ACIA) is to evaluate and synthesize knowledge on climate variability, climate change, and increased ultraviolet radiation and their consequences...Three major volumes will be completed by 2004'. Information from the ACIA can be found on the home page of the Arctic Council at http://www.arctic-council.org/ [7.2.2001]. For recent overviews on the effects of global warming and ozone layer depletion in the Arctic, see Lange 2000, pp. 517-553 and Fabo and Björn 2000, pp. 555-573. For a general review of threats to the Arctic environment, see Stokke 1990, pp. 22-41 and Tennberg 1994, pp. 135-137.

[14] There is a recent example of such a joint action, an intervention made by Ambassador Kari Karanko, Finland, who spoke on behalf of the Arctic Council in the International Negotiating Committee (INC 5) on the Persistent Organic Pollutants (POP) Convention. He stated: 'The Ministers together with the representatives of the indigenous peoples held a roundtable discussion on the threat of contaminants in the Arctic and opportunities for collective action to address the

The Arctic Council has thus far focused its efforts on the coordination of the Arctic states' Arctic policies.[15] The natural field of action of the Arctic Council is intra-Arctic pollution, i.e., pollution which originates in the Arctic and remains there.

The principal problem of intra-Arctic pollution is that of emissions from different kinds of stationary activities.[16] The salient characteristic of the Arctic is that it is sparsely populated,[17] with correspondingly few towns and villages. Accordingly, intra-Arctic pollution tends to be produced comparatively more by point sources, i.e., pollution that can, at least potentially, be traced back to its source, rather than by diffuse pollution caused by a large number of minor polluting activities. Typically, point sources of pollution in the Arctic are profit-oriented industrial activities such as hydrocarbon exploitation, mineral extraction and forestry, which make use of the vast natural resources of the region. Characteristic of these activities is that they are planned to operate for a long period of time and that they cause pollution throughout their life cycle. The thesis does not confine itself to these typical industrial activities,

---

problem...the main findings on persistent organic pollutants made by the Arctic Monitoring and Assessment program are compiled in a fact sheet which is available at this meeting. Madam Chairman, on behalf of the Arctic Council, Finland would like to urge all governments to strengthen efforts to finalize a comprehensive and verifiable convention addressing this important environmental issue at this last session here, in Johannesburg, South Africa'. The intervention can be found on the home page of the Arctic Council at http://www.arctic-council.org/whatsnew.asp [4.4.2001].

15 Relevant documents here are, e.g., CAFF's the Strategic Plan for the Conservation of Arctic Biological Diversity, Co-operative Strategy for the Conservation of Biological Diversity in the Arctic Region, Circumpolar Eider Strategy and Action Plan, International Murre Conservation Strategy and Action Plan, CPAN Strategy and Action Plan; the EPPR's Field Guide for Oil Spill Response in Arctic Waters; PAME's Regional Programme of Action for the Protection of the Arctic Marine Environment from Land-based Activities, Arctic Offshore Oil and Gas Guidelines; Sustainable Development Working-Group's Guidelines for Environmental Impact Assessment (EIA) in the Arctic. These documents can be found on the home page of the Arctic Council at http://www.arctic-council.org/workgrps.asp [15.2.2001].

16 As Young (1997, pp. 267-268) points out: 'A second focus for the efforts of the Arctic Council to promote sustainable development involves large-scale industrial activities (e.g. oil and gas development, hydroelectric power production, the extraction of nonfuel minerals, industrial fishing) taking place in the Arctic...The ongoing effort to develop "Guidelines for Environmental Impact Assessment (EIA) in the Arctic,"...is a commendable step in this direction'.

17 For a study of people living in the Arctic, see AMAP Assessment Report 1998, pp. 141-182. See also Knapp 2000, pp. 459-483.

however; it examines all environmentally harmful stationary activities, including military ones.

It is thus activities exploiting natural resources that form the main focus of this study. They present one of the most serious threats of intra-Arctic pollution at present[18] and, as a consequence, the Arctic cooperation process has provided normative guidance, especially for their planning stage. There are also other kinds of point sources which may cause intra-Arctic pollution. For example, plans exist for the opening of two sea routes, the Northeastern[19] and the Northwestern[20] routes, which would increase the risk of pollution from mobile sources, especially oil tankers.[21] However, it seems that the opening of these routes will not take place in the foreseeable future, at least not on a large scale.[22]

---

[18] As put by Roginko (1995), in his contribution to the International Northern Sea Route Programme (INSROP): 'But the greatest threat to IES [international environmental security] in the Arctic arises from natural resource extraction and harvesting activities'.

[19] There was an international program developing the Northeastern sea route until 1999, the International Northern Sea Route Programme (INSROP), the home page of which is at http://www.fni.no/insrop/defaultINSROP.html [7.2.2001]. For an analysis of the prospects of the Northeastern sea route, see Ostreng 1998.

[20] For a brief overview, see Rothwell 1996, pp. 191-200.

[21] It is precisely the transportation of petroleum hydrocarbons that is the main expected use of the northern sea routes. According to the six-year INSROP project: 'The main conclusion of INSROP's research is that a substantial increase in international commercial shipping is feasible – in economic, technological and environmental terms. The largest and most obvious cargo potential is found in the huge oil and gas reserves in the Russian Arctic – both onshore and offshore – where marine export towards western markets is likely to start up early in the new Century'. The conclusions can be found on the home page of INSROP at http://www.fni.no/insrop/defaultINSROP.html#Overview [7.2.2001]. This has also been stressed by the AMAP Assessment Report 1998, p. 694, which strongly recommends 'that the work of the International Northern Sea Route Programme (INSROP) be supported to assess the feasibility and potential international consequences of using the Northern Sea Route for the transport of oil via tankers', Section 10.8 (13).

[22] The conclusions of the INSROP project also acknowledged: 'In spite of this, the Northern Sea Route has so far not been utilised commercially to a significant degree by non-Russian vessels...The main problems which will need to be solved before the NSR's potential can be fully utilized, are on the political level. The Russian ice-breaker fleet is slowly, but steadily, deteriorating, and continuing economic and political turmoil in Russia makes it difficult to raise the necessary investments towards upgrading and maintaining an adequate infrastructure for the NSR. The economic and political problems have so far also prevented Russia from establishing a stable, competitive tariff regime'. For the source, see the previous footnote.

Another set of human activities posing problems of intra-Arctic pollution are military activities, of which only stationary military activities, such as the building of storage houses for weapons and the construction of processing facilities for radioactive military waste, are considered in this thesis.[23] There are many kinds of military activities that pose pollution threats to the Arctic environment, and stationary activities form only a small part.[24] While military activities certainly present a serious threat to the Arctic environment, the main problems are mostly concentrated in the area of one Arctic area only, the northwestern part of the Russian Federation.

Not all the norms which apply to environmentally harmful stationary activities will be dealt with in this study. In order to make a detailed study possible, only some of these rules have been chosen, i.e., the rules that apply to the planning phase of stationary activities. These norms are called here 'environmental impact assessment (EIA) norms', a term that will be specified below.[25] Typically, the life cycle of stationary activities consists of a planning stage, which ends when the project is granted a licence to operate, an operational phase and a termination phase.[26] All of the procedural rules that apply after a licence has been issued have been excluded from the scope of the thesis.[27] In addition, the thesis excludes international law norms

---

[23] For a review in Finnish of the concrete problems caused by military presence in the Arctic, see Heininen 1991, especially pp. 59-94. See also Heininen 1990, pp. 42-69 and Heininen 1992, pp. 55-71 (in English).

[24] For an analysis, see Gizewski 1993-1994.

[25] See *infra* Chapter IV.

[26] For an analysis, see Lefeber 1996, pp. 35-46.

[27] In Chapter V, 'The Espoo Convention in the Arctic', the approach is a bit broader, including a study on the post-project analysis provided by the Convention. In addition, only those norms applicable to stationary activities which require a prior licence to operate are studied. This criterion excludes procedural norms that establish communication procedures between states with respect to plans of policy or strategy which may cause negative consequences to the area of another state. See, e.g., the 1983 Agreement for Cooperation Relating to the Marine Environment (Article VI,2). It also excludes assessment duties related to diffuse sources of pollution. A good example is the 1979 Convention on Long-Range Transboundary Air Pollution, Article 5 of which provides: 'Consultations shall be held, upon request, at an early stage between, on the one hand, Contracting Parties which are actually affected by or exposed to a significant risk of long-range transboundary

that aim to prevent the causing of pollution by, for example, prohibiting the exercise of a certain environmentally harmful stationary activity.[28]

Although the thesis confines itself in the main to norms of international law, some directives of European Community (EC) law are also discussed; their inclusion is justified in that they are part of an international agreement, the 1992 Agreement on the European Economic Area (EEA), and thus are treaty obligations in international law.[29] Since EC law[30] in general governs only a small part of the Arctic as defined in this thesis - the parts of mainland of Finland and Sweden that are above the Arctic Circle - it will not be considered here.[31] It is also considered unwarranted to use all the various terms related to the European integration process. Since it is only directives of EC and the role of the EC as a treaty partner that are relevant in this study, the author has deemed it practical to use only the abbreviation 'EC' throughout the study although in some cases another term might better capture the reality.[32]

---

air pollution and, on the other hand, Contracting Parties within which and subject to whose jurisdiction a significant contribution to long-range transboundary air pollution originates, or could originate, in connexion with activities carried on or contemplated therein'. Yet, the Convention applies only to air pollution for which 'it is not generally possible to distinguish the contribution of individual emission sources or groups of sources' (Article 1 (b)).

[28] If a stationary activity is planned for the production of a certain item or substance which itself is prohibited, the entire activity is clearly illegal in international law. A good example can be found in the 1992 Convention on the Prohibition of the Development, Stockpiling and Use of Chemical Weapons. A recent convention that prohibits the use of certain chemicals is the 2001 Stockholm Convention on Persistent Organic Pollutants. Another example is where states have prohibited certain stationary activities in a certain area, for instance, mining in the Antarctic Treaty Area, prohibited on the basis of Article 7 of the 1991 Protocol on Environmental Protection to the 1959 Antarctic Treaty.

[29] For further information, see *infra* chapter IV (4.1.1.).

[30] When the legal system of the European integration process is discussed, the term 'European Community law' – EC law – is preferred. This term, rather than, e.g., European law, is chosen to emphasize that although EC law is evidently of *sui generis* character, it is also, in its basis, international law since all the founding treaties are international treaties binding in international law. EC law can most appropriately be viewed as a self-contained regime within international law; see Wyatt and Dashwood 1993, pp. 53-55. The term 'EC law' is also preferred to, e.g., 'European Union law'. Some scholars use the term 'EU law' to emphasize the changed nature of this legal system; see e.g., Kent 1996, p. 8.

[31] As will be shown *infra* in chapter II (1.1.), Greenland is not part of the EC.

[32] The terminology used when referring to the inter-governmental organization behind the European integration process is rather complex. The most inclusive term would be the European Union (EU).

The second principal aim of the thesis is to provide an analytical perspective on the legal status of the Arctic cooperation process and the norms adopted within it. Doubts have been voiced as to the legal nature of the normative instruments establishing the constitutional rules of Arctic cooperation and the environmental protection norms adopted within this process.[33] With respect to the 1991 Declaration and Strategy for the Protection of the Arctic Environment, the majority of scholars studying the issue consider these norms binding, but it is not clear in what way; some scholars opine that the norms are politically binding, while most seem to treat them as softly binding, without, however, specifying what they mean by the concept.[34] There are also those who view the cooperation as operating outside of international law altogether, especially in regard to the Declaration Establishing the Arctic Council.[35] There is thus a great need to examine the legal status of Arctic cooperation comprehensively.

This term was introduced in the 1992 Maastricht Treaty on European Union (TEU) and refers to the three main pillars of the Union: the first is the European economic integration process, which now includes the cooperation based on the three founding treaties, i.e., the 1951 Treaty on European Coal and Steel Community (ECSC), the 1957 Treaty on European Economic Community (EEC; amended as the 1992 Treaty Establishing the European Community through the TEU) and the 1957 Treaty on European Atomic Energy Authority (EURATOM), that were designated as the European Community (EC) by the Treaty on European Union; the second pillar is Common Foreign and Security Policy; and the third Police and Judicial Cooperation in Criminal Matters. Since only the environmental directives of the EC are of interest in this thesis, it is considered appropriate to refer to the European Community (EC). Before the Treaty on European Union, it was common to use the term 'European Communities (EC)' when referring to the three inter-governmental organizations acting through the same institutions. It should also be noted that when a reference is made to directives enacted before the Treaty on European Union, the EEC is their legal basis. In order to avoid any terminological confusion, the abbreviation 'EC' is used throughout the thesis. There are doubts as to whether the EU can be defined as an international legal person, capable of, e.g., concluding treaties with other international legal subjects. Some scholars reject this status, e.g., Eaton 1994, pp. 215-225, but some take the view that such can be presumed, e.g., Klabbers 1998, pp. 231-253. Since it has been the 'European Communities' or 'European Community' that has been a party to all the treaties studied in this thesis, it is appropriate to refer to the EC when discussing participation in these treaties and no need arises to take a position on the legal personality of the EU.

[33] See *infra* chapter III (1).

[34] See *infra* chapter III (1.1.).

2. Structure of the Thesis

Before proceeding to the principal aims of this thesis as outlined above, it will be necessary to give the reader some background information on the legal status of physical space in the Arctic. EIA norms regulate the planning phase of proposed stationary activities in the Arctic, and the geographical scope of these norms is partly defined by the norms giving international legal subjects rights over physical spaces of the Arctic. Accordingly, chapter II first presents the definition of the Arctic used in this study and proceeds to examine the international legal subjects present there as well as the rights over physical space they exercise. The idea is to provide the reader with an introduction to the international legal context that prevails in the Arctic, a regime that sets limits on the functioning of the EIA norms.

Chapter III takes up the second aim of this thesis, i.e., the study of how the Arctic cooperation process can be qualified from the viewpoint of international law. The chapter starts with a discussion of the historical development of Arctic cooperation and the legal status of its instruments as well as a review of how these have been evaluated by scholars. Thereafter, it goes on to examine in depth the different views of how contractual obligations are created in international law, and what general international law stipulates with regard to this issue. The more detailed view on how a scholar of international law ought to examine the intentionally produced documents is illustrated via a study on three separate problems posed by the written normative instruments adopted in the Arctic cooperation process. The study on Arctic instruments forms the basis for a method by which the bindingness of norms of the applicable instruments will be studied in Chapters IV and V. Finally, some questions relating to the terminology used in this thesis are taken up.

---

[35] See *infra* chapter III (1.2.).

The main aim of the thesis is addressed in chapters IV and V. Chapter IV examines all the EIA norms to be found in the different categories of sources of international law, mainly treaties and customary international law. These are organized into different sections: obligations on the national environmental protection system of the Arctic states, obligations on preventing transboundary pollution affecting the environment of other Arctic states, obligations related to Arctic transboundary watercourses and, finally, obligations applicable in international areas of the Arctic. The most important chapter is clearly the one on EIA norms that apply to transboundary pollution between the Arctic states since most international treaties regulate precisely such situations; these prior transboundary procedures, as they are called, are studied in detail. The most threatening source of potential transboundary pollution in the Arctic, offshore hydrocarbon exploitation, is examined in a section of its own. Finally, conclusions are drawn regarding the present state and future prospects of EIA norms applicable in the Arctic. The chapter also analyzes a case that has taken place in the Finnish Arctic area, the plan to construct the so-called Vuotos reservoir and dam, which has triggered EIA norms of varying kinds and thus merits a detailed study.

Chapter V examines in detail one category of EIA norms, transboundary EIA procedures. It starts by examining the development of the concept of transboundary EIA in the global context and proceeds to analyze the negotiation process leading to the conclusion of the Espoo Convention, which is the most important treaty providing transboundary EIA procedure in the Arctic. Significantly, the chapter also examines how the Convention has been promoted in the Arctic Cooperation process. Thereafter, the scope and content of the instrument are studied in detail, and, finally, conclusions are drawn as to its present importance for the management of situations of potential transboundary pollution between the Arctic states. Chapter VI sums up the findings related to the principal aims of the thesis, the systematization of EIA norms in the Arctic and the legal status of Arctic cooperation process.

II THE LEGAL STATUS OF THE ARCTIC AS A PHYSICAL SPACE

There exists no consensus on what can be regarded as the southernmost boundary of the 'Arctic'. Numerous definitions are on record, and thus there is need to define what the term will mean in this thesis.[36] Since the rules systematized in this thesis - EIA norms - concern the protection of the environment, it may seem most reasonable to rely on those definitions of the Arctic which have a bearing on the Arctic environment and its ecosystems. A fine-grained definition is that made by ecologists, who divide the region into three zones 'on the basis of climate and the presence of certain vegetation types and animals'[37]: the high Arctic[38], the low Arctic[39] and the subarctic[40] (see figure 1). The following map of the area includes the treeline, which is sometimes used in defining the southernmost limit of the terrestrial Arctic.

Another criterion for defining the southernmost limit of the Arctic is the 10°C July isotherm; i.e., the southernmost border is where the average temperature in the warmest month of the year is below 10°C degrees. The map below incorporates this criterion together with other definitions of the Arctic (see figure 2). The problem with these definitions of the Arctic is that they normally exclude the Eurasian part of the region (Scandinavia, the Kola Peninsula and Iceland). For instance, the treeline criterion includes only a minor part of Scandinavia; Finland, Sweden and Iceland are excluded altogether, evidently due to the warming effect of the Gulf Stream in these countries. The 10°C July isotherm criterion also excludes Finland and Sweden but does include one part of Iceland.

---

[36] For a brief overview of these definitions in Finnish, see Kuusisto 2000, pp. 163-167.

[37] AMAP Assessment Report 1998, p. 119.

[38] Ibid., the high Arctic is defined in Section 4.4.1.1.

[39] Ibid., the low Arctic is defined in Section 4.4.1.2.

[40] Ibid., the subarctic is defined in Section 4.4.1.3.

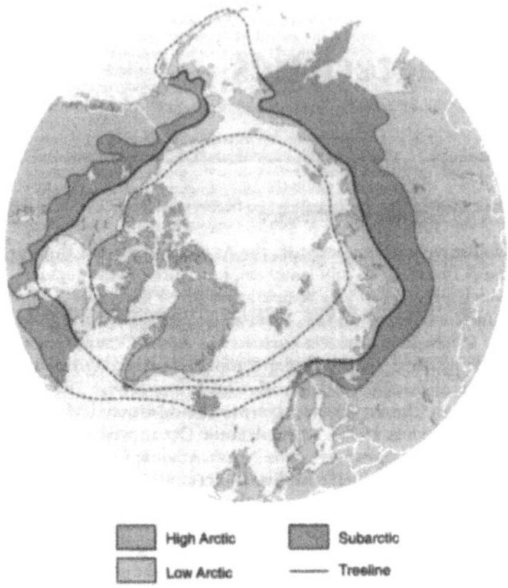

Figure 1 (source: AMAP Assessment Report 1998, p. 10)

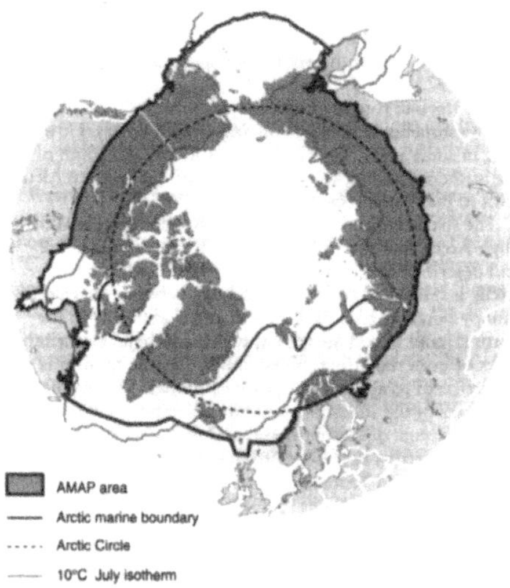

Figure 2 (source: AMAP Assessment Report 1998, p. 9)

One interesting definition is that suggested in an Arctic-wide cooperation treaty proposed by Pharand.[41] In this draft treaty, he suggests that the 'Arctic Region', and thus the area of application of the treaty, should cover 'the area north of 60 North Latitude, including Labrador and the region in northern Quebec known as "Nunavik"'[42]. However, this definition would include areas belonging to the territory of other than the eight Arctic states, e.g., the Shetland Islands, an area part of the United Kingdom (Scotland); it is no wonder that Pharand did not exclude other states than the eight Arctic states from the membership of the 'Arctic Region Council'.[43]

The Arctic-wide cooperation process could have started with fewer states than it did. For example, the Polar Bear Convention of 1973 could have served as a model for general Arctic-wide cooperation although it included only the littoral states of the

---

[41] Pharand's (1991) proposal can be found on the home page of the Canadian Arctic Resources Committee at http://www.carc.org/pubs/v19no2/5.htm [9.2.2001].

[42] Ibid., Article 1. In the commentary to this Article he states: 'The expression "Arctic Region" was chosen instead of "Arctic Basin" because the latter would normally exclude Iceland, Finland and Sweden, which should be included. Of the numerous ways to define the "Arctic Region", the 60th parallel of latitude appears to be the most appropriate. It includes all the areas covered by the tundra or continuous permafrost, except for parts of northern Quebec and of Labrador; hence, the qualified definition. More specifically, the area north of 60° includes virtually all of the relevant regions of the eight Arctic States: Canada: Yukon, N.W.T., most of northern Quebec and the tip of Labrador. Denmark: all of Greenland. Finland: all. Iceland: all. Norway: all of Svalbard and most of mainland Norway (north of Oslo). Sweden: most of the territory (north of Stockholm). U.S.S.R.: roughly the northern half (counting the archipelagos) of the territory, including the numerous rivers emptying in the Arctic Ocean. U.S.A.: virtually all of Alaska. The area includes, of course, all of the Arctic Ocean and surrounding seas, as well as the southern limit of sea ice. By contrast, the Arctic Circle would have left out significant bodies of water, large areas of the tundra and of sea ice'.

[43] Ibid. Article 3, Membership, reads: 'The founding Members of the Council shall be the eight States whose territory projects north of the Arctic Circle...The admitted Members may be States, Governmental and Non-Governmental Organizations, territorial and regional governments. Such States, organizations, and governments are eligible for membership if they have demonstrated a substantial interest in the work of the Council and a capacity to further its purposes. Admission shall be decided by the Assembly upon the recommendation of the Commission. Commentary: To permit participation of all those with sufficient interest, the Council is open to membership of other States (such as France, Germany, Japan, United Kingdom), organizations of States (such as the European Community), NGOs (such as the ICC) and territorial governments (such as Alaska, Yukon, N.W.T., Greenland) and regional governments (such as Chakotskiy, Nunavik and the Nordic Sami Council)'.

Arctic Ocean (The United States (Alaska), the Russian Federation, Canada, Denmark (Greenland) and Norway).[44] Significantly, the Arctic was defined in a different manner when the idea of Arctic-wide cooperation was launched by Finland in 1989. The states who were invited to take part in this cooperation were defined using the criterion of the Arctic Circle (66°32°N), 'which approximates the southern boundary of the midnight sun'[45]. All eight states that were invited to join the cooperation process have areas of territorial sovereignty above the Arctic Circle; seven also have terrestrial areas above this limit.[46] It is thus this boundary that serves best the purposes of this study since it clearly determined participation in the Arctic cooperation process. Hence, the southernmost boundary of the Arctic as defined in this study is the Arctic Circle with the exception of the land territory of Iceland, which is included although it lies below the Circle (see figure 3).

Figure 3 (source: Arctic Centre/University of Lapland)

---

[44] The 1973 Agreement on the Conservation of Polar Bears. There are also other Arctic-specific international treaties, e.g., the 1988 Agreement on Arctic Cooperation and Exchange of Notes Concerning Transit of Northwest Passage between Canada and the United States and the 1992 Agreement between the Government of Canada and the Government of the Russian Federation on Cooperation in the Arctic and the North.

[45] AMAP Assessment Report 1998, p. 9. As stated by Finland as chair of the Arctic Council in the Global Ministerial Environment Forum and the Twenty-First Session of the Governing Council of UNEP (Nairobi, 5-9 February 2001): 'The Arctic Council was founded in 1996 as a circumpolar high-level forum that brings together the Governments of those countries which have a territorial outreach beyond the Arctic Circle'. The speech can be found on the home page of the Arctic Council at http://www.arctic-council.org/unep.asp [16.2.2001].

[46] The Arctic Circle also crosses the northern territorial sea area of Iceland.

Brownlie divides the status of territories in international law into 'territorial sovereignty; territory not subject to the sovereignty of any state or states and which possesses a status of its own (trust territories, for example); the *res nullius* and the *res communis*'[47]. Two of these categories have become outdated: *res nullius*[48], because such areas no longer exist in practice,[49] and mandated and trust territories, which have become obsolete. It would thus seem that in practical terms there exist only two categories of territories in international law: areas of territorial sovereignty and areas belonging to all states (*res communis*).[50] With the exception of the Exclusive Economic Zone (EEZ) and continental shelf, these two categories also cover almost the whole of the Arctic physical space. The EEZ entitles the coastal states to establish sovereign rights for all natural resource exploitation in this water-column, although all states still possess rights of navigation there. Despite the fact that the zone is thus still partly an area of *res communis*, it is also essentially subject to the sovereign rights of the coastal states. Most scholars of the law of the sea classify it as an area *sui generis*, and this is also the position adopted in this study.[51]

The category of 'territorial sovereignty' as used here means only those areas in which a state exercises full sovereignty. 'Areas of jurisdictional competence' denote those

---

[47] Brownlie 1998, p. 105.

[48] *'Res nullius'* means those areas which have not yet been subjected to the territorial sovereignty of any state but which can be so subjected.

[49] The only exceptions here are Elsworth Land and Marie Byrd Land in the Antarctic over which no territorial claims have been made. It should be remembered, however, that there are other parts of the Antarctic continent that are not under regular territorial sovereignty. Although territorial sovereignty over these areas was claimed by seven states, these claims were 'frozen' by the Antarctic Treaty of 1959.

[50] *'Res communis'* refers to areas which cannot be subjected to the territorial sovereignty of any state, for instance, the water-column of the high seas or the deep sea-bed.

[51] See, e.g., Kwiatkowska 1989, pp. 4-9. Kwiatkowska asserts (p. 6) that 'EEZ is...without precedent in international law' and calls it a multifunctional zone. Churchill and Lowe (1999, p. 166) advance that 'the EEZ must be regarded as a separate functional zone of a *sui generis* character, situated between the territorial sea and the high seas'. See also O'Connell 1982, pp. 575-579.

areas where the coastal state exercises extensive exclusive decision-making power[52] and includes continental shelf areas. This study focuses on one specific set of economic activities carried out on the continental shelf, offshore hydrocarbon exploitation, not on the area itself, and thus the categorization of continental shelf areas as areas of jurisdictional competence is justified. Finally, the category of 'international area' as used here denotes only the areas that are important from the viewpoint of this thesis, i.e., the high seas, international air space and the deep sea-bed.

The central part of the Arctic is composed of an ocean, the Arctic Ocean, surrounded by continental land masses and islands. Accordingly, the question of the legal status of the Arctic physical space poses very different problems than in the Antarctic, which is composed of a continent surrounded by seas. It is thus especially important

[52] From the viewpoint of planned environmentally harmful stationary activities there are only two relevant differences in the status of a region: those areas which are subject to the exclusive decision-making power of the Arctic states and those areas which are subject to the decision-making power of all states, i.e., the deep sea-bed. If the question of legal status is viewed solely from this standpoint, it seems that there is no reason to make any distinction between an area under the territorial sovereignty of a state and an area not under the territorial sovereignty of a state but in which the state nevertheless exercises exclusive decision-making power in relation to stationary activities; an EEZ is one such area. As will be shown below, this is due to the fact that essentially the same obligations apply irrespective of whether the stationary activity is carried out in the area of territorial sovereignty or other areas over which states have exclusive jurisdiction. In addition, since all the stationary activities which may cause harmful consequences can only be practiced in the deep sea-bed, there seems to be no reason to study anything other than the legal status of this area. Yet, my contention is that it is relevant to make a distinction between areas of territorial sovereignty and other areas where states exercise exclusive decision-making power in relation to stationary activities, i.e., areas of jurisdictional competence, inasmuch as the obligations of international environmental law are almost always connected with the potential or actual environmental impact caused by human activities. From the viewpoint of environmental impact, areas of territorial sovereignty and areas of jurisdictional competence are quite different. Areas of territorial sovereignty are areas in which the states exercise full powers in all dimensions of the physical space, whereas in areas of jurisdictional competence states only have powers in regard to certain activities, such as fishing, or exploitation of hydrocarbon resources. Therefore, the environmental protection obligations and rights of states also differ in extent in relation to areas of territorial sovereignty and areas of jurisdictional competence. From the viewpoint of environmental impact, it is relevant to study the extent of all international areas in the Arctic, not only the deep sea-bed, since in most cases it is the high seas or international air space that are exposed to environmental pollution.

to determine which rules of the law of the sea are applicable in the Arctic. Another difference to the Antarctic is that all the land areas in the Arctic are under the territorial sovereignty of the eight states in the region; as is well known, the sovereignty claims over the Antarctic continent were 'frozen' by the 1959 Antarctic Treaty.[53] The Arctic has not always been under the territorial sovereignty and jurisdiction of the states in the region. In fact, in the beginning of the 19th century, the Arctic could best be described as a place where most of the physical space was not subject to territorial sovereignty and which thus qualified as an international freedom area, such as the high seas at present.[54] This situation is in stark contrast to the present one, in which much of the Arctic is under the territorial sovereignty or jurisdiction of the eight Arctic states.

Given these basic characteristics of the Arctic physical space, it is important to study those areas in which the eight Arctic states are territorial sovereigns. These include areas that states can subject to their sovereignty under international law, i.e., land areas, internal waters and the sea-area up till the limit of the territorial sea of coastal states, and the associated air space. In order to shed light on the problematic legal status of certain spaces that are claimed to be under the territorial sovereignty of an Arctic state, it will be useful to study claims to land spaces and maritime spaces separately inasmuch as separate sets of rules apply to the acquisition of territorial sovereignty over the two types. Finally, the so-called sector claim is taken up in order to show the peculiar sovereignty claim for the polar areas in general, and the Arctic in particular.

---

[53] Before the 1959 Antarctic Treaty, seven states had presented formal territorial claims to parts of the Antarctic. The claims of three of these states partly overlap. The United States and the former Soviet Union have not recognized these claims. This problematic situation was solved in Article IV of the Treaty: 'nothing in the Treaty be interpreted as: a renunciation by any party of previously asserted rights of or claims to territorial sovereignty in Antarctica; -a renunciation or diminution by any party of any basis of claim to territorial sovereignty in Antarctica; - prejudicing the position of any party as regards its recognition or non-recognition of any other state's right of or claim or basis of claim to territorial sovereignty in Antarctica.' For a recent treatment of the Antarctic Treaty system, see Stokke and Vidas 1996.

The basic principle that states can obtain territorial sovereignty only over a minor maritime belt was called into question after the Second World War with the expansion of coastal state maritime zones.[55] Claims over sea areas increasingly represented sovereignty claims over land area and thus posed a serious challenge to the whole of the law of physical space. This trend was checked, however, by the developments leading to the conclusion of the 1982 United Nations Convention on the Law Of the Sea (hereinafter the LOSC).[56] The Convention hindered the phenomenon of 'creeping jurisdiction' by making new maritime zones functional zones that did not allow establishment of sovereignty over maritime territory and by establishing outer limits for the continental shelf areas of coastal states. It provided for a strong check on the other possible development, i.e., claims of sovereign title over maritime areas, and thus possibly also the areas above and beneath this maritime area. As will be shown below, the Arctic littoral states have made ample use of these new maritime zones.[57] The main problem with these zones has been the need to delimit overlapping zones of states in general, and of the Arctic littoral states in particular.

Another obstacle to the expansion of the territorial sovereignty of states over seas was the continued strong status of the principle of freedom of the high seas, also codified in the LOSC.[58] One novelty introduced by the LOSC was the distinction made

---

[54] See Timtchenko 1996, p. 46.
[55] On the phenomenon of 'creeping jurisdiction' in general, see, e.g., Schachter 1991, pp. 274-281; and Schrijver 1997, pp. 203-222. This phenomenon has been very pronounced in the Arctic waters. See Theutenberg 1984, pp. 1-7, 46-49.
[56] The Convention on the Law of the Sea has been abbreviated in several ways. Some scholars refer to it as 'UNCLOS', or 'CLOS', but these terms cannot be considered the best to describe the Convention itself, because the negotiation conference was generally referred to as 'UNCLOS III' and the term referring to the Convention should be different in nature. I prefer here the term 'LOSC' (Law of the Sea Convention) for this reason. Moreover, it is used in the widely cited textbook on the law of the sea, Churchill and Lowe 1999 "The Law of the Sea", p. 18.
[57] See *infra* chapter II (2).
[58] Part VII of the LOSC.

between the water-column of the high seas and the sea-bed whereby the deep sea-bed can never be subjected to the sovereign title of states but rather is considered a possession of the whole of mankind under the principle of the common heritage of mankind.[59] In addition, the natural resources of the sea-bed are placed under international governance. In the Arctic waters, both the areas of high seas and the deep sea-bed are still of minor significance from the viewpoint of economic activity because of the very difficult natural conditions prevailing in the region.

1. Territorial Sovereignty Over the Physical Space of the Arctic

In areas of territorial sovereignty, states have general governance over physical space. This means legislative, judicial, administrative and enforcement powers which are exclusive in the sense that no other state can exercise these rights in relation to those areas; only international law can limit this freedom of states within such areas. Akehurst argues that the term 'sovereignty' should be replaced by the term 'independence' because the word 'sovereignty' carries 'emotive overtones of unlimited power above the law'[60]. Yet, since the term 'territorial sovereignty' is so widely used in the literature, it is also preferred in this study.

1.1. Territorial Sovereignty Over Land Areas

There are a number of land areas in the Arctic whose legal status is problematic. As a rule, only states can exercise territorial sovereignty over an area, but there are two Arctic areas over which an entity other than a state exercises powers similar in function to those deriving from territorial sovereignty. First, although it is true that the EC exercises many attributes of statehood within the physical area of its member states, international law accords territorial sovereignty only to the individual member

[59] Article 136 of the LOSC reads: 'The Area and its resources are the common heritage of mankind'.

states of this regional organization. It would require finalization of the process of integration to the point where the EC became a full-blown state for international law to accord territorial sovereignty to the EC. Hence, although three of the eight Arctic states are members of the EC - Finland, Sweden and Denmark - they all possess territorial sovereignty in the eyes of international law.[61]

Second, certain indigenous peoples exercise broad powers over certain physical spaces. One example is Greenland, whose legal status is still evolving. The primarily Inuit population of Greenland has a semi-autonomous relation, known as Home Rule, with Denmark, which is still in charge of the island's external relations with other states.[62] In fact, there are some signs that the population of Greenland may want to acquire full independence in time.[63] Another important case is the establishment of the Nunavut Territory in Canada in 1999, which placed the primarily Inuit population of 26 000 in charge of a considerable number of policy issues in a vast area.[64]

---

[60] Akehurst 1991, pp. 15-16.

[61] Denmark is not dealt with here because Greenland, the only part of Denmark that qualifies as part of the Arctic as defined in this thesis, withdrew from the EC following a referendum organized in 1982. The 1982 Treaty amending, with regard to Greenland, the Treaties establishing the European Communities.

[62] It is interesting that the agreement between Denmark (Greenland) and Iceland on maritime boundaries was negotiated and signed not only by Denmark's representatives but also by Greenland's home rule representatives. Moreover, the Agreement on Cooperation in Research, Conservation and Management of Marine Mammals in the North Atlantic was signed in 1992 by Norwegian, Icelandic, and, most importantly, not by Danish representative, but by the representatives of the Home Government of the Faroe Islands and the Home Rule Government of Greenland.

[63] According to Rehof 1992, p. 29: 'The question of formal independence of Greenland has, however, recently been clarified in a statement by the Danish Prime Minister delivered at a Nordic Council meeting on February 26, 1991. In his statement, the Danish Premier recognized the right of the Greenlanders and the Faroese to opt for (full) independence, should this desire be clearly expressed by the Greenland and Faroe Islands populations'. In fact, the Faroe Islands will gradually become an independent state by the year 2012. See the Finnish newspaper Lapin Kansa, 22 April 2001, p. 9.

[64] Information on the Nunavut territorial arrangement is provided at http://npc.nunavut.ca/eng/nunavut/ [29.4.2001]. On the history of the eastern Arctic Inuit since the Second World War, see Duffy 1988. For an examination of the political structure of the Nunavut arrangement, see Hicks 1999, pp. 21-53.

Nunavut's autonomous relation with Canada is analogous to that of Greenland with Denmark in that Canada is responsible for the Territory's external relations.

The legal status of the Svalbard Islands is problematic to define. The Islands were considered *res nullius*, i.e., they were not subject to any state's sovereignty, before the signing of an international treaty in 1920.[65] The starting point in general international law in such cases is that only effective occupation establishes valid title, but this possibility was excluded once the legal status of the Islands was determined by international treaty. According to Article 1 of the Svalbard Treaty, all the contracting states recognize that Norway has full and absolute sovereignty over the Islands. Hence, Norway was accorded full sovereignty over the Islands, but not the usual kind of territorial sovereignty. First of all, the Islands were demilitarized and neutralized, meaning that Norway is not allowed to locate any military forces there.[66] Second, all the contracting states' ships and nationals were given non-discriminatory rights to practice peaceful and economic activities on the Islands. In addition, it seems that since the legal status of the Islands was determined by an international treaty, amendments to this legal status must be made by the same means. In this perspective, the legal status of Svalbard is clearly not of the normal type and can thus be considered of sui generis character.

In order to study certain particular claims put forward by the Arctic littoral states, it is important to explain how states can establish title over land area in international law. With one marginal exception,[67] all land space in the Arctic is under the territorial

---

[65] The 1920 Treaty Relating to Svalbard.

[66] Ibid. The demilitarization of the Svalbard Islands is regulated by the Preamble and Article 9 of the Svalbard Treaty. See especially Article 9, which states that the Islands 'may never be used for warlike purposes'.

[67] The only exception is the dispute between Denmark and Canada over the sovereignty of Hans Island. See *infra* chapter II (3).

sovereignty of one of the eight Arctic states: Canada[68], the Russian Federation[69], the United States[70] and the five Nordic states[71]. From this viewpoint, it would not seem necessary to study what the norms are that apply to the acquisition of territorial sovereignty rights over land space, i.e., establishing title over land. Yet, because of

---

[68] The territorial sovereignty of Canada from the British Empire has been qualified by Akehurst (1991, p. 147): 'Acquisition of independence by a colony through a gradual process of constitutional development (as happened in the case of Australia, Canada and New Zealand) can be regarded as a sort of 'quasi-prescription''. The transfer of sovereignty over the Arctic areas from the British Empire to the Canadian Dominion took place already in 1880. At least by the beginning of the Second World War, there were no longer any claims by other states to the Arctic islands of Canada. See Franckx 1993, pp. 71-74.

[69] The Soviet Union was terminated through the Alma Ata Declaration of 21 December 1991: 'with the establishment of the Commonwealth of Independent States, the Union of Soviet Socialist Republics ceases to exist'. The Declaration is reproduced in 31 I.L.M. 148 (1992). Although the scholarly circles of international law have debated whether the Russian Federation has succeeded to the Soviet Union or continued its legal personality, the latter position prevails in practice. For a thorough analysis, see Långström 2000, pp. 723-826. Since the Soviet Union had already confirmed its sovereign title over all the islands lying north of its mainland, the Russian Federation enjoys the same rights. See also Franckx 1993, pp. 142-144.

[70] The United States bought Alaska from Russia through the 1867 Treaty between the United States and Russia concerning the Cession of Alaska of 30 March 1867. In 1959, Alaska became the forty-ninth state admitted to the Union.

[71] Roughly speaking, for almost three centuries after the Kalmar Union (1523-1809) the Nordic areas were governed by the Danish-Norwegian King from Copenhagen (Denmark, Norway, the Faroe Islands, Iceland and Greenland) and by the King of Sweden-Finland in Stockholm. This status quo was changed when the Napoleonic Wars started in the beginning of the nineteenth century. First of all, after the 1808-1809 war between Sweden and Russia, Sweden had to cede Finland and the Åland Islands to Russia. The border was established by the Hamina Agreement in 1809 as following the Tornio and Muonio rivers and still serves as the border between Finland and Sweden. As part of a coordinated attack against Napoleon in 1813 Sweden was able to compel the King of Denmark to cede Norway to Sweden in 1814. Importantly, Sweden established its rule only over the mainland of Norway, not over the old Norwegian possessions of Faroe Islands, Iceland and Greenland. The next round of changes took place in the beginning of the twentieth century, when Norway withdrew from the Union with Sweden in 1905, which was confirmed through a referendum organized later that year. Sovereignty over Greenland was decided by a judgment of the Permanent Court of International Justice (PCIJ) in 1933. The Court decided to favor Danish sovereignty over Greenland to Norwegian sovereignty. Finland and Iceland became independent after the First World War, although Denmark still continued to handle the foreign policy of Iceland until 1944, when the Icelandic Republic was established. Finland became an Arctic coastal state after the First World War because one of its districts agreed in Tarto (1920) with the Soviet Union included the county of Petsamo, a coastal area on the Barents Sea. Petsamo was lost to the Soviet Union in the Second World War, with the new border agreed in 1944 and confirmed in the Paris Peace Treaty. The Faroe Islands and Greenland have both increased their independence from Denmark, and the Faroe Islands even have a program to become an independent state by the year 2012. For brief accounts of Scandinavian history, see Wendt 1981, pp. 11-30 and Fitzmaurice 1987, pp. 1-37. For a more comprehensive study of the history of Scandinavia, see Derry 1979.

the overwhelming presence of different types of ice formations in the Arctic waters, the question poses itself whether ice can be equated to land where obtaining title is concerned. As argued by Brownlie, the modern doctrine of the criteria for obtaining title to land space is based primarily on the concept of effective occupation.[72] In the past, this concept was understood more 'in terms of settlement and close physical possession'[73]; nowadays, the legal facts are more important, i.e., whether a certain state has claimed sovereignty over certain area and whether it has established verbal acts of administration to this effect. This is particularly true in remote areas such as the Arctic, as was determined in the *Eastern Greenland* case.[74] As there are only limited requirements for establishing effective occupation over remote land areas such as the Arctic, one might ask if a state could obtain territorial sovereignty over ice as well. With most of the Arctic waters covered by ice of differing types, this question takes on essential importance where claims to territorial sovereignty are concerned.

---

[72] Brownlie 1998, pp. 136-142.
[73] Ibid, pp. 138-139.
[74] This was established in the judgment of the PCIJ in *the 1933 Eastern Greenland case (Legal Status of Eastern Greenland, Judgment of April 5th, 1933 (PCIJ Series C, No. 62; Series A./B., Fascicule No. 53, p. 46)* : 'In many cases the tribunal has been satisfied with very little in the way of actual exercise of sovereign rights, provided that the other state could not make out a superior

It is patently hard to distinguish land and sea in the Arctic, especially the high Arctic, where permafrost land areas transform gradually into shelf ice[75] and/or fast ice[76] and then break into pack ice[77] (see figure 4). The main problems relate to the legal status of fast ice and especially shelf ice since these are the most permanent forms of ice which extend into sea areas. Although some commentators have referred to the fact that indigenous peoples have used fast ice for certain activities, and that in view of its physical properties it could be equated to land, no Arctic state has made a territorial claim to this ice.[78]

Shelf ice is the most permanent form of sea ice. It has been defined as 'a floating ice sheet of considerable thickness attached to a coast'[79]. Shelf ice is clearly closest to land area from a functional viewpoint. Since an ice shelf may extend for many miles into the sea, the question arises whether it is the outer edge of the ice shelf or the outer edge of the land area which should form the basis for the measurement of baselines. This question was explicitly left out of the United Nations Conference on the Law of the Sea (hereinafter 'the UNCLOS III') negotiations for fear that such a provision would re-open the discussions of the legal status of the Antarctic.[80] It would thus seem that there exists no principled objection to claiming title over permanent forms of ice but that, in practice, this is unlikely to take place. However, as will be

---

claim. This is particularly true in the case of claims to sovereignty over areas in thinly populated or unsettled countries'.

75 The definitions are obtained from the Illustrated Glossary of Snow and Ice by Armstrong & Roberts & Swithinbank 1973. An 'ice shelf' is defined (p. 26): 'A floating ice sheet of considerable thickness attached to a coast. Ice shelves are usually of great horizontal extent and have a level or gently undulating surface. They are nourished by the accumulation of snow and often by the seaward extension of land glaciers. Limited areas may be aground'. An 'ice sheet' is defined (p. 26): 'A mass of ice and snow of considerable thickness and large area. Ice sheets may be resting on rock or floating'.

76 Ibid. 'Fast ice' is defined (p. 14): 'Sea ice which remains fast along the coast, where it is attached to the shore, to an ice wall, to an ice front, or over shoals, or between grounded icebergs. Fast ice may extend a few m or several hundred km from the shore'.

77 Ibid. 'Pack ice' is defined (p. 31): 'any area of sea, river or lake ice other than fast ice, no matter what form it takes or how it is disposed'.

78 For an analysis, see Boyd 1984, pp. 98-152, especially pp. 140-144.

79 See supra footnote 75.

shown below, ice-covered regions do accord coastal states a right to establish special zones of protection.[81]

At present, there exist only two cases in which ice has been equated with land. In the first, Canada has drawn its northern straight baselines in such a way that '8 basepoints on the northern coast of Ellesmere Island are located on the outer edge of ice shelves'[82]; this is a straightforward case of ice being equated with land. In the second case, Denmark has used the ice shelf of Flade Isblink as a coast line from which its territorial sea is measured. Denmark has defended this solution by insisting that it is impossible to determine where the land mass under Greenland's ice shelf ends.[83]

---

[80] Churchill and Lowe 1999, p. 33.
[81] See *infra* chapter II (2).
[82] Franckx 1993, p. 98.
[83] Timtchenko 1996, pp. 118-119.

Figure 4 (source: Finnbarents)

1.2. Territorial Sovereignty Over Maritime Areas

For a long time international law has accorded coastal states two separate maritime areas of territorial sovereignty, internal waters and territorial sea. The basic difference between these areas is that innocent passage of foreign vessels is allowed in a territorial sea but not, as a rule, in internal waters. Sovereignty over the two types of maritime areas follows from the operation of international law: territorial sovereignty over coastal land areas gives rise to the concomitant rights and duties to establish the zones. In principle, states cannot exercise sovereignty over other maritime areas. One exception, however, is that long and peaceful use by one state of a certain maritime area belonging to the high seas may consolidate into sovereignty over that area, as is the case with historic bays.[84]

It is important for a coastal state to have as large an area of internal waters as possible, as all of its other maritime zones are drawn from the outer limit of its internal waters, i.e., baselines.[85] One significant consideration, of course, is that vessels of other states have no right of innocent passage in these areas.[86] It is therefore no surprise that international law relating to the drawing of baselines leaves ample room for coastal states to advance their national interests.[87] There are two main

---

[84] Some scholars take the view that it would be a more realistic approach to distinguish between those states that have acquiesced to the usage and those that have protested against it. Title can thus be considered valid between states that have acquiesced, and invalid with regard to those that have protested against it. See, e.g., Churchill and Lowe 1999.

[85] The drawing of the outermost limit of the continental shelf is an exception here since the continental shelf starts from the outer limit of the territorial sea. Yet, if the continental margin does not naturally extend to the outer limit of the EEZ, the outer limit of the continental shelf is also drawn using the baselines. See Article 76 of the LOSC.

[86] Innocent passage rights are maintained, however, in the waters subject to archipelagic baselines. In addition, if the coastal state has drawn straight baselines over waters which previously were not internal waters of that state, the innocent passage rights of other states are maintained. See Articles 52 and 8 (2) of the LOSC.

[87] The law relating to the drawing of baselines has also been violated quite often. See Churchill and Lowe 1999, pp. 39-40.

bases on which a state can claim a certain maritime area as its internal waters: valid drawing of baselines in international law and sea-areas which are claimed on the basis of historic title.

The principal example of a claim based on historic title is Canada's drawing of straight baselines around its northern islands in 1985, whereby it enclosed all water and ice in that area within its internal waters.[88] This act was controversial from the viewpoint of prevailing norms of customary international law on straight baselines in that it relied on historic title and disregarded the geographical criteria laid down in present-day customary international law. Historic title was accepted by the ICJ as one of the relevant criteria in the *Fisheries* case.[89] However, after that decision both codification conventions on the subject have established geographic criteria as the main standard and economic criteria as a subsidiary one; no mention is made of historic title.[90] Besides, as shown by Scovazzi and Pharand for instance, Canada's past actions do not even prove the historic title invoked.[91]

---

[88] Order Respecting Geographical Co-Ordinates of Points from which Baselines may be Determined, 1985. The Law of the Sea; Baselines: National Legislation With Illustrative Maps 1989, pp. 86-98. Canadian Secretary of State for External Affairs, J. Clark, removed the doubts as to the Canadian position in regard to its Arctic archipelago: 'No previous government, however, has defined its precise limits or delineated Canada's internal waters and territorial sea in the Arctic. This government proposes to do so. An order in council establishing straight baselines around the outer perimeter of the Canadian Arctic Archipelago has been signed today, and will come into effect on January 1, 1986. These baselines define the outer limit of Canada's historic internal waters.'. Reproduced in Franckx 1993, pp. 97-98.

[89] *Fisheries case, Judgment of December 18th, 1951: I.C.J. Reports 1951, p. 116.*

[90] The 1958 Convention on the Territorial Sea and the Contiguous Zone (hereinafter 'the 1958 Territorial Sea Convention', Article 4), and, later, the LOSC (Article 7) confirmed the hierarchy between these bases that was exemplified above.

[91] See Scovazzi 1987, pp. 663-680. Scovazzi distinguishes between two geographic locations: Hudson Bay and the Hudson Strait on the one hand, and the Arctic archipelago of Canada on the other. Scovazzi concludes that Canada can maintain the historic waters claim with respect to the former area but not the latter. Pharand (1988, p. 125) also states: 'the conclusion is that Canada would not succeed in establishing that the waters of the Canadian Arctic Archipelago are historic internal waters'.

Scovazzi is of the view that Canada should be allowed to keep the lines it has drawn, but mainly on the basis of geographic and partly on the basis of economic criteria.[92] From this viewpoint, it could be said that the present lines drawn by Canada are valid in international law, but not on the basis Canada asserts. This is due to the fact that Article 8 of the LOSC requires that if the use of straight baselines encloses areas which were high seas before the lines were drawn, innocent passage rights for other states' vessels are to be maintained.[93] Since the LOSC does not permit reservations, and it is possible that Canada will ratify the Convention, this may cause some problems for Canada. Yet, some scholars have already found it possible for Canada to both ratify the LOSC and maintain its historic title, which would preclude the innocent passage of vessels in the whole of the country's Arctic archipelago.[94] They maintain that the straight baselines can be drawn on a historic basis alone because this action was taken before the LOSC would enter into force for Canada. However, as was argued above, it seems that at the time of Canada's claim, customary law did not allow the drawing of straight baselines on the basis of historic criteria, but only on the basis of geographic criteria complemented by economic criteria. Furthermore, as mentioned above, it is very doubtful whether Canada's claim can be backed up by historical evidence.

The 1958 Territorial Sea Convention and the LOSC contain almost identical provisions on bays, which are based on a geometrical test of whether a certain

[92] Scovazzi (1987, p. 680) concludes: 'La conclusion qui nous semble préférable est donc que le système de lignes droites dans l'Arctique canadien serait dans son ensemble admissible sur la base de la géographie, c'est-à-dire des articles 4 ou 7 des deux conventions de codification du droit de la mer et du droit coutumier correspondant, et non pas sur la base de la confuse théorie des eaux historiques'.

[93] This norm was already adopted in the 1958 Territorial Sea Convention, Article 5 (2).

[94] For an overview of these claims, see Franckx 1993, pp. 105-107. There are scholars who maintain that the baselines were not drawn on an historic basis and that Canada can justify its claim on the basis of criteria in present customary law; see, e.g., McRae 1994-1995. According to McRae, it has essentially been shown that the Northwest passage is not an international strait and that the enclosement of the Canadian Arctic archipelago via straight lines survives the test of customary and conventional law.

indentation can be regarded as a legal bay.[95] However, these rules do not apply to cases where straight baselines are used, or to the claims for a historic bay; in fact, neither the Territorial Sea Convention nor the LOSC regulate the issue of historic bays. Customary law provides certain criteria here, however.[96] Historic title must be established through a long period of effective, open and continuous authority over the bay as a historic bay; the acquiescence of other states is also required.

In the Arctic, Soviet international legal scholars in particular actively advocated the status of a historic bay for all Soviet Arctic seas - the Kara, Laptev, East Siberian and Chukchi Seas - until as recently as the 1980s.[97] Since these seas can hardly be classified as bays, it was no surprise that the Soviet Union claimed only one historic bay officially in 1957, Peter the Great Bay, which met with numerous protests from other states.[98] In 1985, the Decree on Arctic Straight Baselines[99] confirmed that these seas were not considered internal waters because the straight baselines were drawn around the islands or groups of islands that separated the four seas.[100] The Decree also claimed the White Sea, Cheshskii Bay in the Barents Sea and Baidaratskii Bay in the Kara Sea as historic bays.[101] The 1993 Law of the State Boundary of the Russian Federation adopted state legislation on state boundaries in which the concept of

---

[95] Article 7 of the Territorial Sea Convention.
[96] See Churchill and Lowe 1999, pp. 43-45.
[97] Timtchenko 1996, p. 99.
[98] According to Pharand (1988, p. 107), the claim was disputed by many states, especially the United States, the United Kingdom, France and Japan. Pharand (pp. 107-110) also points out the marked difference between governmental policy of the Soviet Union in relation to the concept of historic bays and the broad claims made by Soviet scholars of international law.
[99] See Franckx 1993, pp. 181-186.
[100] According to Franckx (1993, p. 185): 'Without having to argue the international validity of these claims, they imply beyond a shadow of a doubt by their simple existence that this country no longer claims the Kara, Laptev, East Siberian and Chukchi Seas to be internal waters, historically belonging to the U.S.S.R.!'.
[101] Franckx 1993, pp. 181-184.

historic bays was mentioned but the definition was left for the government to decide.[102] It seems that Iceland regards two of its bays as historic bays.[103]

There exist three basic methods in drawing baselines in international law: normal baselines, straight baselines and archipelagic baselines. Both the Territorial Sea Convention and the LOSC provide expressly that the normal baseline is the low-water line along the coast.[104] Hence, if a coast is smooth, a coastal state is obligated to draw the baseline on the basis of the coastline at low tide. Both conventions allow straight baselines if the 'coastline is deeply indented and cut into, or if there is a fringe of islands along the coast in its immediate vicinity'[105]. However, Article 7 (4) of the LOSC requires that '[t]he drawing of straight baselines must not depart to any appreciable extent from the general direction of the coast, and the sea areas lying within the lines must be sufficiently closely linked to the land domain to be subject to the regime of internal waters'. This geographical criterion is supplemented by an economic criterion in Article 7 (5). The method of straight baselines has been much used by the Arctic littoral states but not always in accordance with the criteria in international law.[106] Archipelagic baselines can be drawn only by an archipelagic state as defined in Articles 46 and 47 of the LOSC.[107] No such state exists in the

[102] Timtchenko 1996, pp. 99-100. According to Brubaker (1996 I, pp. 32-33), of the twenty-eight prominent bays he examines, only four meet the traditional length criterion set out for the enclosure of bays in the Territorial Sea Convention and the LOSC. Brubaker (1996 I, p. 33) suggests: 'Russian domestic legislation and other sources indicate that historic claims made include the White Sea...Gulf of Cheshskaia...Gulf of Pechorskii...Gulf of Baidarskaia...Obskii Inlets...and the Enisei Inlets'.

[103] Prescott (1985, p. 260) states: 'The bays called Faxaflói and Breidhafjordhur are closed by straight lines measuring 73 and 40 nautical miles respectively, and it is generally believed that they are considered to be historic bays by Iceland'.

[104] Article 3 of the Territorial Sea Convention and Article 4 of the LOSC.

[105] Article 7 (1) of the LOSC.

[106] See Timtchenko 1996, pp. 98-114.

[107] A state can be considered an archipelago if 'it is constituted wholly by one or more archipelagos'. An archipelago is: 'A group of islands, including parts of islands, interconnecting waters and other natural features which are so closely interrelated that such islands, waters and other natural features form an intrinsic geographical, economic and political entity, or which historically have been regarded as such (Article 46 (b) of the LOSC). If a state considers that it is an archipelagic state, it can draw baselines 'joining the outermost points of the outermost islands and

Arctic since the only archipelagos of the Arctic, Frans Josef Land and the Svalbard Islands, do not form states of their own, but are part of Russia and Norway, respectively.

The concept of territorial sea is of considerable importance for the present study since the outer limit of a state's territorial sea also defines the outer limit of its sovereignty over other dimensions of space as provided in Article 2 (1 and 2) of the LOSC:

> 1. The sovereignty of a coastal State extends, beyond its land territory and internal waters and, in the case of an archipelagic State, its archipelagic waters, to an adjacent belt of sea, described as the territorial sea
> 2. This sovereignty extends to the air space over the territorial sea as well as to its bed and subsoil.

According to the customary law of the sea, the outermost limit of a state's territorial sea is 12 nautical miles (hereinafter 'miles') measured from the baselines. The coastal state is not obligated to extend its territorial sea up to this limit; it may establish a narrower zone. The six Arctic littoral states have all established territorial seas. Their claims range from 3 to 12 nautical miles and are therefore legitimate in international law.[108]

1.3. Sectoral Sovereignty Over All Areas

The sector claim does not fit within the rules of international law since it does not rely on the distinction between land and sea, but, in its broadest expression,

---

drying reefs of the archipelago' (Article 47 (1) of the LOSC). Yet, a state's enclosing the waters falling inside these archipelagic baselines does not mean that they can be considered normal internal waters; in fact, the LOSC ensures that other states maintain their innocent passage rights in such waters (Article 52 of the LOSC). It should be noted here that an archipelagic state may draw normal straight baselines around all of its individual islands as well as archipelagic baselines around the archipelago. It is only on waters which are part of archipelagic waters that some of the high seas freedoms of all states are maintained.

[108] See Rothwell 1996, p. 276 and the notes thereat.

encompasses all physical space which is included within the lines of longitudes extending to the Pole under the claiming state's territorial sovereignty. It is entirely possible to claim territorial sovereignty over a sector extending to the Pole in terms of land space only, but the legal basis for such a claim must be established through effective occupation.[109] All but one of the 'frozen' sovereignty claims to the Antarctic are sectoral claims mainly to land areas, i.e., the Antarctic continent.[110] Claiming sovereignty over a given sector that includes all physical space is an entirely different matter.

The history of the maritime policy of both Russia and Canada indicates that these states have occasionally referred to the sector theory.[111] Yet, as Franckx has shown, the official action of both states in this respect was very modest, whereas the academic community of those states advocated sectoral claims quite strongly.[112] Although some official activity of Canada and the Soviet Union/Russian Federation has occurred during the 20th century, it is clear that neither state claims full sectoral sovereignty.[113] In the words of Franckx: 'Canadian and Soviet governments...have long abolished this theory in its original form, i.e. founding a claim to all land, ice and water expanses in a given sector'[114]. Indeed, this is evident from the fact that, as shown in this chapter, both states have drawn maritime boundaries in accordance with the prevailing law of the sea.

---

109 Brownlie 1998, pp. 148-149.

110 Seven states have made claims to the continent of the Antarctic: the United Kingdom, New Zealand, France, Norway, Australia, Chile and Argentina. The claims of three of these partly overlap, namely, those of the United Kingdom, Chile and Argentina. Norway's proclamation in 1939 referred to 'land lying within this coast and the environing sea' and can thus not be considered a traditional sector claim; see Rothwell 1996, p. 58. Interestingly, one of the reasons why Norway has not made a sector claim to the Antarctic as the other claimants have seems to be its perceived negative consequences for the pending dispute with the Russian Federation. If Norway made such a claim, the Russian Federation could argue that Norway has also made use of the sector claim. For an overview, see Stokke 1996, pp. 384-408

111 A useful summary of the official action of both of these states can be found in Timtchenko 1996, pp. 79-83.

112 Franckx 1993, pp. 143, 152-153, 168-170.

113 See Timtchenko 1996, p. 84.

The sector theory still figures prominently in the disputes over the location of the maritime boundary between Canada and Russia. Elferink argues that the theory has influenced the Arctic maritime policy of Russia in relation to both of its Arctic maritime delimitation disputes, of which that with the United States has already been settled on a provisional basis and that with Norway is still pending. In his view, Russia agreed to base the maritime boundary with the United States on the 1867 Treaty between the United States and Russia concerning the Cession of Alaska, which upholds sectors of longitude as lines of delimitation, partly because this strengthens its position in its ongoing dispute with Norway.[115] There may be a point in Elferink's argument since the equidistance method would have placed more maritime space under Russia's jurisdiction in its maritime border with the United States.

Russia has used the sector theory in two ways in its delimitation dispute with Norway. First, it has argued that, owing to a set of special circumstances, the line should be drawn not in accordance with the equidistance method, but on the basis of the sector of longitude. Second, it seems to treat the sector line as a special circumstance itself inasmuch as the line has been applied in administrative practice in such a way that it has acquired political and psychological significance.[116] Thus, Russia does appear to advance the sector theory but under the constraints of present customary law of the sea. It seems that the majority of commentators do not accept Russia's sector argument in this dispute.[117] Canada argues that the line of delimitation of its maritime boundary with the United States in the Beaufort Sea should follow the 141' W meridian. Yet, unlike Russia, Canada does not justify the use of the sector on

---

[114] Franckx 1993, p. 237.
[115] Elferink 1997, p. 20.
[116] Ibid, pp. 11-12.
[117] See *infra* chapter II (3).

the basis of past practice but on the basis of two old treaties in which, according to Canada, the line was established.[118]

2. Areas of Jurisdictional Competence in the Arctic Waters

Since World War II, the agreement on coastal state jurisdiction over the sea areas falling under the freedom of the high seas has considerably reduced the rights enjoyed by all states in international areas. This phenomenon, called 'creeping jurisdiction', has been very pronounced in the Arctic waters for several reasons. First, the Arctic littoral states have made ample use of the new possibilities of maritime jurisdiction to enclose areas previously considered international areas within their coastal state jurisdiction, a practice which has also resulted in many conflicts between the Arctic states. Second, the Arctic waters are shallow in many places, which allows for extensive claims to the continental shelf and potentially reduces international areas even more. Before studying these different zones of jurisdictional competence in detail, it will be instructive to consider one problematic area of jurisdictional competence, the maritime zones around the Svalbard Islands.

At the time of the conclusion of the Svalbard Treaty in 1920, international law only allowed coastal states to draw a very limited territorial sea, whereby the off-shore economic rights of the contracting states were limited to the territorial sea of the Svalbard Islands only. The phenomenon of creeping jurisdiction has since complicated the interpretation of the Treaty and prompted the question of whether the economic rights given to the other contracting states apply to maritime zones other than the territorial sea. A recent thorough analysis of the problem by Ulfstein advances the position that Svalbard is an island in the meaning of the law of the sea and is thus entitled to all of the maritime zones.[119] Furthermore, it cannot be

[118] See *infra* chapter II (3).
[119] Ulfstein 1995, pp. 418-441.

maintained that the continental shelf which starts from northern Norway includes the Svalbard Shelf since Norway has used Svalbard as the basis for delimiting an equidistance line towards Russia and thus established that the continental shelf around Svalbard is the Svalbard Shelf, not the mainland shelf.[120] The Norwegian scholar is also of the opinion that if Norway decides to claim an EEZ and delineate a continental shelf zone around Svalbard Islands, the non-discrimination rights of the other contracting states would have to be maintained.[121]

To date, Norway has maintained that the wording of the Treaty restricts its application to the territorial sea of Svalbard and that Norway has full jurisdictional competence that excludes other states parties and their nationals in relation to the EEZ and continental shelf zones around the Islands. Another argument Norway has advanced is that its continental shelf starts from its northern mainland and continues through the Svalbard region as far as the continental margin, or, in other words, that the Svalbard Shelf is part of the Norwegian continental shelf proper. The other contracting states have either protested against the Norwegian view or reserved judgment.[122] Norway has not as yet established an EEZ and/or continental shelf zones due to this dispute but it did establish a 200- mile fisheries protection zone around the Islands in 1977.[123] This, too, has been disputed by the Russian Federation.[124]

---

[120] Ibid., pp. 423-425.

[121] Ulfstein (1995, p. 441) argues: 'The general purpose to preserve the former terra nullius economic rights indicates that arts. 2 and 3 should be applied to the continental shelf and the 200-mile zone through an extensive interpretation or an analogy'. He adds: 'anomalies would be created if arts. 2 and 3 applied within but not beyond the territorial sea. The close link between sovereignty to land areas and sovereign rights to maritime zones beyond the territorial sea indicates also that limitations to sovereignty over land should also apply to sovereign rights in maritime zones'. In a similar vein, see Timtchenko 1996, p.134. Some scholars, however, do not make such a direct conclusion. For instance, Hakapää (1990, pp. 72-73) asserts: 'The relevant formulation of the Svalbard Treaty, for instance, is straightforward: it only refers to territorial waters, which in any juridical consideration stand apart from the continental shelf. For an expansive interpretation it would seem necessary to secure the approval of the parties involved'. See also Theutenberg 1984, p. 54.

[122] Ulfstein 1995, p. 422.

[123] Ulfstein (1995, p.439) asserts: 'It has been necessary to regulate fishing activities off Svalbard. A 200 mile fisheries protection zone around Svalbard was established by Norway by Royal Decree

A continental shelf zone is different from an EEZ in that there is no need to proclaim it. The important question is what is, legally speaking, the outermost edge of the continental shelf, since the Arctic waters are naturally very shallow. The criteria for defining the continental shelf were transformed substantially with the adoption of the 1982 LOSC. In the Convention on the Continental Shelf of 1958, the continental shelf was understood to be that continental shelf area that started from the outer edge of the territorial sea limit and ended either at the 200-meter depth limit (distance criterion) or where it was no longer possible to exploit its natural resources (exploitability criterion). This exploitability criterion was restricted in interpretation by the understanding that the shelf was primarily a geological feature.[125]

The exploitability criterion was abandoned in the LOSC and all coastal states have a right to the continental shelf up to a 200-mile limit from the baselines irrespective of geological conditions. In other words, a coastal state is entitled to include the deep sea as part of its continental shelf but only to a distance of 200 miles as measured from the baselines. If the continental margin extends beyond the 200-mile limit, a coastal state has a right to the continental shelf up to 'natural prolongation of its land territory to the outer edge of the continental margin'[126]. Certain absolute rules for the outer edge of the continental shelf are laid down in Article 76 (4 and 5), and while these norms provide a certain outer limit, on balance it can be asserted that the 'limits established by the 1982 Convention allow the inclusion of within national jurisdiction

of 3 June 1977, with effect from 15 June the same year. The zone was established on the basis of the Norwegian Act on the Economic Zone of 17 December 1976, but section 2 of the Royal Decree provides that section 3 of the Act, which prohibits foreign fishing, will not 'for the time being' be applied in the fisheries protection zone. As stated above, Norway wanted in this way to signal that it has the right to establish an EEZ around Svalbard, but has not as yet made use of this right. In the meantime, a fisheries protection zone based on non-discrimination is being applied'.

124 Ibid., p. 446.

125 Brownlie (1998, p. 218) states: 'The 200-metre depth criterion is subject to the exploitability criterion, but the latter is controlled by the general conception of the shelf as a geological feature, and by the principle of adjacency in Article I'.

126 Article 76 (1) of the LOSC.

of substantially the whole of the physical continental margin'[127]. This flexibility in drawing the outer limit of the continental shelf is heightened by the fact that there exists an exception to the outermost limit of the continental shelf in the case of 'submarine elevations that are natural components of the continental margin, such as its plateaux, rises, caps, banks and spurs'[128].

At present, all six Arctic littoral states have defined the criteria by which they determine the outermost limit of their continental shelf zones, but these criteria do not specify the limit of the continental shelf clearly. Denmark (Greenland) and the United States have used the criterion of exploitability and Canada, Norway, Iceland and the Russian Federation[129] that of natural prolongation.[130] It must be kept in mind, however, that since the Arctic waters are generally very shallow, these zones will probably be extensive. This is especially true for the continental shelf in the Russian Arctic.

The rules relating to the Exclusive Economic Zone are nowadays to be found both in conventional law, i.e., the LOSC, and the customary law of the sea. Although some

---

[127] Churchill and Lowe 1999, p. 149. The LOSC does contain certain absolute limits to the continental shelf, defined in Article 76 (5): 'The fixed points comprising the line of the outer limits of the continental shelf on the sea-bed, drawn in accordance with paragraph 4 (a) (i) and (ii), either shall not exceed 350 nautical miles from the baselines from which the breadth of the territorial sea is measures or shall not exceed 100 nautical miles from the 2,500 metre isobath, which is a line connecting the depth of 2,500 metres'. Because of the general shallowness of the Arctic waters, it is possible to apply the 2,500-meter isobath criterion, which gives the Arctic coastal states plenty of room to include the whole natural continental shelf as their legal continental shelf.

[128] Article 76 (6) of the LOSC. Timtchenko (1996, p. 127) argues: 'This provision is very important for the Arctic because there are two huge underwater formations: the Lomonosov Ridge and the Alpha Ridge. The scientific researches have not yet given the absolutely clear answer concerning the origin of the ridges. If they are of continental origin it is necessary to use the above mentioned exception for the definition of the outer limit of the continental shelf of Russia, Canada and Denmark whose continental shelves have these ridges as their portions'. It seems, however, that Article 76 (3) of the LOSC defines the continental margin as not including oceanic ridges.

[129] The Law on the Continental Shelf of the Russian Federation is reproduced in 35 International Legal Materials (I.L.M.) 1498 (1996).

[130] See Rothwell 1996, p. 278 and the notes thereat.

differences exist between these sources, the basic elements are the same.[131] First, the establishment of such a zone is optional, and some coastal states have chosen to establish more limited zones, such as fishing zones. The maximum extent of an EEZ is 200 miles from the baselines where the territorial sea is measured. This zone accords the coastal states broad, exclusive rights with regard to the economic uses of this area as well as marine scientific research and the protection of the environment.[132] Varying systems of economic jurisdiction have been claimed among the Arctic states. Some of the littoral states, such as Denmark and Canada, have restricted their economic jurisdiction to fisheries; others claim 200-mile EEZ zones.[133]

Also important from the viewpoint of this study is the special coastal state competence with regard to ice-covered areas.[134] In the negotiations on the LOSC, Canada, as the main proponent of this zone, succeeded in having a special zone accepted that entitled the coastal states of ice-covered areas to take environmental protection measures in these areas.[135] This norm was meant to apply only to Arctic waters, not the waters surrounding the Antarctic.[136] The main norms relating to the establishment of this zone are regulated in Article 234 of the LOSC:

---

[131] Churchill and Lowe (1999, pp. 161-162) point out that 'what is part of customary international law are the broad rights of coastal and other States as enumerated in Articles 56 and 58 of the Convention. It is much more doubtful whether the detailed obligations in the articles relating to the exercise of coastal State jurisdiction over fisheries, pollution and research have passed or are likely quickly to pass into customary international law'.

[132] Article 56 of the LOSC.

[133] See Rothwell 1996, p. 280 and the notes thereat.

[134] It should be remembered that a state may also proclaim contiguous zones to 'prevent infringement of its customs, fiscal, immigration or sanitary laws and regulations within its territory or territorial sea'. Article 33 of the LOSC. These are not studied here since they are of marginal interest for the problem at hand.

[135] See O'Connell 1984, pp. 1022-1025. Canada adopted the Canadian Arctic Waters Pollution Prevention Act already in 1970. The Act was contrary to the prevailing law of the sea at the time, but Canada succeeded in obtaining international approval for the Act through its incorporation in the LOSC. The Canadian Act applies to areas 100 miles from the baselines and to a much larger set of polluting activities than Article 234, which only applies to vessels.

[136] See Hakapää 1999, p. 114. According to Hakapää (1981, p. 256), during the UNCLOS III 'some delegations also suggested that not only arctic waters but certain others as well – especially those in

> Ice-covered areas
> Coastal States have the right to adopt and enforce non-discriminatory laws and regulations for the prevention, reduction and control of marine pollution from vessels in ice-covered areas within the limits of the exclusive economic zone, where particularly severe climatic conditions and the presence of ice covering such areas for most of the year create obstructions or exceptional hazards to navigation, and pollution of the marine environment could cause major harm to or irreversible disturbance of the ecological balance. Such laws and regulations shall have due regard to navigation and the protection and preservation of the marine environment based on the best available scientific evidence.[137]

This provision sets out the following criteria by which a coastal state is entitled to establish such a zone: 'Particularly severe climatic conditions' exist in this area; 'Presence of ice covering such areas for most of the year' creates 'obstructions or exceptional hazards to navigation'; and 'Pollution of the marine environment could cause major harm to or irreversible disturbance of the ecological balance'. These criteria are mainly overlapping since if ice is present in an area for most of the year, the other conditions are normally also fulfilled.[138] If these conditions are fulfilled, the coastal state is entitled to establish an environmental protection zone but 'within the limits of the exclusive economic zone'[139].

---

tropical regions – should be subject to very special coastal competence'. Yet, after criticism from maritime states, the competence was confined to ice-covered areas only.

[137] Article 234 of the LOSC. Franckx refers to the problems of interpretation of this Article. He argues that the use of the term 'where' can be understood in two senses. First, it refers to 'the area where the coastal state can enact additional laws and regulations, or, secondly, it 'has a meaning similar to "when" in which case it defines the circumstances required before the coastal state can enact such legislation'. See Franckx 1993, p. 96.

[138] According to Hakapää (1981, p. 257): 'The other qualifications of the ice-covered areas are of a supplementary nature. The very fact that certain waters are covered by ice "for most of the year" in itself evidences both a "particularly severe" climate and the presence of "obstructions and exceptional hazards to navigation". Similarly, as discussed before, the climatic conditions in the areas concerned imply that pollution may "cause major harm or irreversible disturbance of the ecological balance".'

[139] Article 234 of the LOSC. In strict literal interpretation, this might be interpreted as covering only the area of the EEZ, not the territorial sea and internal waters, since the EEZ starts from the outer limit of the territorial sea. Yet, as pointed out by Hakapää, a more reasonable interpretation is that the environmental protection zone covers all marine areas inside the outermost limit of the EEZ

Article 234 also specifies the measures that are possible in the special zone. First, the laws and regulations in the special environmental protection zone must be directed at vessels, and these measures must be non-discriminatory in that they do not make distinctions between vessels based on the flag they are flying. Second, the laws and regulations must only be aimed at 'prevention, reduction and control of marine pollution'[140]. Finally, these laws and regulations must also 'have due regard to navigation'[141] and use the best scientific evidence in the protection of the marine environment. Thus, within the zone the coastal state can interfere with the innocent passage rights of all ships sailing those waters. However, these laws and regulations cannot be used as a pretext for preventing foreign shipping in areas where innocent passage rights exist; a coastal state must base any control measures on scientific evidence.

In the Arctic, Canada and Russia have established such zones. In 1970, Canada enacted the Arctic Waters Pollution Prevention Act, which established a 100-mile pollution prevention zone off its northern coast.[142] It should be noted that the Act applies to a broader set of polluting activities than Article 234, which applies only to vessels.[143] The Soviet Union incorporated Article 234 as part of the 1984 Edict on the Economic Zone.[144] Interestingly, Canada and Denmark (Greenland) agreed on several measures to protect the environment of Nares Strait, Baffin Bay and Davis Strait in the 1983 Agreement for Cooperation Relating to the Marine Environment between Denmark (Greenland) and Canada. As stated in the preamble to the Agreement, the

since a literal interpretation would give rise to the extraordinary situation that the coastal state would be entitled to take stricter environmental protection measures in its EEZ area than in its territorial waters. See Hakapää 1999, p. 115.

140 Article 234 of the LOSC.

141 Ibid.

142 Bill C-202. The Bill is reprinted in 9 I.L.M. 543 (1970). For an analysis of the Bill with regard to its legality in international law, see Utton 1974, pp. 140-153.

143 O'Connell 1984, pp. 1022-1025.

144 See Franckx 1993, pp. 179-181.

Agreement also implements Article 234 of the LOSC.[145] In a similar vein, the Strategy adopted by the eight Arctic states in 1991 encourages them to take joint action in order to 'strengthen recognition of the particularly sensitive character of ice-covered parts of the Arctic Ocean'[146].

3. The Overlapping Claims of States to Physical Space in the Arctic

In some parts of the Arctic waters, the phenomenon of creeping jurisdiction has resulted in overlapping jurisdictional spaces among the Arctic littoral states. As will be shown below, in some cases the Arctic states have been able to resolve these differences through an international agreement on the location of the boundary, or some joint development scheme, but in others the disputes are still pending. In fact, the Arctic states have managed to reach agreement on the location of their maritime boundaries in most instances, and the Arctic waters are now largely governed by delimitation agreements between the Arctic states.[147]

A good example here is Jan Mayen Land, an island belonging to Norway. Although the island is very small, 53 kilometers in length and 15-20 kilometers in width, and is only used as a meteorological station, according to the law of the sea it is entitled to the same maritime spaces as all islands.[148] Norway drew a 200-mile EEZ zone adjacent to its mainland coast in 1977, but did not extend this zone to Jan Mayen.

---

[145] The preamble to the Agreement provides: 'Taking into account the developments at the Third United Nations Conference on the Law of the Sea, in particular the text on "Ice-covered Areas"'. See also Articles VII (ii) and Article X (2).

[146] Section 7 (iii) of the Strategy. See *infra* chapter III (1.1.).

[147] The agreements are, in chronological order, the following: Norway - Soviet Union (the Russian Federation) 1957 (partial); Canada - Denmark (Greenland) 1974; Iceland – Norway (Jan Mayen) 1981; Soviet Union (the Russian Federation) - the United States (Alaska) 1990; Denmark (Greenland) – Norway (Jan Mayen Land) 1995; Denmark (Greenland) – Iceland 1998.

[148] According to Article 121 (3) of the LOSC 'rocks' do not qualify as 'islands' in the juridical meaning: 'Rocks which cannot sustain human habitation or economic life of their own shall have no exclusive economic zone or continental shelf'. Jan Mayen Land has only a population of some 10-20 persons, who run a meteorological station.

When it was found that waters near Jan Mayen support capelin resources, however, Norway had an interest in extending this zone to include the island. Since this zone overlapped with that of Iceland, the two states entered into negotiations and concluded an agreement in 1980; interestingly, because of Iceland's dependence on its fisheries industry, Norway allowed it to have a full 200-mile EEZ.[149]

The parties established a joint organ to manage the capelin resources. However, they could not agree on the delimitation of their overlapping continental shelf claims and referred the dispute to a conciliation commission. According to Article 9 of the 1980 agreement, the commission was to make unanimous recommendations on the basis of a broad set of considerations. The governments were not bound by these recommendations but were obligated to pay reasonable regard to them. The commission reported back to the governments in 1981 and recommended that the boundary of the continental shelf should follow the line of the EEZ, except where the most promising potential for hydrocarbon exploitation existed.[150] In the same year, the parties reached an agreement on the basis of the commission's recommendation and established a complex arrangement for developing the hydrocarbon resources jointly.[151]

Since the distance between Jan Mayen Land and Greenland could not support their claims of a 200-mile fishing zone, Norway and Denmark also had to enter into negotiations. Denmark had taken notice of the concessions made by Norway to Iceland and accordingly, in the beginning of the 1980s, claimed a full 200-mile fishing zone, and, later on, a 200-mile continental shelf zone as well. Norway was not

---

[149] Some scholars have found it hard to accept such a solution. For instance, Prescott (1985, p. 267) states: 'it is a puzzle why Norway was so unselfish in this matter'.

[150] The commission's Report is reproduced in 20 I.L.M. 797 (1981). For an analysis of the complex joint arrangement as provided by the Chairman of the conciliation commission, see Richardson 1988, pp. 443-459.

[151] The 1981 Agreement on the Continental Shelf between Iceland and Jan Mayen. For a brief overview of the dispute, see Anderson 1993, pp. 1755-1760.

ready to make the same concession to Denmark as it had to Iceland, however, and argued that a median line should be drawn between the two opposite areas. In 1988, Denmark filed an application with the International Court of Justice (ICJ) to initiate proceedings against Norway in order to clarify these jurisdictional boundaries.

The Court was of the opinion that an equitable solution was what international law required in these cases, normally a median line. However, the Court also found that there existed some special circumstances requiring that the median line be adjusted, e.g., the coast of Greenland is strikingly longer than that of Jan Mayen Land in the area concerned. In addition, it ruled that Greenland must be assured access to the best fishing resources in the area. The Court did not accept as special circumstances differences in human population or economy between the two islands or the fact that Norway had made concessions to Iceland. Ultimately, the Court decided to draw a delimitation line of both the continental shelf and fishing zone roughly midway between the full 200-mile line asserted by Denmark and the median line pursued by Norway. The line has been drawn on the basis of three separate areas; in the first, the southernmost area and the principal fishing grounds the line between the median line and the 200-mile line advocated by Denmark is divided evenly; in the other two, the line has been drawn such that Norway has received two-thirds of the areas concerned.[152] In 1995, the parties concluded an agreement which was based on the judgment of the Court.[153]

Another recent agreement is that concluded on 11 November 1997 between Greenland (Denmark) and Iceland between the representatives of both Greenland

[152] Maritime Delimitation in the area between Greenland and Jan Mayen, Judgment, I.C.J. Reports 1993, p. 38. For an analysis of the case, see Decaux 1993, pp. 495-513.

[153] The 1995 Agreement between the Kingdom of Denmark and the Kingdom of Norway relating to the Delimitation of the Continental Shelf in the Area between Jan Mayen and Greenland and of the Boundary between Fisheries Zones in the Area. As pointed out by Anderson (1998, p. 2519): 'In the Agreement of 18 December 1995, the parties carried out geodetic calculations and adjusted slightly three of the Court's four points, but without changing the basis of the judgment'.

Home Rule and the Danish government, on the one hand, and Iceland on the other. The parties agreed to delimit their overlapping continental shelf zones and fishery zones on the basis of the median line between the relevant stretches of coast of Greenland and Iceland.[154] Since one of the areas under consideration is also claimed by Norway (Jan Mayen), the Agreement requires that 'the delimitation point A has been established in association with the kingdom of Norway and is to be confirmed in bilateral agreements with the Government of the Kingdom of Norway'[155]. The Agreement also contains rules on the exploitation of the hydrocarbon resources that can be exploited on both sides of the continental shelf boundary. The Agreement between Greenland (Denmark) and Iceland as well as the agreements by the two parties with Norway entered into force on 27 May 1998.

Denmark (Greenland) and Canada already settled the location of their maritime boundaries in 1973.[156] The boundary line is primarily based on the equidistance principle, although some adjustments have been made because of islands and the configuration of the coast. The boundary does not, however, cover all overlapping claims between the two states. First, as will be shown below, since both parties claim sovereignty over Hans Island, the maritime areas around this island are not covered by the boundary agreement. Second, the agreement does not settle the states' overlapping claims in the Lincoln Sea since it applies only as far as 82° 13' North.[157]

---

[154] The 1998 Agreement Between the Government of the Republic of Iceland on one hand, and the Government of the Kingdom of Denmark with the Council of Greenland on the other on Delimitation of the Continental Shelf and Fisheries Zones in the Area Between Iceland and Greenland.

[155] Ibid., Article 1.

[156] The 1974 Agreement between the Government of the Kingdom of Denmark and the Government of Canada Relating to the Delimitation of the Continental Shelf between Canada and Greenland.

[157] For a brief analysis, see Alexander 1993, pp. 371-378.

The Soviet Union and the United States settled the location of their maritime boundary in 1990.[158] Although the original convention by which Russia ceded Alaska to the United States was interpreted as not providing for a maritime boundary, its basic method of demarcation was chosen as the method for drawing maritime boundaries between the two states. The line starts from the North Pacific Ocean and runs through the Bering Sea to the Bering Strait. It moves north along a meridian through the Chukchi Sea and then towards the Arctic Ocean to the extent permitted by international law. This single line covers all water column and continental shelf areas. Russia has not yet ratified the agreement as required by the agreement in Article 7, but, through an exchange of notes, the parties have agreed to abide by the terms of the 1990 agreement as of 15 June 1990 pending its entry into force.

Some examples can be cited of states' exercising joint jurisdictional competence over maritime areas in the Arctic. For example, Norway has agreed with two other Arctic littoral states on the establishment of joint maritime zones. As discussed above, Jan Mayen Land has engendered conflicts, although these have now been resolved. Iceland and Norway established a conciliation commission to resolve the location of their continental shelf boundary. The commission recommended that the continental shelf boundary follow in general the EEZ line of these states with the exception that a joint development zone was to be established for the area where the hydrocarbon potential was estimated to be the greatest. This zone was established through the Agreement on the Continental Shelf between Iceland and Jan Mayen in 1981.[159]

Norway and the Russian Federation have not been able to settle their dispute over their maritime boundaries in the Barents Sea. Since there exists a lot of fishery potential in this area, Norway and the Soviet Union concluded a provisional

---

[158] The 1990 Maritime Boundary Agreement Between the United States of America and the Union of the Soviet Socialist Republics. For a brief analysis, see Verville 1993, pp. 447-452, for a more extensive one Elferink 1994, pp. 255-274.

[159] See *supra* note 151.

agreement on fishing in this zone, the so called grey zone, which will not prejudice any final agreement on the maritime boundary. The grey zone comprises 67,500 square km, of which 41,500 lie in the disputed area. This agreement establishes national fishing quotas and certain conservation measures.[160] As was reviewed above, in 1990 the United States and the Soviet Union were able to conclude a comprehensive maritime boundary agreement covering all maritime zones. Article 3 of the Agreement provided an interesting jurisdictional solution: in the parts of the sea areas where the EEZs of the states overlap, both states have EEZ jurisdiction; in other words, there is joint jurisdiction of the parties in these areas.

There are also a number of pending controversies over physical space in the Arctic. Canada and the United States dispute the location of their maritime boundary in the Beaufort Sea. Canada argues that the line should follow the 141°W line of longitude in accordance with the original 1825 boundary treaty between Great Britain and Russia. This treaty provided that the line of demarcation was to follow the meridian of 141°W to the frozen ocean. The United States is of the opinion that the line should be drawn using the equidistance method, which would place it to the east of a line based on 141°W. This dispute will not be easily resolved because of the high potential for hydrocarbons in the contested area.

Norway and the Soviet Union/ the Russian Federation have taken steps to establish their maritime boundary in the Barents Sea but only to a very modest extent. In 1957 the two countries agreed on a territorial sea boundary at the mid-point of the closing line of the Varanger Fjord.[161] This required an agreement not to extend their territorial seas beyond that point before agreement was reached on other maritime

[160] The 1978 Agreement between Norway and the Soviet Union on a Temporary Practical Arrangement for Fishing in an Adjacent Area in the Barents Sea with Attached Protocol.
[161] The 1957 Agreement between Norway and the Soviet Union Concerning the Sea Frontier in the Varangerfjord of 15 February 1957 and Protocol of 29. For a brief analysis, see Anderson 1993, pp. 1781-1784. For a thorough analysis of the entire dispute, see Elferink 1994, pp. 232-248.

boundaries. This minor breakthrough has not been followed by other agreements although some progress has been made.[162] The parties were able to agree on a joint fisheries zone in 1977, the so-called grey zone referred to above. A controversy is also pending on the delimitation of the EEZ as well as continental shelf boundaries although agreement seems to exist that only one line need be drawn, i.e., a single line between the continental shelf and the EEZs. The problem hinges on whether the basis of the boundary should be a sector line, as advocated by Russia, or an equidistant line, as argued by Norway.[163] The parties seem to have made some progress recently.[164]

---

[162] According to Elferink (1997, p. 10): 'Since 1988 considerable progress has been made in the informal consultations over the delimitation of the boundary. In June 1991, during a press conference in Oslo, Mr. Gorbachev and the Norwegian Prime Minister Brundtland stated that the delimitation had been resolved for three quarters. This seems to refer to the fact that three quarters of the area under dispute had been delimited on a provisional basis. Reportedly, the area that still needs to be delimited is situated in the southern part of the Barents Sea...Recent meetings between the parties have not led to substantial progress and delimitation of the southern part of the Barents Sea remains the main unsettled issue'.

[163] According to Elferink (1997, pp. 11-16), the basic juridical difference between these states lies in their position on which method should be used to draw the boundaries. Norway argues that Article 6 of the Continental Shelf Convention of 1958 should be interpreted as meaning that the equidistance method should be applied unless special circumstances exist. Since Norway takes the view that this is not the case here, the equidistance method should be applied. Russia, on the other hand, argues that the boundary line should follow the sector line. It also refers to the Continental Shelf Convention of 1958 but does not see any presumption that the equidistance method should be applied unless special circumstances exist: it takes the view that the equidistance method and special circumstances are of similar value, and that special circumstances cover factors other than geography. In other words, Russia considers the law of the sea as requiring only that an equitable solution be reached by the parties. The sector theory is advanced by Russia in two ways. First, because of the existence of a whole set of special circumstances, the line should be drawn not in accordance with the equidistance method but on the basis of sector line. From the legal point of view, however, there seems to be no support for this approach since 'none of these circumstances is in any way related to the sector line'. Indeed, it is hard to argue that coastal configuration, ice conditions, population size, etc. would lead to an equitable delimitation corresponding exactly to a fixed meridian. In addition, 'considering that a sector line leads to an encroachment on Norwegian maritime zones it seems difficult to accept that its application leads to an equitable solution'. Second, Russia seems to treat the sector line as a special circumstance itself in that it has been applied in administrative practice in a way that has given it political and psychological significance. This would require that 'state practice can provide evidence that a certain line is accepted by the parties as a maritime boundary through a process of acquiescence or estoppel, or the conduct of the parties, although not amounting to a situation reflecting acquiescence or estoppel, indicates what they consider to be an equitable solution'. Elferink sees no evidence that Russia applied the sector line as a maritime boundary prior to the negotiations or that Norway accepted such claims. However, he acknowledges that geographic circumstances in the region might justify some

The problems of this maritime boundary dispute are amplified by the fact that there are several different areas to be delimited. The maritime boundary starts from the countries' mainland frontier in Varanger Fjord, on which the parties have agreed. In the Barents Sea, the first part of the maritime boundary is influenced by the mainland coasts, which are adjacent, and the second by the opposing coasts of the Norwegian mainland and Svalbard as well as the Russian archipelago of Novaya Zemlya. One of the islands of Svalbard, Hope Island, may also affect the maritime boundary. It should be noted that the uncertainty surrounding the right of Norway to establish an EEZ and continental shelf zone without giving non-discriminatory rights to other contracting states makes the dispute even more complicated. The final part of the potential delimitation goes via Svalbard Island and the Russian island of Franz Joseph Land. This line is much influenced by two small islands, Kvitoya (Norway) and Viktoria (Russia). To the north, the boundary extends as far as the outer limits of the states' EEZs and the continental shelves.[165]

A potential dispute over areas of jurisdictional competence exists between Norway (Svalbard) and Denmark (Greenland) as the distance between the maritime zones of these islands is less than 400 miles. This is unlikely to materialize, however, as delimitation of maritime zones between Svalbard and Greenland will only become

---

modifications of the equidistance line in favor of Russia. For a comprehensive analysis of the dispute, see Elferink 1994, pp. 232-248.

[164] In the Declaration made by the Prime Minister of Norway and the Prime Minister of the Russian Federation, which was signed on 29 September 2000 in Oslo, it was provided: '6. Delimitation In The Barents Sea. The governments of our two countries agree on the importance of further progress towards a solution as regards delimitation in the Barents Sea. A new round of talks at senior-official level on delimitation in the Barents Sea was held in Oslo on 6 September 2000. It was noted that the process started in 1998 for determining common methods and consistent geodetic data is nearing completion. This will provide a framework for taking small, balanced steps with a view to achieving an agreement on delimitation in the Barents Sea. It was agreed that in the further talks, a broad-based approach would be followed, including the various aspects of the delimitation process such as the fisheries and possible cooperation in the petroleum sector. The next round of talks will be held in Moscow at the end of the year'. The English version of the Declaration can be found at http://odin.dep.no/ud/engelsk/p10002187/032001-990361/index-dok000-b-n-a.html [6.2.2001].

possible when Norway has resolved its dispute over economic rights with some of the contracting states of the Svalbard Treaty.

4. International Areas

As was mentioned above, in the beginning of the 19th century most of the Arctic as defined here could be considered as an international area not subjected to any state sovereignty. By the middle of the twentieth century, all land space was already firmly under the territorial sovereignty of the Arctic states, and it was thus mainly the Arctic Ocean and other Arctic waters, as well as the international air space, whose status remained international. After World War Two, however, the status of the Greenland, Barents, and Norwegian Seas in particular as parts of the high seas was much diminished by expanding coastal state jurisdictional spaces. It is thus essentially the core of the Arctic Ocean whose status can still be regarded as 'purely' international if the normal principles of the law of the sea are applied.

The great majority of commentators share the opinion that the normal principles and rules of the law of the sea apply to the Arctic Ocean.[166] Another debate in the scholarly world is whether the Arctic Ocean should be regarded as a semi-enclosed sea as defined in Article 122 of the LOSC. If it were so qualified, the littoral states would have more obligations to cooperate in regard to environmental protection

---

[165] See Elferink 1997, pp. 6-7.

[166] For a general review of the law of the sea applicable in the Arctic Ocean, see Boczek 1986, pp. 154-189. He concludes (p. 187): 'despite its unique geophysical features, this ice-covered marine space is subject to the same universal legal regime of the oceans as that which has now been codified and developed in the United Nations Convention on the Law of the Sea of 1982'. For similar arguments, see Kish 1973, pp. 26-28, Pharand 1982, pp. 257-277; in Finnish, see Hakapää 1995, pp. 87-105. For an early theory advocating that the Arctic's physical space be divided between the Arctic coastal states on the basis of 'meridian limits running up to the North Pole', see Lakhtine 1930, pp. 703-717.

there.[167] However, the Arctic Ocean seems not to qualify as a semi-enclosed sea as defined in Article 122.[168]

There exist two main categories of international areas: international freedom space, and international governance space. International freedom space denotes that space in which states are in principle free to operate unless constrained or prohibited by international law.[169] As mentioned above, such space has considerably diminished in regard to the water column due to creeping coastal state jurisdiction; it has lost much of its significance in the areas of exclusive jurisdictional competence of coastal states, although some freedoms have been left intact. In the Arctic waters, those areas over which the Arctic littoral states have not established an EEZ or fishery zones can be considered freedom areas as defined here. Another international freedom space, the international air space, is larger than the area of the high seas due to the fact that the coastal states have enlarged their maritime jurisdiction. Jurisdiction over air space does not follow these lines, but, rather, in accordance with general international law, the outermost line of states' territorial seas; the air space which is not above the

---

[167] See Article 123 of the LOSC.

[168] Article 122 of the LOSC reads: 'For the purposes of this Convention, "enclosed or semi-enclosed sea" means a gulf, basin or sea surrounded by two or more States and connected to another sea by a narrow outlet or consisting entirely or primarily of the territorial seas and exclusive economic zones of two or more coastal states'. There is a wide outlet from the Arctic Ocean to the Atlantic Ocean at the Greenland Sea and this outlet is not under the EEZs of the coastal states. For a view that the Arctic ocean cannot be qualified as a semi-enclosed sea, see Harders 1997, pp. 57-62 and p. 232. Even if the Arctic Ocean could be defined as a semi-enclosed sea, it is not clear whether this would entail environmental protection obligations for the littoral states, and to what extent. The entire obligation is written in a manner that calls into question whether an agreement has been reached by the parties. Vukas (2000, p. 40), however, opines: 'Taking into account the motives for establishing the category of enclosed or semi-enclosed seas, and the adopted definition, the characteristics of the Arctic ocean would seem to justify its being considered as an enclosed or semi-enclosed sea'. In addition, he argues (2000, p 42): 'It cannot be said that the commitment of states under Article 123 is entirely devoid of legal force. Although states are not obligated to coordinate their activities, it can be claimed that acts systematically rejecting any negotiations on the protection and preservation of the marine environment of an enclosed or semi-enclosed sea would represent a contravention of the Convention'. For a review of scholarly proposals, see Franckx 1993, pp. 240-243.

Arctic states' land areas or above their territorial sea areas is thus part of international air space, where the freedom regime prevails.

In the LOSC, the deep sea-bed, i.e., the sea-bed beyond any coastal state's continental shelf area, was declared part of the common heritage of mankind, an area of international governance. The conception of the deep sea-bed (hereinafter the Area) as part of the common heritage of mankind was implemented in the LOSC by creating an inter-governmental organization, the International Sea-Bed Authority (ISA), which represents potentially all states of the world, but at the moment only those that have ratified the Convention.[170] As a result of the definition of the Area in the LOSC as part of the common heritage of mankind, the provisions relating to the Area in the LOSC define exclusive authority over the resources of the Area for the international community. If it is understood that the Area was part of the high seas before the LOSC this conception is difficult to reconcile with the contractual basis of the LOSC in general and the creating of obligations for third states through an international treaty in particular.

There evidently existed different theories as to the legal status of the Area before the LOSC. Some maintained that the natural prolongation of coastal states' land territory into the sea could be extended as far as possible albeit limited by similar rights of other coastal states. Others, advocated by the Western states in particular, held that the continental shelf had natural geological limits and that the sea-bed beyond that point had always been covered by the high seas regime of freedom. Since the end of the 1960s, the majority of states seem to have adopted the position that the Area is

---

[169] There are certain basic obligations applicable also in the freedom spaces, for instance, the principle to have reasonable regard to the rights of other states. See Article 87 of the LOSC. For an analysis, see Brownlie 1998, pp. 235-249.

[170] The International Sea-Bed Authority (ISA) consists of the legislative branch, the Assembly, and the executive branch, the Council. In addition, it has its own body for exploiting the resources of the deep sea-bed, the Enterprise.

part of the common heritage of mankind.[171] This is reflected in what is known as the 'Moratorium' Resolution[172] adopted by the United Nations General Assembly (UNGA) in 1969 and especially in the 1970 Declaration of Principles Governing the Sea-Bed and Ocean Floor and the Subsoil Thereof.[173] Although there might have been different theories as to the legal status of the deep sea-bed before 1970, it seems that the UNGA Declaration changed the course of the debate permanently. At present, a great majority of states accept the status of the deep sea-bed as an area belonging to the common heritage of mankind. As will be shown below, the problems are related mainly to the practical implementation of this principle. Although creeping jurisdiction has diminished the Area considerably in the Arctic waters, much of the Arctic Ocean can still be considered part of it (see figure 5).

[171] For a general treatment, see Schachter 1991, pp. 291-296.
[172] UNGA resolution 2574 (XXIV), 15 December 1969. The text is reproduced in 9 I.L.M. 419 (1970).
[173] UNGA resolution 2749 (XXV), 17 December 1970. The text is reproduced 10 I.L.M. 220 (1971).

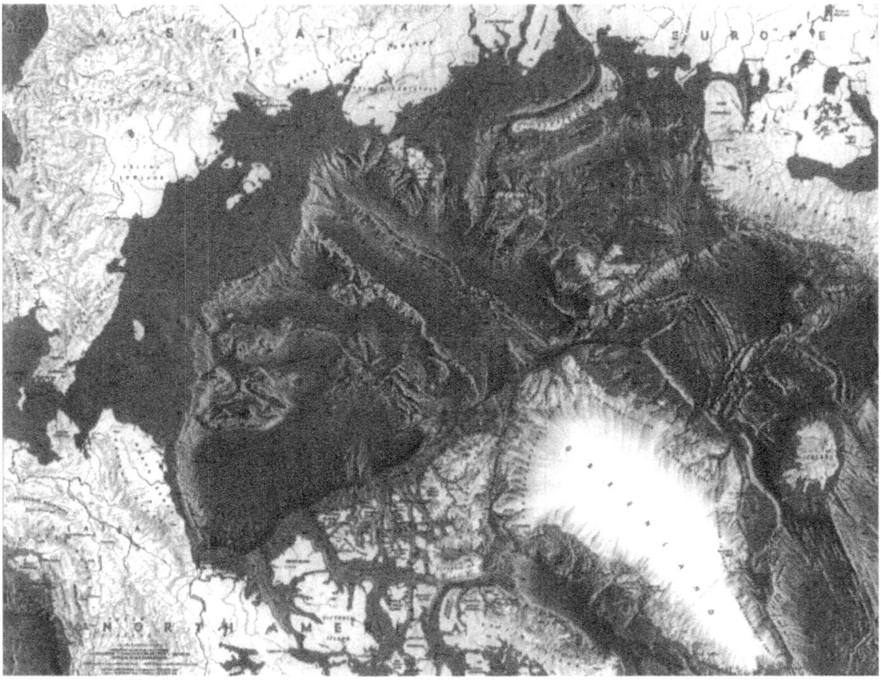

Figure 5. Arctic Ocean floor (source: National Geographic (supplement), Vol. 177, No. 1, January 1990)

III INTENTIONAL NORM-MAKING IN INTERNATIONAL LAW: THE CASE OF ARCTIC COOPERATION

The present chapter has two main goals, to examine the legal status of Arctic cooperation and, on this basis, to elaborate a method by which intentionally produced instruments should be evaluated in international law. The first of these aims is the one of the principal focuses of the thesis set out in the introductory chapter. It is essential that it be taken up here not only to clarify the confusion regarding the legal status of the norms and instruments adopted in this cooperation but also to furnish the reader with basic information on the nature of Arctic cooperation and its international legal possibilities, which will be used in the concluding chapter of this thesis.

The discussion of the second, broader, goal begins with an examination of how general international law regulates intentional norm-making, mainly the conclusion of agreements[174], the aim of which is to clarify the considerable confusion as to how the intentionally adopted instruments and norms should be evaluated in international law. As the written normative instruments (hereinafter 'instruments')[175] adopted in Arctic cooperation have proven difficult to qualify in international law, they serve here as an ideal basis for elucidating the principal problems related to all of the international instruments studied in this thesis. This study of Arctic instruments forms the basis for the method used to determine the bindingness of the norms studied in chapters IV and V.

The chapter starts with a review of the historical development of Arctic cooperation and the legal status of its instruments as well as how these have been evaluated by

---

[174] The concept of international agreement subsumes both written and verbal agreements, with 'treaty' normally used to refer to the former.

[175] The term 'instrument' as used in this thesis refers to the outcome of negotiations between international legal subjects in which a set of norms in written form are adopted.

scholars. Thereafter, it examines in depth the different views of how contractual obligations are created in international law, and what general international law stipulates with regard to this issue. The more detailed view on how a scholar of international law ought to examine the intentionally produced instruments is illustrated via a study on three separate problems posed by the instruments adopted in the Arctic cooperation process.

1. The Arctic Cooperation Process

The initial idea of Arctic-wide cooperation was launched in 1987 in Murmansk by former Soviet Secretary-General Michail Gorbachev. The Soviet leader proposed that the Arctic states could initiate cooperation in various fields, one of these being protection of the Arctic environment.[176] This idea was concretized in part when Finland convened a conference of the eight Arctic states - Canada, Denmark, Finland, Iceland, Norway, Sweden, the Russian Federation and the United States - in Rovaniemi in 1989 to discuss the issue of protecting the Arctic environment. After two additional preparatory meetings in Yellowknife, Canada, and Kiruna, Sweden, the eight Arctic states, as well as other actors met again in Rovaniemi in 1991 to sign the Rovaniemi Declaration, by which they adopted the Arctic Environmental Protection Strategy (hereinafter 'the Strategy').[177] In the Declaration, the states resolved to 'adopt the Arctic Environmental Protection Strategy and commit [themselves] to take steps towards its implementation and consider its further

[176] Gorbachev proposed that a nuclear-weapon-free zone be declared in northern Europe; naval activity be limited in the seas adjacent to northern Europe; peaceful cooperation be the basis for utilizing the resources of the Arctic; scientific study of the Arctic has great significance for all mankind; the countries of the North cooperate in matters of environmental protection; the Northern Sea Route be opened by the Soviet Union to ice-breaker escorted passage. For an analysis, see Rothwell 1996, pp. 229-231.

[177] The history of the negotiation process is described in Rothwell 1996, pp. 229-242 and Tennberg 1998, pp. 53-61. A brief presentation can be found in Caron 1993, pp. 378-379.

elaboration'[178]. The Declaration and the Strategy in fact form a single instrument (hereinafter these two instruments together are referred to as 'the AEPS document').

1.1. The First Phase of the Arctic Cooperation Process: AEPS Cooperation

The Strategy of the AEPS document is not structured like a normal international treaty, which contains a preface outlining the principles of the regime and a set of provisions guiding the behavior of the states parties in the direction of the principles. Instead,

> The Strategy is comprised of a number of component parts, beginning with a statement of objectives. These objectives establish the broad direction in which the eight Arctic countries are intending to move. The objectives are accompanied by statements of principle which are designed to guide the actions of Arctic countries individually and collectively, as they move toward achievement of the objectives. The Strategy also describes the problems and priorities which the eight Arctic countries agree need to be addressed at this time.[179]

The Strategy is more goal-oriented than a typical international environmental treaty. The goal of protecting the Arctic environment determines the means adopted, whereas in the case of standard international environmental treaties the emphasis is on the means by which the behavior is directed towards the values expressed in its principles. The Strategy contains very ambitious objectives, such as 'to identify, reduce, and, as a final goal, eliminate pollution'[180], and the principles are laid down in

---

[178] The 1991 Declaration on the Protection of the Arctic Environment and the 1991 Arctic Environmental Protection Strategy.

[179] Preface to the Strategy.

[180] Ibid., Section 2 (1v). The Section on objectives reads as follows: 'i)To protect the arctic ecosystem, including humans; ii) To provide for the protection, enhancement and restoration of environmental quality and the sustainable utilization of natural resources, including their use by local populations and indigenous peoples in the Arctic; iii) To recognize, and to the extent possible, seek to accommodate the traditional and cultural needs, values and practices of the indigenous peoples as determined by themselves, related to the protection of the Arctic environment; iv) To

very ambitious terms.[181] Six pollution problems are chosen as priority areas: persistent organic contaminants[182], oil pollution[183], heavy metals[184], noise[185], radioactivity[186] and acidification[187]. After briefly describing the six problems, the Strategy goes on its Section four to study the international mechanisms for the protection of the Arctic environment in regard to each problem. Section 5 then takes up the action plans of the Strategy to combat each of the problems.

The Strategy also contains obligations requiring the Arctic states to establish working groups specialized in certain aspects of pollution problems in the region. It establishes the Arctic Monitoring and Assessment Program (AMAP)[188], whose primary objective is 'the measurement of the levels of anthropogenic pollutants and the assessment of their effects in relevant component parts of the Arctic environment'[189]. Section 7, entitled 'Protection of the Arctic marine environment', is very much linked to international law. The opening paragraph emphasizes the importance of the LOSC in the protection efforts of the Arctic states and the remaining paragraphs emphasize the need to adhere to the existing international

---

review regularly the state of the Arctic environment: v) To identify, reduce, and, as a final goal, eliminate pollution'.

181 Ibid., Section 2 (2).

182 Ibid. The problem of persistent organic contaminants is regulated in Sections 3 (1), 4 (1) and 5 (1).

183 Ibid. The problem of oil pollution is regulated in Sections 3 (2), 4 (2) and 5 (2).

184 Ibid. The problem of heavy metals is regulated in Sections 3 (3), 4 (3) and 5 (3).

185 Ibid. The problem of noise is regulated in Sections 3 (4), 4 (4) and 5 (4).

186 Ibid. The problem of radioactivity is regulated in Sections 3 (5), 4 (5) and 5 (5).

187 Ibid. The problem of acidification is regulated in Sections 3 (6), 4 (6) and 5 (6).

188 Ibid., Section 6. The first action of the AMAP is to distinguish 'human-induced changes from changes caused by natural phenomena'. This 'will require estimates and regular reporting by the Arctic countries'. 'The eight Arctic countries should therefore agree to establish an Arctic Monitoring and Assessment Program (AMAP) to fulfill these monitoring objectives'. The Strategy recommends that the AMAP be implemented through the AMAP task force and a small secretariat. The task force should first review the existing programs in order to avoid duplication of effort. AMAP should first concentrate on persistent organic contaminants and on heavy metals and radionuclides, and 'monitor ecological indicators to provide a basis for assessments of the status of Arctic ecosystems'. The 'State of the Arctic Environment' Reports will be delivered regularly to Arctic countries.

189 Ibid., Section 6 (5).

environmental treaties[190] and to take joint action in order to 'strengthen recognition of the particularly sensitive character of ice-covered parts of the Arctic ocean'[191]. In addition, the Arctic states are encouraged to become parties to relevant 'international instruments connected with the protection of the marine environment'[192] and to apply the existing standards according to the conditions of the Arctic.[193]

Section 8 of the Strategy, 'Emergency Prevention, Preparedness and Response' (EPPR)[194], does not establish a separate working group for this purpose, but requires that the Arctic states 'take steps to convene a meeting of experts to consider and recommend the necessary system of cooperation'[195]. The Section also frequently emphasizes the importance of existing international treaties.[196] Finally, Section 9 establishes cooperation on conservation of Arctic flora and fauna: 'The eight Arctic countries will cooperate for the conservation of Arctic flora and fauna, their diversity,

---

190 For instance, Section 7 (ii) reads: 'To this end the Arctic countries agree to...take measures as soon as possible to adhere to the strictest relevant international standards within the conventions, to which the countries are parties, regarding discharges irrespective of origin'. Section 7 (v) requires the Arctic countries to '[j]ointly support the appropriate initiatives of international organizations in developing mandatory standards in order to improve the protection against accidental pollution affecting the marine environment, and actively ensure application of such standards'.

191 Ibid., Section 7 (iii).

192 Ibid., Section 7 (iii).

193 Ibid. Section 7 (i) reads: 'the Arctic countries agree to i) Apply the principles concerning the protection and preservation of the Marine Environment as reflected in the 1982 United Nations Convention on the Law of the Sea, and, in accordance with the continuing development of international environmental law, to further strengthen rules in order to protect the Arctic'.

194 The Strategy acknowledges that although most of the pollution threatening the Arctic comes from the south, there 'has been an increase in development activities and shipping within the Arctic'. These activities pose a danger of accidental pollution. The Strategy identifies a large number of treaties applicable in this field. First, the Arctic countries agree to review the existing arrangements. 'They will also take steps to convene a meeting of experts to consider and recommend the necessary system of cooperation'. Enumerations of these actions follow with these, including management measures and cooperation in studying the problem of pollution.

195 Section 8 (i) of the Strategy. The rest of the Section outlines the possible elements of this cooperation.

196 Ibid., Section 8 (2). The Section summarizes a number of existing relevant international treaties. According to 8 (i), the Arctic states 'will take steps to review existing bilateral and multilateral arrangements in order to evaluate the adequacy of the geographical coverage of the Arctic region by cooperative agreements'.

and their habitats'[197] and 'agree to establish a mechanism for furthering the following aims in close cooperation with Arctic indigenous peoples'[198], i.e., a separate working group with a mandate.[199] Each Arctic country is required in very strong terms to notify the others via the national agency responsible for coordinating the cooperation.[200] The final paragraphs of the Section emphasize the national conditions under which the cooperation established is to be continued.[201]

The Declaration provides for a general obligation to continue AEPS cooperation: 'We agree to hold regular meetings to assess the progress made and to coordinate actions which will implement and further develop the Arctic Environmental Protection Strategy'[202]. The forms of this environmental cooperation are specified in the Strategy. Section 10, paragraph 5, enumerates the terms of reference of AEPS cooperation, providing the cooperation process with a broad mandate for protecting the Arctic environment.[203] Paragraphs 3 and 4 provide for the participation of other than Arctic states; the three international organizations of the Arctic indigenous

[197] Ibid. The rest of the paragraph (i) concentrates on outlining the potential mandate of the cooperation on conservation of Arctic flora and fauna.
[198] Ibid., Section 9 (v).
[199] Ibid. The mandate of the working group is set out in the rest of paragraph (v): 'a) Promoting and facilitating exchanges of information and personnel as provided for in this Strategy; b) Making recommendations with respect to the priorities, the orientation and the nature of research and monitoring programs of the Arctic countries; c) Proposing strategies for enhanced conservation of Arctic species and their habitats; and d) Regularly compiling and disseminating information on activities regarding the conservation of Arctic flora and fauna'.
[200] Ibid. Section 9 (vii) reads: 'By October 1991 each Country will identify to the others its national agency designated to coordinate the cooperation envisaged by this section'.
[201] Ibid. Sections 9 (viii) and (ix) read: 'The countries agree that the terms and conditions of the cooperation and exchanges provided for in this section will be subject to the laws and regulations of the Countries; ix) Each country will make its best efforts to provide resources adequate to carry out its responsibilities under this section. It is understood that the ability of each country to carry out activities is subject to the availability of funds, and that countries will seek to ensure long-term funding for necessary projects'.
[202] Ibid., Section 10 (1).
[203] Ibid. Section 10 (5) reads: 'The Meetings on the Arctic Environment shall serve to: i) identify and coordinate actions to implement and further develop the Arctic Environmental Protection Strategy; ii) initiate cooperation in new fields relevant to the environmental protection of the Arctic; iii) make necessary recommendations in order to protect the Arctic environment; iv) improve

peoples are accorded the legal status of observers,[204] and the criteria for observer status in the AEPS cooperation are presented.[205] The decision-making procedure is not made explicit in the Strategy; the parties only agree that 'the date and venue of the next meeting will be agreed upon at the preceding meeting'[206].

The AEPS document does not have the appearance or format of a regular international environmental treaty. The names of the two instruments comprising the document, 'Declaration' and 'Strategy', are not common as names of international treaties. The Declaration is signed only by the representatives of the eight Arctic states, and the AEPS lacks the other standard formalities of international treaties, such as provisions regulating the instrument itself. In addition, the AEPS has been drafted in a specific manner in that it is structured in terms of sections and paragraphs, not provisions. The normative guidance of the AEPS document is expressed in most places through language which is of a general and/or vague nature.[207] On the other hand, there are also some parts of the AEPS that convey the prescribed pattern of behavior quite clearly, especially in regard to norms requiring the Arctic states to continue cooperation.[208] In the main, the AEPS document provides general discussion before proceeding to normative guidance. However, in the preface of the Strategy it is provided that:

---

existing environmental regimes relevant to the Arctic; and v) assess and report on progress on actions agreed upon'.

204 Ibid. Section 10 (4) reads: 'In order to facilitate the participation of Arctic indigenous peoples the following organizations will be invited as observers: the Inuit Circumpolar Conference, the Nordic Saami Council and the U.S.S.R. Association of Small Peoples of the North.'.

205 Ibid. Section 10 (3) reads: 'The decision to invite observers should be based on a pragmatic and functional evaluation of their involvement in and contribution to Arctic environmental questions.'

206 This was agreed upon in the Declaration: 'We agree to meet in 1993 and accept the kind invitation of the Government of Denmark and the Home Rule Government of Greenland to hold the next meeting in Greenland'.

207 See especially Sections 1 and 2, in which the norms are expressed mainly in very general language, or Sections 5 (5) and 5 (6), in which all the requirements are expressed using the modal 'should'.

208 See, e.g., Section 5 (2v) which reads: 'The Arctic Countries agree to undertake joint actions in relevant international fora to further strengthen recognition of the particularly sensitive character of

> The implementation of the Strategy will be carried out through national legislation and in accordance with international law, including customary international law as reflected in the 1982 United Nations Convention on the Law of the Sea.[209]

The AEPS document has been analyzed by scholars in various ways, no doubt in part because it does not look like a regular international environmental treaty.

Scholars frequently refer to Arctic instruments as examples of soft law instruments without giving a basis for this characterization.[210] Some have considered the issue more rigorously, however. For instance, Samson starts from the premise that the AEPS is 'nothing more than an agreement to improve cooperation for environmental monitoring'[211]. Although he concedes that it is an agreement, he states:

> Perhaps the weakest point of the AEPS lies in the fact that it is not legally binding, and is not even part of a larger international Arctic regime. All attempts to introduce regulatory elements failed in both Rovaniemi and Nuuk due to a lack of political will. Some NGOs call for the AEPS to be upgraded to the status of a legally binding treaty.[212]

Samson asserts that the AEPS is not an international treaty because the political will to produce one was lacking. He sums up his opinion by stating that 'the AEPS is "soft law", and it specifically mentions the importance of other such instruments, namely, the Brundtland Report (1987)'[213]. He defines the concept of soft law as 'texts which

---

ice-covered parts of the Arctic Ocean'. Section 10 (1): 'the eight Arctic countries agree to hold regular meetings on the Arctic environment'.

209 Preface to the Strategy, paragraph 8.

210 See, e.g., Heininen et al. (1995, p. 84): 'The Declaration of 1991 and the AEPS are only recommendations, but not binding international agreements'. Tennberg (1998, p. 70) argues: 'The Arctic regime is "soft"; it is based on producing knowledge about the state of the Arctic environment and guidelines that are not legally binding or ratified'.

211 Samson 1997, p. 69.

212 Ibid., p. 69.

213 Ibid., p. 77.

are adopted by states without being legally binding'[214]. Thus, in his view the Arctic states have opted to not create an international treaty but a soft law instrument, which is not legally binding. Samson's conclusion is probably influenced by the views of an American officer involved in Arctic Affairs whom he interviewed.[215]

Caron argues simply: 'though the strategy is "soft" in that its nature as a plan of action is aspirational, it represents the most far-reaching effort to date by the Arctic countries to protect the Arctic environment'[216]. It seems that Caron deduces the softness of the AEPS from the general and indeterminate nature of the norm-texts of the document. Yet, he does not define what he means by the term 'soft'. The same applies to Timtchenko, who opines: 'The documents of the First Ministerial Meeting in 1991 as well as of the Second one in Nuuk, Greenland, September 16, 1993 did not establish an effective regime of environmental protection in the Arctic. They may be regarded as examples of 'soft law''[217].

Franckx asserts that 'strictly legally speaking, a non-binding declaration was signed on 14 June in which the parties agreed to cooperate in monitoring pollution of the Arctic environment and to protect the region's fauna and flora'[218]. Franckx thus does not view the AEPS document as an example of soft law, but adheres to the traditional binary mode of analysis and deems it non-binding altogether. Bloom provides a third view on the legal status of the AEPS document: 'The AEPS, along with a Declaration on the Protection of the Arctic Environment, was approved by Arctic ministers at Rovaniemi, Finland, in June 1991 as a political - but not a legal - commitment to

---

214 Ibid., p. 76.

215 Ibid. Samson (p. 77) states: 'According to Ray Arnaudo, Chief of Polar Affairs Office at the US Department of State, the fact that the AEPS consists of "soft law" has so far not hampered progress, and may be a reason for it success. Had a convention been attempted in the first instance, the ratification procedure would probably still be going on'.

216 Caron 1993, p. 379.

217 Timtchenko 1996, p. 273. In a similar vein, Kolodkin et al. (1993, p. 287) state: 'It is apparent that the documents adopted in Rovaniemi are recommendatory by their nature, reflecting the typical "soft law" approach'.

establish a more comprehensive structure for cooperation'[219]; however, he does not define what he means by political commitment. In sum, whereas Caron and Timtchenko consider the AEPS document 'soft', Franckx judges it as non-binding, and Bloom sees it as politically binding.

A more thorough analysis is that advanced by Rothwell, who proceeds from the definition of international treaty provided in the Vienna Convention.[220] He points out that although the 'AEPS is in written form and concluded between states, there is doubt as to whether it was intended to be governed by international law'[221]. First of all, he studies whether the Strategy contains any of the customary formalities of international treaties and notes that it neither contains any ratification procedure nor is intended to be registered with the United Nations. On the other hand, he observes that 'commentators on the law of treaties emphasise that a treaty may take many forms and be concluded under a great many differing circumstances'[222], and adds that 'it is also essential that the parties to the instrument intended to create international legal relations'[223], supporting this argument with Rosenne's argument on how treaties are created.[224] Rothwell also draws attention to the provision in the preface of the Strategy providing that the implementation of the Strategy is to be carried out through national law and in accordance with international law. This evidently does not convince him since he concludes that 'the language which is used throughout is not such that legal rights and obligations are created or imposed upon the states

---

[218] Franckx 1993, p. 247.
[219] Bloom 1999, pp. 712-713.
[220] Rothwell 1996, pp. 238-242. See also Koivurova 1999, pp. 143-161.
[221] Rothwell 1996, p. 239.
[222] Ibid., p. 240.
[223] Ibid.
[224] Rosenne (1989, p. 87) asserts: 'if the participants in that negotiation, for whatever reason, do not intend the arrangement which comes out of the negotiation, however formal that arrangement might look, to be an agreement covered by international law, it will not be a 'treaty' for the purposes of international law, whatever else it might be, and however much of a 'commitment' it might express'.

parties'[225]. On the other hand, he acknowledges that 'there are numerous instances where the state parties do make commitments to pursue various courses of action to protect the Arctic environment, and there is contemplation of joint action being taken to deal with the identified Arctic environmental problems'[226].

The conclusion Rothwell arrives at is that the Strategy can most appropriately be classified as an example of soft law:

> even if the AEPS is properly classified as an example of soft law, this does not mean an absence of legal rights and obligations for the parties to it, but rather different consequences for the enforcement of those soft law rights and obligations.[227]

It can be inferred that for Rothwell 'softness' means no more than the indeterminate/general nature of the provisions, not the legal status of the instrument itself; he considers that legal rights and obligations are created but, because of the indeterminate/general nature of the provisions, views the enforcement of these provisions as differing from the enforcement of more clearly delineated ones. This approach comes very close to that of this thesis. In fact, the difference seems to be only of a terminological nature. The instruments Rothwell calls 'soft law instruments' will be discussed in detail below and referred to as 'informal arrangements'.[228] Indeed, most commentators seem to treat these instruments as binding in some manner, but the terms used vary.

As laid down in the Strategy, the Arctic states were free to decide the date and place of their next meeting. It was agreed that the first ministerial-level follow-up meeting would be held in 1993 in Nuuk, Greenland, where a Declaration on Environment and

---

225 Rothwell 1996, p. 239.
226 Ibid.
227 Ibid., p. 240.
228 See *infra* chapter III (4).

Development[229] was signed by the representatives of the eight Arctic states. The 1993 Nuuk Declaration was complemented by a Report, which recorded in a more detailed fashion the agreements concluded by the ministers of the Arctic states; these two instruments form a whole similar to that formed by the 1991 Declaration and the Strategy comprising the AEPS document.

The institutional structure which was laid down in the AEPS document was extended somewhat at the Nuuk meeting. As was mentioned above, only two working groups for cooperation were established in the AEPS document, the AMAP and the CAFF. However, the AEPS laid a foundation for EPPR and the working group for the protection of the Arctic marine environment (PAME). The Nuuk Report states: 'Ministers noted with satisfaction that the eight Arctic Countries have embarked upon cooperative action on emergency prevention, preparedness and response'[230] and thus acknowledges the EPPR working group. The meeting in Nuuk also established the PAME working group and a new task force on sustainable development.[231] Other institutional developments at Nuuk included the meetings of the Senior Arctic Affairs

---

[229] The 1993 Nuuk Declaration on Environment and Development in the Arctic. The name of the Declaration clearly derives from the Declaration on Environment and Development adopted at the Rio Conference on Environment and Development organized a year earlier. The Nuuk Declaration acknowledges the Rio Conference as follows: 'Recognizing the importance of applying the results of the United Nations Conference on Environment and Development to the Arctic region'. The Declaration further affirms Principles 2 and 22 of the Rio Declaration.

[230] See p. 12 of the Nuuk Report.

[231] Paragraph 2 (3 and 4) of the Nuuk Declaration. The working groups mainly consisted of representatives of the respective governmental ministries of the Arctic states as well as representatives of the indigenous peoples, who were given financial support by the Danish government at the Nuuk Meeting to establish a secretariat for the purpose of participating in the working groups. Annex to the Report of the Nuuk meeting reads: 'Statement by Mr. Svend Auken, Minister for the Environment of Denmark: The Danish Government and the Greenland Home Rule Government have noticed the recommendations of the indigenous peoples' organizations regarding a special program area within the AEPS to address all issues related to the participation of indigenous peoples...Government of Denmark, in cooperation with the Greenland Home Rule Government, is pleased to announce that we can support this recommendation not only verbally, but also by offering to establish a small Secretariat for this purpose in Denmark'. Later in the Annex it is added that this 'statement is reproduced for information purposes and is not part of the report'.

Officials (SAAO). This form of contact was developed informally since it was not explicitly provided for in the AEPS document.[232]

The work of the five working groups on the Arctic environmental protection, and meetings of the SAAOs were given guidance by the Report. In the document, the ministers both provide guidance as to the kind of environmental protection work the groups were expected to carry out and laid down the institutional rules for these working groups, such as the procedure for deciding who is to act as the host of a certain working group. At this phase of AEPS cooperation, the Report mainly consisted of guidelines for the working groups as to what they should concentrate on. More principled issues of environmental protection were also taken up in the Declaration.[233]

The Nuuk Declaration and Report are very similar in form to the AEPS document. The Declaration is signed by the ministers of the Arctic states, which adopted the Report through the Declaration.[234] The Nuuk Declaration and Report both contain normative guidance expressed in a general and/or indeterminate manner[235] but also

---

[232] The SAAO activity is described in the Nuuk Report: 'The Ministers also requested Senior Arctic Affairs Officials to hold regular consultative meetings, at least once a year, to monitor and review progress in the implementation of the AEPS' (See p. 16 of the Nuuk Report).

[233] The transparency of environmental information and the participation of both the public and indigenous peoples in environmental decision-making were supported (Paragraphs 6 and 7 of the Nuuk Declaration). A precautionary approach to development with environmental implications was promoted 'including prior assessment and systematic observation of the impacts of such development' as was the adoption by the Arctic states of internationally transparent environmental impact assessment procedures, thus supporting implementation of the provisions of the Espoo Convention (paragraph 8). Also, the importance of 'prior and timely notification and consultation regarding activities that may have significant adverse transboundary environmental effects' was underlined (paragraph 9). Finally, the states agreed to 'undertake to consider the development of regional instruments concerned with the protection of the Arctic environment' (paragraph 11).

[234] Paragraph 2 of the Nuuk Declaration.

[235] For instance, paragraph 5 of the Declaration reads: 'We support the achievements of the United Nations Conference on Environment and Development, and state our beliefs that the Principles of the Rio Declaration on Environment and Development have particular relevance with respect to sustainable development in the Arctic'. Paragraph 9 reads: 'We underline the importance of prior and timely notification and consultation regarding activities that may have significant adverse transboundary environmental effects, including preparedness for natural disasters and other

use language that conveys the prescribed direction of behavior very clearly. In fact, it seems that the language used in the Declaration and Report of the Nuuk is considerably clearer with respect to normative content than that in the AEPS document.[236] Only two scholars, to the knowledge of the present author, have taken a position on the legal status of the agreements recorded in these documents and both are of the opinion that the Nuuk Declaration is an example of soft law.[237]

The second follow-up meeting of AEPS cooperation took place in Inuvik, Canada, in March 1996, where the Inuvik Declaration was adopted.[238] The Declaration signed there was adopted at the time when the negotiations on the establishment of the Arctic Council were underway.[239] In the Declaration, the Arctic states also adopted a Report, and the Declaration and the Report together can be considered a single normative document.[240]

The Inuvik Declaration and Report concentrated on setting priorities for the work of AEPS cooperation organs. The task of SAAOs was detailed by providing that 'the priorities are directing the AEPS process and providing integration, policy and management direction to the AEPS Programmes and the AEPS Secretariat'[241]. It is interesting that already at this phase the SAAOs are given instructions to prepare for

---

emergencies that are likely to produce sudden harmful effects on the Arctic environment or its peoples'.

236 See, e.g., paragraph 2 of the Declaration, which reads: 'establishing a working group to assess the need for further action or instruments to prevent pollution of the Arctic marine environment'. Paragraph 3 states: 'We will in particular cooperate to strengthen the knowledge base and to develop information and monitoring systems for the Arctic region'. Paragraph 11 provides: 'We undertake to consider the development of regional instruments concerned with the protection of the Arctic environment'.

237 Samson 1997, p. 69 and Timtchenko 1996, p. 273.

238 The 1996 Inuvik Declaration on Environmental Protection and Sustainable Development in the Arctic.

239 For example, paragraph 15 of the Declaration states: 'We are fully committed to the earliest possible establishment of the Arctic Council'.

240 Ibid., paragraph 2.

241 Ibid., paragraph 6.

the establishment of the Arctic Council.[242] Priorities were also assigned to the AMAP working group[243], Conservation of Arctic Flora and Fauna (CAFF)[244], PAME[245] and EPPR[246]. The Declaration provided that the already established task force on sustainable development was to become a working group.[247] By the time of the Inuvik Declaration, most of the work in the sphere of environmental protection had been transferred to the working groups. Yet, these working groups had drawn up few environmental protection instruments,[248] and the number of clear commitments which

---

242 Ibid. The last sentence reads: 'The SAAOs, with the assistance of the Permanent Participants, will also undertake to develop revised Terms of Reference for SDU and an initial workplan for the Arctic Council's sustainable development work, to be presented for discussion to the Arctic Council Senior Arctic Officials'. Hence, the plans for the Arctic Council were clearly well advanced since the terms 'permanent participant' and 'Senior Arctic Officials' are used in the documents establishing the Council.

243 Ibid. The most important priorities for the AMAP were: 'publishing in early 1997, the State of the Arctic Environment Report...preparing two short papers on i) the feasibility of developing sub-regional cooperative oil-related monitoring and assessment activities, and ii) the state of knowledge concerning organotins in the marine environment'.

244 Ibid. The priorities for the CAFF were: 'continuing the development of the Circumpolar Protected Area Network (CPAN); assisting countries with the implementation of the Circumpolar Murre Conservation Strategy and Action Plan; and developing a draft Arctic strategy relating to the goals of the Convention on Biological Diversity'.

245 Ibid. The most important priorities for PAME were: 'to address the development of an Arctic Regional Programme of Action on marine pollution resulting from land-based activities; to develop, in cooperation with other AEPS programmes, guidelines for offshore petroleum activities; to collect information on current and future shipping activities and their associated environmental effects; and to maintain an overview of the effectiveness of existing international arrangements relevant to the objectives of the PAME programme area'.

246 Ibid. The priorities for EPPR were: 'completing the Arctic Guide for Emergency Prevention, Preparedness and Response; contributing to the development of preventative, mitigating and response measures for oil and gas accidental releases in the Arctic; continuing to involve indigenous peoples in accident prevention and response; preparing an analysis of the effectiveness of existing accident reporting systems; refining the Risk Analysis on Environmental Threats to the Arctic, including an assessment of the need for future action; and analysing the adequacy and effectiveness of the existing international agreements and other arrangements in the Arctic within EPPR's area of expertise'.

247 Ibid. In Paragraph 6 it is provided: 'we agree to establish a Working Group on Sustainable Development and Utilization (SDU)'. The priorities for the SDU were: 'to continue to cooperate with the current terms of reference of the Task Force on SDU and with specific direction from the SAAOs, pending the expeditious creation of the Arctic Council'.

248 One example is the Circumpolar Murre Conservation Strategy and Action Plan prepared by the CAFF.

the Arctic states assumed in the Declaration was rather limited. Some commitments were expressed in indeterminate and/or general language,[249] some quite clearly.[250]

1.2. The Second Phase of the Arctic Cooperation Process: the Arctic Council

The Arctic Council was established in an informal manner in September 1996 in Ottawa, Canada. The Arctic states signed an instrument called the Declaration Establishing the Arctic Council and issued a joint communiqué to explain the newly created body.[251] The establishment of the Arctic Council amended the forms of Arctic cooperation previously based on the AEPS document. The AEPS document and the Declaration establishing the Arctic Council (hereinafter 'the constitutive instrument') partly overlap in scope and neither provides rules specifying which is to prevail should provisions of the two conflict. This conflict was avoided by making AEPS cooperation the second main field of action of the Arctic Council.

The constitutive instrument substantially extends the terms of reference of Arctic cooperation. As was shown above, Arctic cooperation based on the AEPS document concentrated only on protecting the Arctic environment. Now, the terms of reference

[249] See, e.g., paragraph 12 of the Inuvik Declaration, which states: 'We endorse the EPPR recommendation that where significant accidents occur which directly or indirectly cause, or threaten to cause, environmental or health effects in the Arctic, all Arctic Countries shall, through their National Contact Points, be notified promptly and receive further information as appropriate'; or paragraph 11, which states: 'We support the continuing negotiations and collaboration in relevant international fora to integrate the AEPS with local, regional, circumpolar and global environmental protection activities'.

[250] Ibid. See, e.g., paragraph 13, which states: 'We agree to cooperate with the Russian Federation in searching for investments required for implementing programmes and projects aimed at the conservation and restoration of the traditional habitats of the indigenous peoples of the North of the Russian Federation'; and paragraph 14, which adds: 'We agree to ensure implementation of the priorities as listed in the present Declaration and to make every effort to provide and maintain the necessary resources to enable each country and indigenous peoples to participate fully in the activities of the AEPS'.

[251] The 1996 Declaration on the establishment of the Arctic Council. On the history of the negotiation process, see Rothwell 1995 II, pp. 302-304 and Scrivener 1999, pp. 51-58. The Joint Communiqué can be found on the home page of the Arctic Council at http://www.arctic-council.org/f_documents.asp [18.2.2001].

of the Arctic Council were defined as 'common Arctic issues, in particular issues of sustainable development and environmental protection in the Arctic'[252]. Hence, in principle, all issues which are defined as 'common' can be dealt with by the Arctic Council. This yields a very large mandate for the Council since the 'common issues' can include almost any international policy issue; however, in a footnote the instrument provides that 'the Arctic Council should not deal with matters related to military security'[253]. Environmental cooperation is now included as a principal focus within the mandate of the Council,[254] with the four working groups of AEPS cooperation continuing under the umbrella of the Arctic Council.[255] The second 'pillar' of the mandate is cooperation on sustainable development.[256]

The constitutive instrument amends and greatly elaborates the rules on participation. It provides for three categories of participants: members, permanent participants and observers. The eight Arctic states are members; the three organizations which represent the indigenous peoples of the Arctic are permanent participants[257]; and the criteria for observers are laid down.[258] In addition, the instrument specifies the criteria for the status of permanent participants and the decision-making procedure for determining that status.[259] It is also important that the indigenous peoples' secretariat

---

[252] Article 1 (a) of the constitutive instrument.
[253] Ibid., footnote at p. 3.
[254] Ibid., Article 1 (b).
[255] Ibid. Article 1 (b) reads: 'The Arctic Council is established as a high level forum to...b. oversee and coordinate the programs established under the AEPS on the Arctic Monitoring and Assessment Program (AMAP); Conservation of Arctic Flora and Fauna (CAFF); Protection of the Arctic Marine Environment (PAME); and Emergency Prevention, Preparedness and Response (EPPR)'.
[256] Ibid. Article 1 (c) reads: 'The Arctic Council is established as a high level forum to...c. adopt terms of reference for, and oversee and coordinate a sustainable development program'.
[257] Ibid. Article 2 of the constitutive instrument enumerates the following as permanent participants: 'The Inuit Circumpolar Conference, the Saami Council and the Association of Indigenous Minorities of the North, Siberia and the Far East of the Russian Federation'.
[258] Ibid. Article 3 of the constitutive instrument reads: 'Observer status in the Arctic Council is open to: a) non-Arctic states; b) inter-governmental and inter-parliamentary organizations, global and regional; and c) non-governmental organizations that the Council determines can contribute to its work'.
[259] Ibid. Article 2 (2) reads: 'Permanent participation is equally open to other Arctic organizations of indigenous peoples with majority Arctic indigenous constituency, representing: a. a single

established under the AEPS is to continue under the framework of the Arctic Council.[260]

The decision-making procedure which had developed in AEPS cooperation is made explicit in the constitutive instrument. In Article 7 it is provided that: 'Decisions of the Arctic Council are to be by consensus of the Members'. In Article 2, 'member' is defined as including only the eight Arctic states. This decision-making by consensus is to be undertaken only after 'full consultation'[261] with the permanent participants, i.e., the organizations of the Arctic indigenous peoples. Although these permanent participants do not have formal decision-making power, they are clearly in a position to exert much influence in practice on the decision-making of the Council.

As the Arctic Council has only recently been established, only few comments on its legal status have been made.[262] The Conference of the Arctic Parliamentarians stated in its 1998 Salekhard Declaration:

> the Arctic Council is a mutual declaration by governments rather than an international treaty, that it can provide opportunities for co-operation and collaboration with other bodies to their mutual benefit, and that it thus can help Arctic governments address common needs.[263]

Heininen is of the view that the Arctic Council can only be treated as a *de facto* regime, not a full-blown international treaty such as that regulating environmental protection in the Antarctic.[264] Young notes: 'The council has no organizational

indigenous people resident in more than one Arctic State; or b. more than one Arctic indigenous people resident in a single Arctic state'. Decisions by the Arctic states on whether this criterion is fulfilled must be unanimous. Article 2 also states: 'the number of Permanent Participants should at any time be less than the number of members'.

260 Ibid., Article 8.

261 Ibid., Article 2.

262 In addition to those studied *infra*, see Koivurova 1999, pp. 143-160.

263 Paragraph 25 of the Salekhard Declaration. The Declaration can be found on the home page of the Standing Committee of the Arctic Parliamentarians at http://www.grida.no/parl/ [8.2.2001].

264 Heininen 1999, p. 220.

capacity, no resources of its own, and no authority to make decisions that bind its members'[265].

One commentary that has been particularly influential is an article written by an officer involved in the Arctic cooperation on the part of the United States, Evan Bloom, who works as a Legal Adviser to the Bureau of Oceans and International Environmental and Scientific Affairs. When the United States acted as chair of the Arctic Council and was also responsible for maintaining its home page,[266] Bloom's article was made available on that site.[267] Bloom feels that the United States had difficulty in establishing a new inter-governmental organization[268] and sees this as part of the larger framework of the external policy of the United States:

> The establishment of the Arctic Council as a forum without legal personality, and thus not as an "international organization" as that term is understood under international law, was an objective of the United States and is consistent with a tendency in recent American diplomatic practice to seek an informal cooperative structure when that structure is adequate for the purposes of the issues involved.[269]

Bloom thus sees the Council as not being an inter-governmental organization, with the consequence that the Council has no legal personality in international law. Bloom also gives a straightforward answer to the question whether the Council's decisions are binding on member states:

---

265 Young 2000 (chapter 4, paragraph 1). For similar arguments, see Archer and Scrivener 2000, p. 614.

266 See Article 32 of the Rules of Procedure.

267 Bloom's article was also published in the American Journal of International Law 1999, pp. 712-722.

268 Ibid., p. 714.

269 Ibid., p. 721.

> The Rules of Procedure, which apply to all bodies of the Council, specify how meetings are run and how decisions are taken. They are relatively lengthy and detailed for an informal body which has no legal authority to bind its members.[270]

Although Bloom provides no explicit reasons to support his conclusions, he appears to assume that states can decide not only that the decisions of inter-governmental organizations are legally non-binding but that the entire organization is operating outside of international law. Bloom refers to the Organization on Security and Cooperation in Europe (OSCE) as an example of such an organization. He also seems to accept that states can choose to allow their agreements to be regulated by different normative orders.[271]

The integration of AEPS cooperation under the Arctic Council did not take place immediately. The final AEPS ministerial meeting took place in Alta, Norway, in June 1997, where the Alta Declaration was adopted.[272] The format which had been used in the three earlier ministerial conferences of the AEPS cooperation was changed in this fourth ministerial meeting. In the earlier meetings, the signed Declaration also adopted a Report, which itself contained explanations of what the ministers had agreed upon. In Alta, the signed Declaration also 'adopt[ed] the SAAOs recommendations'[273]. The SAAO Report now contained recommendations for the

---

270 Ibid., p. 718.

271 Bloom (1999, p. 721) argues: 'Of course, there is nothing to prevent states from making a political, as opposed to a legal, commitment to provide resources; the United States has done so on a significant scale with respect to the Arctic Council. However, if a state loses support within its government for a particular effort, or if funds are not made available by legislative authorities, funding is easily terminated without legal consequence'.

272 The 1997 Alta Declaration on the Arctic Environmental Protection Strategy. The preface of the SAAO Report to the Ministers states: 'This is the final SAAO Report to Ministers before the integration of AEPS activities into the newly established Arctic Council, whose first meeting will take place in the fall of 1998' (Paragraph 4 of the Preface to the Alta Report of the SAAO's).

273 Ibid., paragraph 9.

decision-making by the ministers; it did not explain what the ministers had agreed upon in the earlier meetings.

The Alta meeting differed from the earlier AEPS meetings in that the environmental protection working groups produced instruments for the representatives of the Arctic states to adopt. A good example is paragraph 3 of the Declaration:

> We receive with appreciation the "Guidelines for Environmental Impact Assessment (EIA) in the Arctic" and the "Arctic Offshore Oil and Gas Guidelines" developed under the AEPS, and **agree** that these guidelines be applied. (original emphasis)

There are also other examples of normative instruments adopted by the working groups of the AEPS cooperation and supported by the representatives of the Arctic states.[274] The Declaration recommends that the Arctic Council cooperate with those international organizations 'which are pursuing complementary objectives, such as the Barents Council'[275]. The vigorous participation of the Arctic states in many international environmental protection efforts is encouraged.[276]

---

274 Ibid. See, e.g., paragraph 4 which reads: '**We receive** with appreciation the "Arctic Guide for Emergency Prevention, Preparedness and Response" and **encourage** the Arctic countries to make use of the Guide as appropriate'; or paragraph 9 (2): 'We also endorse the further development and implementation of the "Circumpolar Protected Areas Network (CPAN) Strategy and Action Plan", the implementation of the "International Murre Conservation Strategy and Action Plan", and the "Circumpolar Eider Conservation Strategy and Action Plan"'. (original emphasis).

275 Ibid., paragraph 14.

276 Ibid. See, e.g., paragraph 15, which reads: '**We agree** that the Arctic countries should take joint action to promote the early completion of the on-going work on an "International Code of Safety for Ships Operating in Polar Waters" (Polar Code) under the auspices of the International Maritime Organisation (IMO)'; or in paragraph 16: '**We call** for full implementation of the International Atomic Energy Agency (IAEA) Convention on Nuclear Safety; the rapid finalisation of the IAEA Joint Convention on the Safety of Spent Fuel Management and on the Safety of Radioactive Waste Management, and the application of the current IMO irradiated Nuclear Fuel Code. In addition, **we recognize** the importance of the on-going co-operation to provide for the early completion of the facilities needed for the implementation of the ban on dumping of radioactive waste at sea adopted by the London Convention of 1972'. (Original emphasis). See also paragraph 17.

The first meeting of the Arctic Council took place in Iqaluit, Canada, in September 1998. The Iqaluit Conference adopted the format which had been used in Alta whereby the Senior Arctic Officials (SAO)[277] provide a Report containing recommendations for the ministers of the Arctic states, which the ministers then approve and issue in the form of a signed Declaration.[278] At the first meeting of the Arctic Council, the practical aspects of the functioning of the Council figured prominently. For example, the Arctic Council Rules of Procedure and Arctic Council Terms of Reference for a Sustainable Development Program were adopted since these are required by the constitutive instrument of the Arctic Council.[279] The Rules of Procedure provide clear rules on how the Arctic Council and its cooperative bodies are to work together in practice.

Ministerial meetings are held biennially and meetings of SAOs are convened between these meetings as provided by the constitutive instrument.[280] The details of how to organize these meetings are provided in the Rules of Procedure.[281] The Rules of

[277] The name "Senior Arctic Affairs Officials (SAAO)" was changed in the constitutive instrument of the Arctic Council to "Senior Arctic Officials (SAO)". According to the rules of procedure adopted by the Iqaluit Declaration: 'Each Arctic State shall designate a SAO, and each Permanent Participant shall designate a representative, to act as focal point for Arctic Council activities, and shall inform the other Arctic States and Permanent Participants of the designation through the Host Country. 22. The Host Country shall provide the chairperson for the SAO meetings, subject to the concurrence of the Arctic States represented at the SAO meeting'.

[278] The 1998 Iqaluit Declaration.

[279] Article 1 c. of the constitutive instrument, which reads: 'The Arctic Council is established as a high level forum to...adopt terms of reference for, and oversee and coordinate a sustainable development program'. Article 6 of the constitutive instrument requires: 'The Arctic Council, as its first order of business, should adopt rules of procedure for its meetings and those of its working groups'.

[280] Article 4 of the constitutive instrument states: 'The Council should normally meet on a biennial basis, with meetings of senior arctic officials taking place more frequently, to provide for liaison and co-ordination. Each Arctic State should designate a focal point on matters related to the Arctic Council'.

[281] Part II of the Rules of Procedure, 'Arctic Council Meetings', paragraphs 3-25. Following are the most important provisions. Especially important is the designation of the host country, which is in charge of organizing the meetings (paragraphs 16 and 32). Rules of procedure also specify the procedures relating to regular contacts between the ministerial meetings. The SAOs are required to meet at least twice yearly 'at the call of the Host Country' (paragraph 25). Paragraph 3 reads: 'Six of the Arctic States shall constitute a quorum for purposes of holding a Ministerial or Senior Arctic

Procedure also provide rules for the functioning of other cooperative activities, for instance, the establishment of working groups, task forces and other bodies by the Arctic Council.[282] The Terms of Reference for the Sustainable Development Program do not really lay down any substance for the program[283] but do set out a procedure by which the proposals for cooperative activities are processed.[284] The Iqaluit Declaration granted observer status to several entities.[285] The Conference also accepted the Aleut International Association as a permanent participant in addition to those already accepted in the constitutive instrument.

Official (SAO) meeting'. Paragraph 5: 'In accordance with the Declaration, the category of Permanent Participation is created to provide for active participation and full consultation with the Arctic Indigenous Representatives within the Arctic Council. This principle applies to all meetings and activities of the Arctic Council'. Paragraph 11: 'The Host Country, an Arctic State, or other subsidiary bodies may undertake communications on Arctic Council matters with other international fora as may be agreed to in advance by the Arctic States'. Paragraph 12: 'During the discussion of any matter, a representative of an Arctic State or Permanent Participant may rise to a point of order and the point of order shall be decided immediately by the chairperson in accordance with these Rules'. Paragraph 15: 'Ministerial meetings shall be held biennially, or at such other times and locations as may be agreed upon by the Arctic States in consultation with the Permanent Participants. Paragraph 23: 'The SAOs shall receive and discuss reports from working groups, task forces and other subsidiary bodies and shall coordinate, guide and monitor Arctic Council activities in accordance with the decisions and instructions of the Arctic Council'.

282 Ibid. Part IV, Implementation of Cooperative Activities. The other sections of Part IV are Secretariat Support Functions, Indigenous Peoples' Secretariat. Part V, Other Matters, contains the following issues: Additional Permanent Participants, Observers, Experts, Languages, Public Communications and Documents of Meetings, Communications with the Arctic Council and Amendment.

283 Paragraph 1 of the Terms of Reference only provides: 'The goal of the sustainable development program of the Arctic Council is to propose and adopt steps to be taken by the Arctic States to advance sustainable development in the Arctic, including opportunities to protect and enhance the environment, and the economies, cultures and health of indigenous communities and of other inhabitants of the Arctic, as well as to improve the environmental, economic and social conditions of Arctic communities as a whole'.

284 Ibid., paragraphs 2 and 3.

285 The following entities were accepted in paragraph 3 of the Iqaluit Declaration: The Federal Republic of Germany, the Kingdom of the Netherlands, Poland, the United Kingdom of Great Britain and Northern Ireland, the Nordic Council, the Northern Forum, the United Nations Economic Commission for Europe (UN-ECE), the United Nations Environment Programme (UNEP), the International Arctic Science Committee (IASC), the Standing Committee of Parliamentarians of the Arctic Region (SCPAR), the World Wide Fund for Nature (WWF) and the International Union for Circumpolar Health (IUCH).

The Iqaluit Declaration adds little in the field of environmental protection. Certain instruments concluded by the environmental protection working groups of the AEPS cooperation are welcomed,[286] and guidance is given to these working groups.[287] In addition, the Arctic states are encouraged to take joint action in international environmental protection fora.[288] Most of the provisions in the Declaration are indeterminate and general although some specific norms are also provided.[289]

In October 2000, the second ministerial meeting of the Arctic Council took place in Barrow, Alaska, where a Declaration was signed[290] and chairmanship of the Council was transferred from the United States to Finland.[291] By this time, the Barrow Declaration could note with pleasure that the 'Arctic Council…has successfully taken

---

[286] Ibid. See, e.g., paragraph 23 of the Iqaluit Declaration: '**Welcome** the Field Guide for Oil Spill Response in Arctic Waters, as a source of information on how to deal with oil spills in the Arctic; and **take note** of both the report of Phase I Analysis of Communication and Notification Systems in Place for Arctic Risks and the Revised Environmental Risk Analysis of Arctic Activities'; or paragraph 24: '**Welcome with appreciation** the Regional Programme of Action for the Protection of the Arctic Marine Environment from Land-based Activities (RPA)'. (original emphasis).

[287] Ibid. See, e.g., paragraph 16: '**Recognize** the need to continue to identify actions to address the pollution sources identified in the AMAP Report, and **instruct** SAOs to continue to develop an overall plan of action complementary to existing legal arrangements and the Regional Programme of Action for the Protection of the Arctic Marine Environment from Land-based Activities'; or paragraph 22: **Endorse** the Emergency, Prevention, Preparedness and Response (EPPR) Work Plan set forth in EPPR's Strategic Plan of Action; **ask** the working group to initiate new projects and activities as indicated in the chapters'. (original emphasis).

[288] Ibid. For instance, paragraph 19: '**Strongly welcome** the establishment of an Intergovernmental negotiating Committee under the auspices of United Nations Environment Programme to work towards the conclusion of a global agreement on POPs by the year 2000, and **encourage** the Arctic States to act together to assist the early conclusion of such a global agreement'. (Original emphasis).

[289] Ibid. For instance, Paragraph 8: 'We **request** that Canada and the Arctic Monitoring and Assessment Programme cooperate in reviewing knowledge on the impacts of environmental contamination on the health and development of children and youth, under the direction of SAOs'; or, paragraph 18: '**agree** to work vigorously for the early ratification and implementation of the Protocols on the elimination or reduction of discharges, emissions and losses of Persistent Organic Pollutants (POPs) and of Heavy Metals under the framework of the United Nations Economic Commission for Europe Convention on Long-Range Transboundary Air Pollution'. (original emphasis).

[290] The 2000 Barrow Declaration on the Occasion of the Second Ministerial Meeting of the Arctic Council. See also the minutes of the Barrow meeting on the home page of the Arctic Council, and the plan for the Finnish term as Chair of the Council (2000-2002) at http://www.arctic-council.org/whatsnew.asp [3.3.2001].

on the responsibilities of the Arctic Environmental Protection Strategy'[292] although some problems remained as to the coordination of activities.[293] New permanent participants were also accepted.[294] The Barrow Ministerial Declaration welcomed the SAO report and 'adopt[ed] the recommendations contained in that Report'[295].

The Arctic states adopted the Arctic Council Action Plan to Eliminate Pollution in the Arctic (ACAP) and 'determin[ed] that ACAP...will be a basis for developing and implementing actions under the Council's auspices with respect to pollution prevention and remediation'[296]. Importantly, it was decided that an *ad hoc* steering group would be established 'awaiting a review of the structure of the Arctic Council organization'[297]. Another organ established was the Arctic Climate Impact Assessment Programme (ACIA), a joint project of AMAP, CAFF and International Arctic Science Committee (IASC), with a Steering Committee to coordinate its work.[298] The task of the project was to 'evaluate and synthesize knowledge on climate variability and change, increased ultraviolet radiation, and their consequences for Arctic ecosystems and societies'[299]. The Declaration welcomed the opportunity presented for the Arctic states to bring the 'Arctic issues to the attention of the global community through the preparatory processes associated with the ten year review of

---

[291] Ibid., paragraph 25.
[292] Ibid., Preface.
[293] Ibid. Paragraph 19 reads: '**Take note** that as the Arctic Council has evolved and taken over the structures established under AEPS, some overlap of functions has occurred among the new and existing institutional structures of the Council'. (original emphasis).
[294] Ibid. Paragraph 22 reads: '**Approve** the Arctic Athabascan Council and the Gwich'in Council International as Permanent Participants in the Arctic Council and **welcome** those organizations to the Arctic Council'. Emphasis included in the document. (original emphasis).
[295] Ibid., paragraph 5.
[296] Ibid., paragraph 2 (2).
[297] Ibid., paragraph 2(3).
[298] Ibid., paragraph 3.
[299] Ibid.

Agenda 21'[300]. The meeting also endorsed the results of the environmental protection working groups of AMAP[301], CAFF,[302] EPPR[303] and PAME[304].

In conclusion, it can be seen from the foregoing that the Arctic cooperation process has become more and more institutionalized with time. It started with very vague and general cooperation provisions in the AEPS document; the rules on participation, decision-making procedures and the mandate of the process became much more specific with the establishment of the Arctic Council; and the process has since become more detailed in its organizational structure. The Arctic cooperation process seems to be in line with the recent practice of concluding new kinds of cooperative arrangements that do not neatly fit in with the traditional concepts of treaty and inter-governmental organization.[305]

From the perspective of law, the main problem is the legal status of the cooperation process in general and of the instruments adopted in the process in particular; indeed, a considerable variety of views have been presented on this issue. The AEPS document seems to be qualified by the majority of scholars as binding in some manner, but it is not clear in what manner. The available comments on the legal status of the Arctic Council do not view it an inter-governmental organization for the purposes of international law.

[300] Ibid., paragraph 7.
[301] Ibid., paragraph 8.
[302] Ibid., paragraph 9.
[303] Ibid., paragraph 10.
[304] Ibid., paragraph 11.
[305] See *infra* chapter VI.

2. The Creation of Contractual Rules in International Law

Before the 1950s there were no major problems in determining what was a legal agreement and what was not.[306] The vast majority of scholars of international law agreed that verbal agreements, treaties in simplified form, and, of course, formal treaties were agreements binding in international law. In other words, there was no requirement of form for an agreement and there were few exceptions to the basic principles; a good example is a gentlemen's agreement, which did not bind states legally but was seen as binding statesmen morally.[307] This concept of agreement was firmly based on the principle of *pacta sunt servanda*, which is the pivotal principle of all legal systems in general and international law in particular: if an agreement between states as to future behavior is reached, the agreement is to 'be performed by them in good faith'[308] whatever the form in which it is recorded. If such *pacta* are not reached, no obligation or obligations are created. In this conception, there is no reason to distinguish between agreements of oral or written form in determining whether a legally binding agreement has been reached.

---

[306] See Klabbers 1996 I, pp. 4-6, pp. 66-68 and the notes thereat. Klabbers refers (p. 5) to the Harvard Draft Convention on the Law of Treaties, published in 1935, which 'specifically excluded exchanges of notes and exchanges of letters from its scope, albeit on the understanding that these instruments were nevertheless legally binding'. He adds (p. 5): 'debates in the 1950s and 1960s revolved around the notion of 'treaties in simplified form', leading to the conclusion that there are no fundamental differences between such 'treaties in simplified form' and 'regular' treaties as far as their legally binding force is concerned'. Klabbers also refers to studies of diplomacy at the time in which the idea of having agreements governed in any other way than a legally binding instrument was inconceivable.

[307] According to Klabbers (1996 I, p. 6): 'statesmen could sometimes make declarations of policy, jointly or separately, thereby outlining their plans for the future without those declarations being given normative force. These declarations, then, were simply treated as non-binding'.

[308] Article 26 of the Vienna Convention.

There still exists virtual unanimity among scholars of international law that formal instruments are international treaties and that oral agreements are binding in international law. The problems arise in regard to informal instruments, i.e., instruments that do not contain the regular formalities of treaties. Increasingly, the approach has been to treat informal instruments simply as non-binding in international law but binding in some other manner, e.g., as non-legal agreements[309], soft law instruments[310] and/or morally/politically[311] binding instruments.[312] This standpoint seems to fit in with the basic principles of treaty-making because these scholars operate on the basis of a binary mode of analysis: either an instrument is a treaty or it is not. They accept that states are free to create international treaties binding in international law, but even where states do not, the instruments they adopt still contain normative guidance and must thus be somehow binding on them. This approach faces considerable problems.[313]

---

[309] Schachter (1977, pp. 296-297) states: 'International lawyers generally agree that an international agreement is not legally binding unless the parties intended it to be. Put more formally, a treaty or international agreement is said to require an intention by the parties to create legal rights and obligations or to establish relations governed by international law. If that intention does not exist, an agreement is considered to be without legal effect ("sans portée juridique"). States are, of course, free to enter into such nonbinding agreements, whatever the subject matter of the agreement'. A recent detailed argument is that presented by Ahlström (1999, pp. 59-60), who argues that states are entitled to conclude 'non-legally binding agreements' or 'non-legal agreements'.

[310] For a recent proposal to this effect, see Hildengard 1999, pp. 499-515.

[311] For a recent argument in Finnish, see Tupamäki 1999, pp. 109-120.

[312] There is no reason why a scholar could not argue that all of these different normative orders exist. According to Virally, states can choose the 'pocket' in which they place their agreements. Virally 1990, pp. 135-145.

[313] If a scholar argues that these norms are binding in some other manner than in international law through the *pacta sunt servanda* principle, s/he must also explain how these binding norms relate to the rules and principles of international law, since in practically every instance of binding norms there already exist norms of international law regulating the same issue. International law contains collision rules, such as *lex posteriori* or *lex specialis*, or norms of higher status, such as *jus cogens*, but only to resolve conflicts between different sources of international law, not conflicts between norms originating from different types of normativity. If it cannot be shown that there exist orders independent of international law, international law can only apply its own rules and principles, which do not recognize the existence of other, competing normative orders. It is also necessary to point out that if it can be proven that these normative orders are independent of international law, it should be possible to show the source from which they derive their binding force. An independent general principle of *pacta sunt servanda* would have to be established in moral, political and soft law; for otherwise it would be hard to argue that the norms contained in such instruments are

If it is accepted that states are free to conclude agreements, then the phases preceding an agreement, written or oral, cannot be binding in international law. The purport of the principle of freedom of contracting is precisely to safeguard states from being bound in international law before an agreement is reached. In order for this approach to be convincing it must be shown that states are entitled to enter into their agreements as part of independent normative orders that contain their own rules for how political agreements, soft law agreements, and the like are created; otherwise international law deems that all such norms are non-binding in international law. Do such independent normative orders exist in international society?

Since the soft law thesis[314] has assumed such an influential status in international environmental law, it is taken up here.[315] The problems of having a moral[316] or

---

binding. For a scholarly proposal to this effect, see Hildengard 1999, pp. 499-515. Hildengard argues that there exists a principle of pacta sunt servanda in soft law as well, not only in hard law.

[314] The concept of soft law is problematic because there seems to be no accepted definition of the term. Some scholars conceive of it as including all normative material that does not fall within the sources mentioned in Article 38, paragraph 1 of the ICJ Statute; see, e.g., Francioni 1996, pp. 167-178. Some view it as mainly restricted to intentional norm-making, i.e., the conclusion by international legal subjects of soft law instruments; see e.g., Shelton 2000, pp. 1-18. It is this latter view which is adopted in the present consideration of the concept of agreement.

[315] Klabbers (1996 II, p. 170) provides an appropriate explanation of why this soft law thesis has become so successful among scholars of international law: 'International law has always been the subject of a two-pronged attack: Austinian domestic lawyers think that international law is not law at all, whereas self-proclaimed political realists claim that even if it were law, it is irrelevant anyway. The soft law thesis provides something of an answer to both prongs of the attack: it gives the impression that we actually talk about law, that it actually, just like real law, creates commitments and gives rise to expectations, but that it is also capable of accurately reflecting politics; soft law can bow gracefully to overriding political demands'.

[316] Some scholars have postulated the existence of an independent international moral order and/or international political order. Perhaps the best known proposal is that made by Virally (1990, pp. 135-145). He asserted that states could choose to have their agreements governed by a number of social orders, which could even apply simultaneously. Some problems appear when postulating that a moral order exists also in international society. Even if we can postulate such an order with states as its subjects, an agreement governed by that order is untenable in that moral subjects can hardly create moral obligations intentionally. Indeed, the basic meaning of the concept of morality lies in its unintentional character, which derives from the separate acts of moral subjects and responses to them. All moral subjects can, of course, try to influence public morality as states can try to influence the development of customary law, but they cannot dictate what that morality will be through mutual agreement. Hence, even if states included a separate provision in an instrument

political[317] order to govern international agreements have been demonstrated adequately elsewhere. To the knowledge of the present author, no scholar of international law has taken up the task required for the soft law thesis to be accommodated within the framework of treaties, that is, to show that there exists an independent order to which states can submit their agreements. The first question to be asked is how such a revolutionary change in international law could come about, a change that would mean that not only a new source but also an entirely new level of international law had emerged; international law would now have two separate and independent orders. Any attempts to document this change clearly require hard evidence. First, one must ask if is it the case that when states negotiate informal instruments they are intentionally making soft law. In response, it must be acknowledged that there is no informal instrument on record in which the states have expressly provided that the instrument is a soft law instrument. The soft law status of an instrument must thus be inferred from its informality, or from the real intent of

stipulating that the instrument and the norms in it were morally binding, these norms could not be moral norms for the simple reason that morality cannot be dictated by will. For a detailed analysis, see Klabbers 1996 I, pp. 143-151.

317 There are some who argue that states can have their agreements governed by an international political order and in at least one case states have provided explicitly that an instrument is politically binding. See the Document of the Stockholm Conference on Confidence- and Security-Building Measures and Disarmament in Europe, Article 101 of which provides: 'measures adopted in this document are politically binding'. The document can be found in 26 I.L.M. 191 (1987). If such an independent political order existed, it would mean that we would first need to study whether the states intended that a treaty be created, and if they did not, we could conclude that the instrument was politically binding. The problem here is that the concept of politics is limited to the intentional actions of individuals or other subjects of the community. However, if all agreements are the product of intentional action, i.e., politics, surely they are also always politically binding; states could not choose to have their agreements governed by the international political order, because these would always, irrespective of states' will, be politically binding. But does there even exist an international political order? Everyone agrees that international politics is very important in international relations, but this idea pertains to the intentional behavior of states, not to a social order called 'the international political order' which has its own rules for agreements between international legal subjects. If such an order did exist, it would represent a revolutionary change in international society similar to that implied by the existence of a separate order of international soft law. All international treaties governed by international law are also necessarily politically binding since the creation of international treaties is the result of political interaction between states. States cannot intend that their agreement be governed by an international political order, because none exists. For a detailed analysis, see Klabbers 1996 I, pp. 152-156.

states negotiating the instrument, determined for instance, by interviewing those who participated in drafting the instrument.

Even where evidence could be found that some state negotiators thought that they were concluding a soft law instrument, it is clear that this would not be enough. In order for a new source of international law to come into being, not to mention a new level, there must be solid evidence of states' accepting such a source or level. However, it has been shown that there are no judicial decisions in which the soft law thesis has been accepted.[318] Acceptance would have required that a tribunal give value to a normative instrument to the effect that the states had intended it to be governed by a separate international soft law order.[319] It is not enough to show that an informal instrument has been treated as legally relevant, since, as will be shown below,[320] all kinds of non-binding instruments have differing degrees of legal relevance. On balance, although the soft law concept enjoys great popularity in scholarly circles of international environmental law, it has not become an independent legal order to which states could subject their agreements.[321]

As it seems impossible to argue that states can conclude agreements that are binding in normative orders other than international law, another kind of approach may prove more fruitful: states can simply decide that their agreement is non-binding altogether, i.e., not binding in any normative order. This proposal does not have to rely on any other normative order to accord bindingness to instruments that simply are non-binding. Could it be argued that state practice has changed the pacta sunt servanda rule to accord states the power to make an agreement non-binding altogether?

---

[318] Klabbers 1996 II, pp. 167-182.

[319] As shown by Klabbers, the cases that are normally cited as supporting the soft law thesis do not really support that thesis. Tribunals have given legal value to informal instruments which have been labeled soft law instruments by the scholarly world but only as part of the normal law-making processes of international law, for instance, as elements of the development of customary international law. See Klabbers 1996 II, pp. 167-182.

[320] See infra chapter III (4.).

To the knowledge of the present author, there are only a few instruments to be found in the state practice relating to treaties in which states have agreed on future behavior and provided explicitly that the document is non-binding.[322] Does the accepted definition of an international agreement provide states with this option? The definition of international treaty in Article 2 (1a) of the Vienna Convention on the Law of Treaties reads as follows:

> an international agreement concluded between States in written form and governed by international law, whether embodied in a single instrument or in two or more related instruments and whatever its particular designation.[323]

Especially important in this definition is the expression 'governed by international law', which can be read as suggesting that states are entitled not to have their agreements governed by international law. However, the intention of the drafters of the provision was clearly not to introduce a power of choice for states, but to distinguish between agreements governed by international law and agreements governed by national law as well as to distinguish between intent to agree and situations in which states fail to reach an agreement.[324]

It may even seem absurd to study whether the states may have changed the *pacta sunt servanda* principle through state practice. Lowe submits that there are certain

[321] For a detailed treatment, see Klabbers 1996 I, pp. 157-164 and Klabbers 1996 II, pp. 167-182.

[322] The 2000 Memorandum of Understanding Between the Government of Norway and the Government of the Russian Federation on Cooperation on search and rescue as well as Warning of Serious Incidents provides in paragraph 7: 'This Memorandum does not modify existing legal obligations, nor impose new ones, on the Government of the Kingdom of Norway and the Government of the Russian Federation'. See also the instrument adopted in the 1992 Rio Conference on Environment and Development, the Non-Legally Binding Authoritative Statement of Principles for a Global Consensus on the Management, Conservation and Sustainable Development of All Types of Forests. The document is reproduced in 31 I.L.M. 882 (1992).

[323] The Vienna Convention also makes it clear that this definition serves the purposes of the Convention and does not affect the legal status of oral agreements (Article 3).

juridical concepts which are not amenable to change by international legal subjects, one of these being the *pacta sunt servanda* principle.[325] The ultimate problem of an approach allowing states to decide that an agreement is non-binding altogether is that it is hard to imagine that international law would contain a principle of general application delegating such discretion to them. It is untenable that states would have the power to decide that an agreement with a view to future adherence is not an agreement but something else. *Pacta sunt servanda* confers power on states to establish obligations binding on them; they have only the option of agreeing with other states or not. When the *pacta* have been entered into, they are *servanda* in international law. There is no possibility for states to agree that an agreement they have reached on how they behave towards each other in the future is not really an agreement at all. It is, of course, another matter if states only lay down outlines for future policy which are clearly understood as not requiring any follow-up measures. In such cases, especially after the decision of the ICJ in the *Qatar v. Bahrain* case,[326] it is perfectly understandable for states to lay down an explicit provision of non-

---

[324] See Klabbers 1996 I, pp. 55-63. For a different interpretation, see Ahlström 1999, pp. 132-137.

[325] Lowe (1983, p. 211) argues: 'International law as a coherent system must contain some such rule as *pacta sunt servanda*, whether or not such a rule has attracted the consent of states, because without the rule it would not be possible to speak of an international legal system'. It should be noted that this principle not only makes the accepted treaty norms binding, to be performed in good faith, but also acts as a power-conferring norm. As provided by Lavalle (1982, pp. 20-21): 'The exact purport of the rule *pacta sunt servanda* is not to lay down a norm of conduct, but, as has been demonstrated, to empower states to create, by concluding treaties, obligations binding on themselves. It therefore belongs to the class of power conferring rules, one distinguishing feature of which is that any such rule, not being mandatory in nature, is susceptible of being or not being applied, but not of being observed or breached'. Lavalle's (1982, p. 28) argument is the same as Lowe's, namely, that 'there are compelling reasons for considering that *pacta sunt servanda*, conceived as the basic rule of customary law of treaties, is not, strictly speaking, a rule of customary international law'. This assertion seems straightforward when it is understood that *pacta sunt servanda* is a power-conferring norm and not merely a norm requiring that treaty obligations be performed in good faith. Klabbers (1996 I, p. 130) observes: 'As Thirlway has argued in a slightly different context, some of international law's rules relate not so much to conduct as to concepts. Thus, Thirlway notes that international law possesses some criteria for statehood, and the concept of statehood has some typical characteristics. For example, a state cannot do away with all its territory and still remain a state...Whereas rules of conduct may be altered by agreement, legal concepts such as 'state' or 'internationally wrongful act' are, if not entirely immune to change, relatively resistant to such deliberate change. In the same vein, it is doubtful whether states can simply conclude treaties and nevertheless agree that those treaties are not treaties, but something else'.

bindingness in order to make it clear that the provisions in the document are not to be construed as binding in the future.

3. Problems Related to the Legal Status of Instruments Adopted in the Arctic Cooperation Process

It has been argued that the structure within which intentional norm-making takes place in present international society[327] is still the one based on the principle of *pacta sunt servanda*: states are free to decide whether to agree or not, but as soon as *pacta* have been reached, they are *servanda* in international law. However, this general framework cannot be applied to an analysis of normative instruments without further specification. It will thus be useful to study here the specific content of the concept of agreement in relation to problems occasioned by the instruments adopted in the Arctic cooperation process.

3.1. Problem 1: The Legal Status of Informal Instruments

In general terms, an agreement is reached if communication manifesting a 'meeting of minds' has taken place between the parties as to future behavior. This meeting of minds follows from the intent of states to conclude such an agreement. Indeed, it is

---

[326] See *infra* chapter III (3.1.).

[327] Although the states have only one means of intentionally creating international legal obligations, i.e., concluding international agreements, certain means of intentional norm-making come close to legislative procedures of national legal systems. A good example is an UNGA resolution, which is not formally binding but has a strong effect on the development of customary international law. For an analysis, see Falk 1966, pp. 782-791. Another means in international law quite close to the legislative procedures of national legal systems is the conclusion of near-universal treaties, which provide a strong incentive for the development of customary international law. In *the North Sea Continental Shelf cases of 1969 (North Sea Continental Shelf, Judgment, I.C.J. Reports 1969, p. 3)* the ICJ opined in paragraph 73: 'With respect to the other elements usually regarded as necessary before a conventional rule can be considered to have become a general rule of international law, it might be that, even without the passage of any considerable period of time, a very widespread and representative participation in the convention might suffice of itself, provided it included that of States whose interests were specially affected'.

generally accepted that the essential element in creating an international agreement is the intent of international legal subjects to produce one. Some scholars have even argued that no matter how formal an instrument might be, it is not an international treaty if the intent of states indicates to the contrary.[328]

This approach entails numerous problems. First, one must show what the intent of states is. It is hard enough to show the intent of a single individual, let alone a complex legal entity such as a state or, in the case of multilateral treaties, a group of states.[329] It is thus no wonder that the conclusions as to the non-legally binding nature of a certain instrument tend to rely heavily on a lack of formalities rather than the views of state negotiators or the 'real' views of the participating governments. Second, and more importantly, if there exists an agreement or agreements between states documented by the text of an instrument, it is hard to prove that no agreement was intended. In the case of written agreements, the final intent of states can be determined from the text of the instrument, and if this shows that an agreement has been reached, the *pacta sunt servanda* principle applies.

An important consideration is the presence or lack of formalities. Broad agreement prevails on the importance of formalities since many provisions in instruments between states that regulate the instrument itself, e.g., provisions on its registration and provisions providing for a ratification procedure, indicate strongly that the instrument is intended to be an international treaty binding in international law. The same can be said of the title of an instrument. If an instrument is called a 'convention', 'treaty' or 'agreement', it will be understood as binding in international law. Yet, the reverse does not seem to be true; i.e., the informality of the instrument

[328] See e.g., Rosenne 1989, p. 87. In a similar vein, Fawcett (1953, pp. 381-400) argued that the intent of states is decisive. He opined that normally only those instruments would be treaties the disputes of which were settled by a judicial organ with binding decision-making power.
[329] Klabbers 1996 I, pp. 65-72.

does not mean that it is non-binding in international law.[330] Brownlie's view, to be elaborated below, is that: 'There are no substantive requirements of form, and thus, for example, an agreement may be recorded in an exchange of letters or the minutes of a conference'[331].

The concept of an international agreement is regulated in general international law, over which the ICJ has much influence. The impact of the ICJ in the development of the rules governing the creation of contractual obligations is greater than its influence on the formation of most of the rules of customary law. Although the pronouncements of the ICJ are only a subsidiary source of international law according to the doctrine of the sources of international law,[332] their relevance is pivotal not so much for the so-called primary rules, which prescribe rules of behavior for states, as for the so-called secondary rules, which define how primary rules are developed.[333] The rules which define how contractual obligations are created are just such secondary rules of international law on which the ICJ can have much influence. It is thus useful to study the approach of the ICJ in respect of the creation of contractual obligations in general and the legal status of informal instruments in particular. The most important cases of the ICJ in this respect are the *Aegean Sea Continental Shelf* case between Greece and Turkey and the landmark case *Maritime Delimitation and Territorial Questions* between Qatar and Bahrain.[334]

---

[330] As provided by Elias (1974, p. 13): 'The only legal difference between formal and informal treaties lies mainly in the method of their conclusion and in the manner of their entry into force'.
[331] Brownlie 1998, p. 610.
[332] The accepted formulation of the sources of international law is contained in Article 38 of the Statute of the ICJ. According to this Article, '...judicial decisions and the teachings of the most highly qualified publicists of the various nations' are subsidiary means for the determination of rules of law.
[333] See Hart 1961, pp. 77-97.
[334] See Klabbers (1996 I, pp. 182-219) on other decisions of the ICJ relevant to the concept of agreement in international law. The two cases examined here are similar as to the nature of the agreement; in both, the Court had to decide whether an alleged agreement on submitting the dispute to the ICJ recorded in an informal instrument was an international agreement binding in international law.

In the *Aegean Sea Continental Shelf* case,[335] Greece alleged that what is known as the Brussels Communiqué of 31 May 1975 contained a provision stipulating that the parties would submit the dispute to the jurisdiction of the ICJ. The Communiqué was 'issued...directly to the press during a press conference'[336] held at the conclusion of the meeting of prime ministers of the two states. It contained no signature or initials. The Turkish Government considered it 'evident that a joint communiqué does not amount to an agreement under international law'[337].

The Court first stated unequivocally that a lack of formalities is not decisive as to whether an agreement has been reached, seeing no problems in proceeding with the Joint Communiqué even though it did not even contain signatures or initials.[338] It then made clear the method by which it was to study the issue:

> In determining what was indeed the nature of the act or transaction embodied in the Brussels Communiqué, the Court must have regard above all to its actual terms and to the particular circumstances in which it was drawn up.[339]

The Brussels Communiqué stated that the prime ministers

> decided [ont décidé] that those problems should be resolved [doivent être résolus] peacefully by means of negotiations and as regards the continental shelf of the Aegean Sea by the International Court at The Hague.[340]

---

[335] *Aegean Sea Continental Shelf, Judgment, I.C.J. Reports 1978, p.3.*
[336] Ibid., paragraph 95.
[337] Ibid.
[338] Ibid. Paragraph 96 reads: 'On the question of form, the Court need only observe that it knows of no rule of international law which might preclude a joint Communiqué from constituting an international agreement to submit a dispute to arbitration or judicial settlement (cf. Artc. 2,3 and 11 of the Vienna Convention on the Law of Treaties). Accordingly, whether the Brussels Communiqué of 31 May 1975 does or does not constitute such an agreement essentially depends on the nature of the act or transaction to which the Communiqué gives expression; and it does not settle the question simply to refer to the form – a communiqué – in which that act or transaction is embodied'.
[339] Ibid.
[340] Ibid., paragraph 97.

The rest of the document defined the framework for further meetings of the parties in respect of the dispute.[341] Greece argued that this provision constituted not only an undertaking to negotiate but directly conferred jurisdiction on the ICJ.[342] In addition to contending that the Joint Communiqué was not an agreement under international law, Turkey asserted that the intention of the provisions of the Brussels meeting was only to draw up a plan for further negotiations.[343]

The Court carefully studied the context in which the Brussels Communiqué was drawn up and concluded after examining the negotiation history of the dispute that it was very unlikely that an agreement had been reached by the parties on submitting the dispute directly to the Court. The Court opined that from the beginning Turkey was studying the possibility of negotiating a special joint submission agreement to the ICJ. It considered that the Brussels Communiqué was to be read in light of this and that it thus did not provide direct jurisdiction for the Court.[344] The Court did not pronounce directly on whether the Brussels Communiqué itself was a legally binding agreement.[345] However, several judges in their separate opinions and one judge in his

[341] Ibid. Paragraph 97 reads: 'They defined the general lines on the basis of which the forthcoming meetings of the representatives of the two Governments would take place. In that connection they decided to bring forward the date of the meeting of experts concerning the question of the continental shelf of the Aegean Sea and that of the experts on the question of air space'.
[342] Ibid. Paragraph 98 reads: 'These words, it says, are words of "decision" and of "obligation" indicative of mutual commitment on the part of the Prime Ministers to refer the dispute to the Court. Specifically, it claims that the "agreement" embodied in the Communiqué "is more than an undertaking to negotiate" and directly "confers jurisdiction on the Court"'.
[343] Ibid. Paragraph 99 provides: 'it is clear that there was no commitment to submit to the Court without a special agreement because the following paragraph said in this connection that the two Prime Ministers had decided to accelerate the meeting of the experts concerning the question of the continental shelf of the Aegean Sea'.
[344] Ibid., paragraphs 100 to 107.
[345] Ibid. Paragraph 108 reads: 'Having concluded that the Joint Communiqué issued in Brussels on 31 May 1975 does not furnish a basis for establishing the Court's jurisdiction in the present proceedings, the Court is not concerned, nor is it competent, to pronounce upon any other implications which that Communiqué may have in the context of the present dispute. It is for the two Governments themselves to consider these implications which that Communiqué may have in the context of the present dispute'.

dissenting opinion proceeded directly or indirectly from the premise that the Communiqué was a legally binding agreement,[346] and it can be argued that the majority in fact embraced this view. According to Thirlway:

> Although the Court does not say so, it appears to have proceeded on the assumption that the communiqué was an agreement, to conclude that it was not an agreement of the kind claimed by Greece.[347]

In the Maritime Delimitation and Territorial Questions between Qatar and Bahrain,[348] Qatar took the position that a provision contained in signed minutes was an international agreement between Qatar and Bahrain to submit the dispute to the ICJ. During the annual meeting of the Gulf Co-operation Council of Arab States in December 1990, Qatar announced that it was ready to accept the so-called Bahraini formula. This 'formula'[349] was initiated by Saudi Arabia in its capacity as mediator in a Tripartite Committee established to resolve a longstanding territorial dispute between Qatar and Bahrain.[350] Following the Gulf Co-operation Council meeting, the

---

[346] Ibid. Judges Lachs (pp. 49-53), Nagendra Singh (pp. 46-48) and Tarazi (pp. 55-61) expressed in their separate opinions that there existed an agreement to negotiate a compromis. At least Lachs and Tarazi derived this obligation from the Communiqué itself. *Ad hoc* Judge Stassinopoulos simply proclaimed in his dissenting opinion that the Communiqué was an international agreement (pp. 72-83).

[347] See Thirlway 1991, p. 15. Vierdag (1996, p. 153) is of the same opinion: 'It is to be understood that the Court withheld effect from the communiqué solely as an instrument conferring jurisdiction to the Court to entertain the claim, but that its holding did not in any way affect whatever other commitments may be inferred from it for the two governments that issued it'.

[348] *Maritime Delimitation and Territorial Questions between Qatar and Bahrain, Jurisdiction and Admissibility, Judgment, I.C.J. Reports 1994, p. 112 (Qatar v. Bahrain case).*

[349] Ibid., paragraph 18 provides the 'formula': 'The Parties request the Court to decide any matter of territorial right or other title or interest which may be a matter of difference between them; and to draw a single maritime boundary between their respective maritime areas of seabed, subsoil and superjacent waters'.

[350] The Tripartitite Committee was established in December 1987. The King of Saudi Arabia put forward proposals for the parties, which were accepted by letters from the two heads of states. These proposals included four points, of which the first was: 'All the disputed matters shall be referred to the International Court of Justice, at The Hague, for a final ruling binding upon both parties, who shall have to execute its terms'. The Committee ended its mission after six meetings, the last of these held on 6-7 December 1988. The Bahraini formula was the outcome of the following event: 'Then, on 26 October 1988, following an initiative by Saudi Arabia, the Heir Apparent of Bahrain, when on a visit to Qatar, transmitted to the Heir Apparent of Qatar a text'.

foreign Ministers of Bahrain, Qatar and Saudi Arabia signed minutes in which they agreed on certain issues, one of which was the submission of the dispute to the ICJ if no negotiated solution had been reached within a certain time.[351] The original text of the minutes was in Arabic, and the English translations of the parties differed in some important particulars.[352] When the period for a negotiated solution lapsed, Qatar instituted proceedings before the ICJ against Bahrain on the basis of the exchanges of letters of 1987 by which the Tripartite Committee was established and the signed minutes of 1990. Both parties agreed that the exchanges of letters in 1987 were international agreements binding in international law. Bahrain, however, maintained 'that the 1990 Minutes do not constitute a legally binding instrument'[353].

First of all, the Court observed 'that international agreements may take a number of forms and be given a diversity of names'[354] and cited Article 2 (1a) of the Vienna Convention to support its position. It then cited the relevant passages studied above from the *Aegean Sea Continental Shelf* case providing, in essence, that no requirements of form are set by international law for agreements and that the Court must, above all, study the actual terms of the alleged agreement and the context in which it was drawn up.[355] After studying the 1990 Minutes, the Court concluded that the provisions contained in the signed minutes:

---

[351] Paragraph 19 of the *Qatar v. Bahrain* case.

[352] Ibid. The difference was that Qatar's translation seemed to suggest that either party was entitled to make application to the Court unilaterally whereas the Bahraini translation suggested that the application was to be made by the parties acting jointly. Qatar's translation read as follows: 'After the end of this period, the parties may submit the matter to the International Court of Justice in accordance with the Bahraini formula'. The corresponding sentence in the translation used by Bahrain read: 'The two parties may, at the end of this period, submit the matter to the International Court of Justice in accordance with the Bahraini formula'.

[353] Ibid., paragraph 20. In paragraph 22 the position of Bahrain is specified: 'Bahrain however maintains that the Minutes of 25 December 1990 were no more than a simple record of negotiations, similar in nature to the Minutes of the Tripartite Committee; that accordingly they did not rank as an international agreement and could not, therefore, serve as a basis for the jurisdiction of the Court'.

[354] Ibid., paragraph 23.

> Enumerate the commitments to which the Parties have consented. They thus create rights and obligations in international law for the parties. They constitute an international agreement.[356]

Of greatest importance here is that the ICJ links the concepts of 'law' and commitment' through the word 'thus'. Hence, the ICJ is of the opinion that whenever a commitment has been entered into by states, this commitment is a legal one.

Bahrain maintained 'that the signatories of the Minutes never intended to conclude an agreement of this kind'[357], referring to the following statement made by its Foreign Minister: 'at no time did I consider that in signing the Minutes I was committing Bahrain to a legally binding agreement'[358]. The Foreign Minister also stated that at the time of the signing of the Minutes he was aware that he was not authorized, for constitutional reasons, to sign a legally binding agreement, but was 'prepared to subscribe to a statement recording a political understanding'[359]. The ICJ rejected this argument and concluded:

---

355 Ibid.

356 Ibid., paragraph 25.

357 Ibid., paragraph 26.

358 Ibid.

359 Ibid. Bahrain also presented the argument that the subsequent conduct of the parties showed that no international agreement was concluded. According to paragraph 28 of the judgment '[Bahrain] points out that Qatar waited until June 1991 before it applied to the United Nations Secretariat to register the Minutes of December 1990 under Article 102 of the Charter; and moreover that Bahrain objected to such registration'. The Court pointed out that although registration is necessary for invoking the agreements before the organs of the United Nations 'Non-registration or late registration, on the other hand, does not have any consequence for the actual validity of the agreement, which remains no less binding upon parties' (paragraph 29).

> [the Court] does not find it necessary to consider what might have been the intentions of the Foreign Minister of Bahrain or, for that matter, those of the Foreign Minister of Qatar. The two Ministers signed a text recording commitments accepted by their Governments, some of which were to be given immediate application. Having signed such a text, the Foreign Minister of Bahrain is not in a position subsequently to say that he intended to subscribe only to a "statement recording a political understanding", and not to an international agreement.[360]

This passage could be read to mean that only texts recording commitments which contain norms of immediate application can be legally binding agreements, although this interpretation is hard to maintain in light of the text of the minutes. The only norm of immediate application was the norm providing for the continuance of Saudi Arabia's good offices. Indeed, the norm providing for the jurisdiction of the Court was not one of immediate application but a conditional obligation triggered only after a certain date.[361] The Court's decision in *Qatar v. Bahrain* clarifies the concept of agreement in such unequivocal terms that it has already been hailed as a landmark case that will have a direct effect on state practice.[362]

In order to understand the Court's method for ascertaining when an agreement has been concluded, it is useful to study the basic framework of thought within which the

[360] Ibid., paragraph 27.

[361] For an analysis of this question, see Klabbers 1996 I, pp. 212-216.

[362] As stated by Vierdag (1996, pp. 165-166): 'These holdings are pronounced by a majority of fifteen judges, one judge dissenting, surprisingly not the judge *ad hoc* chosen by Bahrain. They will soon appear to have a decisive impact on inter alia the prevalent practice of drawing up all sorts of informal arrangements between states, governments, ministries, state agencies and the like, the possible legal significance of which is commonly ignored or left ambiguous. It is to be hoped that the Court's *dicta* will heighten the awareness of the legal consequences that this practice may have'. Chinkin (1997, pp. 224-225) views the decision as causing negative developments: 'The decision has accordingly been hailed as 'monumental' in that it opens the way to holding states bound by commitments however informally given. On the other hand such an approach might cause misgivings in that it could make states reluctant to agree to even the most imprecise and unofficial formulations, for example in diplomatic exchanges, for fear of their subsequently being held to have incurred legal rights and obligations. This could impede the regular flow of communications between states and inhibit negotiations, unless unequivocal wording disavowing intent was

Court operates. In both of the cases studied, the Court did not focus on the abstract legal nature of the informal instrument itself but on the wording of the instrument, i.e., what had or had not been agreed by the parties. This approach seems to be in stark contrast to the method of many international legal scholars who start from the presumption that the informality of the instrument signals non-bindingness; the ICJ starts from the presumption that the instrument is binding. According to Klabbers, the Court operates on the presumption that an informal instrument is binding in international law unless otherwise shown.[363] This presumption of bindingness is supported by an extensive study of cases in which the legal status of international written instruments has been decided by international, European and national judicial bodies, and is accepted in this study as describing most adequately the current concept of agreement in international law.[364]

It is, of course, possible to rebut the presumption of bindingness by showing that an instrument is not intended to be binding. To illustrate the point, in cases where an instrument is declared as aiming for a formal convention with, for instance, a ratification procedure, all the agreements reached before the conclusion of a convention containing such a procedure are clearly non-binding.[365] If an agreement is reached between the parties on some issue in these cases, and this is recorded in the minutes of the meeting, but no treaty with ratification procedure is subsequently concluded, no agreement has been concluded because the agreement reached was merely part of the negotiating phase of the treaty process; or, if such negotiations have produced only a draft agreement, or some other informal instrument, the

---

incorporated at all stages'. Ahlström sees the decision as negative but not one having any significant effects on state practice. Ahlström 1999, pp. 169-170.

[363] Klabbers 1994, pp. 267-387.

[364] See Klabbers 1996 I, pp. 165-245. A similar view is represented in the scholarly world by, e.g., Danilenko 1993, pp. 54-58; Reuter 1995, pp. 23-24; Vierdag 1996, pp. 144-165; and Brownlie, 1998, p. 610. For a recent criticism of this approach, Aust 2000, p. 42.

[365] Another example by which to rebut the presumption of bindingness is that of an instrument designated as a declaration which truly contains only declarations outlining the general future goals of the parties and/or describing the existence of certain states of affairs.

intention to be bound by that instrument is still lacking.[366] To establish non-bindingness, it must clearly be shown that a treaty with, for instance, a ratification procedure was the ultimate aim. Hard evidence is required; interviews of negotiators of some states who report they had that feeling in the beginning of negotiations are not sufficient. Appropriate evidence might be, for instance, extensive publicization of the aims of the states in the beginning of the negotiations.

It can be concluded that all the informal instruments studied in this thesis are presumptively binding in international law unless the contrary can be shown. There are certain cases worth studying here in which it might be possible to show the lack of intent. The most difficult cases relating to the Arctic cooperation process are those in which the environmental protection working groups have produced instruments. One might ask whether these are binding in themselves or should perhaps be seen as part of a preparatory process leading to final acceptance by the ministerial conferences of the Arctic states. This question must be postponed here, however, because if AEPS cooperation and the Arctic Council can be considered inter-governmental organizations, it follows that all the decisions of their organs are non-binding; the presumption exists that if the constitutive instrument of an inter-governmental organization does not provide an express mandate for binding decision-making power for the organs of the organization, the organs' decisions are non-binding[367] This issue is studied in more detail in conjunction with the problem 3 below.

---

[366] It should still be kept in mind that states have certain obligations even before ratifying a treaty. According to Article 18 of the Vienna Convention on the Law of Treaties, a state that has signed a treaty is obligated not to defeat the object and purpose of a treaty prior to the treaty's entry into force.

[367] See Schermers and Blokker (1995, p. 141 (§209)), who state: 'A rule of thumb is that, while states are free to act as long as this is in accordance with international law (which may prescribe or

3.2. Problem 2: International Treaties with Provisions of Doubtful Legal Status

The starting point here is the one arrived at the end of the previous chapter, namely, that all informal and formal instruments are presumptively binding in international law. It becomes essential to study an instrument provision by provision to determine whether an agreement has been reached by the parties. This problem pertains to all informal and formal instruments, since in most cases at least some of the provisions prompt doubts as to whether an agreement has been reached on a particular issue.[368] As was shown above,[369] the problem also pertains to the instruments of Arctic cooperation process, which contain many kinds of expressions indicating how the parties are to behave.

It is important to first define what an agreement is because, as was shown by the *Aegean Sea Continental Shelf* case, it is clear that not all texts of treaties convey that an agreement has been reached by the parties. The ICJ uses the word 'commitment' in *Qatar v. Bahrain* and provides no other clues as to what this broad term means. Klabbers refers generally to the importance of 'the actual words used in the document'[370] but this does not suggest any method by which to distinguish between agreement and non-agreement either. The present author[371], and some others[372], have suggested that the main distinction relates to the difference in the use of modals such

proscribe state activities), international organizations are competent to act only as far as powers have been attributed to them by the member states'.

368 Some scholars call these 'soft obligations'. For instance, Baxter (1980, p. 549) argues: 'there are norms of various degrees of cogency, persuasiveness and consensus which are incorporated in agreements between States but do not create enforceable rights and duties. They may be described as "soft" law, as distinguished from the "hard" law consisting of treaty rules which States expect will be carried out and complied with'.

369 See *supra* chapter III (1).

370 Klabbers 1994, p. 384.

371 Koivurova 1999, pp. 146-147.

372 See, e.g., Corten and Schaus (1989, p. 252), who argue: 'L'emploi du terme "shall" – par opposition au "should" contenu dans divers textes à portée déclarative – a bien pour sens ordinaire l'instauration d'une obligation'.

as 'shall', which signifies bindingness and 'should', which denotes non-bindingness. In addition, the use of very general and/or vague language may show that no agreement has been reached by the parties.[373]

There are scholars, for instance, Aust, who argue that states customarily use certain modals and expressions to denote bindingness and non-bindingness in treaty-making. He asserts:

> The actual practice of states is to indicate their intention to conclude a treaty by consciously employing terminology such as 'shall', 'agree', 'undertake', 'rights', 'obligations' and 'enter into force'. When they do not intend to conclude a legally binding instrument, but rather an MOU [memorandum of understanding], instead of 'shall' they use a less imperative term, such as 'will'. Terms like 'agree' or 'undertake' are avoided.[374]

In Appendix G of his textbook, Aust presents a comparative table of expressions that are normally used in MOU's and in legally binding treaties.[375] Aust evidently knows something about the customary practice of states, working as he does at the foreign ministry of Great Britain,[376] but it is highly questionable whether such custom can be extracted from the practice of states since customary law of treaties cannot regulate the matter. To be sure, the wording of instruments normally is checked by the parties' legal experts, but it still is very hard to generalize from the experience of drafting certain treaties that the same rules apply to all treaties. Aust readily admits that his perspective is decidedly British.[377] A better view is provided by Bernhardt, who states:

---

[373] Koivurova 1999, p. 147.
[374] Aust 2000, p. 27.
[375] Ibid., p. 404.
[376] Ibid., p. 1.
[377] Ibid. In the introduction, Aust states: 'Although I have tried to avoid using too many British precedents, many examples are drawn from personal knowledge and experience'.

> The view expressed by some authors that vagueness of treaty provisions leads – or at least can lead – to the conclusion that these provisions are not legally binding, seems incorrect. As far as it is clear that the parties intended to conclude a treaty, it is binding in its entirety, even if part of the content is phrased in broad or unclear terms. Vague provisions leave the parties an especially broad margin of discretion, but this does not mean that they are without legal significance and binding substance.[378]

Bernhardt's view can be taken as a starting-point here and refined through deontic logic.[379] Norms can generally be divided into three basic categories: 'prohibitions', 'orders' and 'permissions'.[380] A state can be prohibited from doing X (course of conduct). This course of conduct cannot be ordered or permitted at the same time. 'Order' entails an obligation to do X. Accordingly, there cannot be a prohibition to refrain from doing X, but X is also permitted if it is ordered. Hence, permission to do X is always present when X is ordered. Yet, it may be that express permission to do X is issued. In this case, permission to do X means that X is neither prohibited nor, importantly, ordered. This means that one has a right to do X, or not to do X. What if there is no prohibition or order to do X on the one hand but no express permission to do X either? This state of affairs is called by some 'free choice permission'[381] or 'free act'[382]. In such situations, the subject is not only free to do X or not to do X, but X is only one course of action which can be chosen or not chosen.

---

[378] Bernhardt 1984, p. 461.

[379] For an in-depth treatment of this issue in international law, see Horn 1988, pp. 70-73.

[380] Hilpinen (1981, p. xi) provides a basic definition of the logical approach: 'The articles included in this collection represent what may be called the ***standard modal approach*** to deontic logic (the logic of normative concepts), in which deontic logic is treated as a branch of modal logic, and the normative concepts of ***obligation, permission (permissibility)*** and ***prohibition*** are regarded as analogous to the 'alethic' modalities ***necessity, possibility*** and ***impossibility***'. (original italics).

[381] Ibid. Hilpinen (p. xiv) states: 'In the standard approach to deontic logic, the concept permission (permissibility) is defined simply as the absence of prohibition. The feature of the standard system has been criticized on several grounds; for example, it has been argued that the standard definition of permissibility excludes the possibility of incomplete or open normative systems, that is, systems which leave the normative status of some acts or states of affairs completely undetermined...the most striking argument against the standard analysis of permissibility is provided by the 'paradox of free choice permission''.

[382] Ross 1974, p. 164.

The EIA norms studied below fall into the category of 'order'. They prescribe obligations such as the duty to establish an EIA procedure or an obligation to inform other states of potentially significant transboundary impacts. However, there exists a category of norms which falls between the categories of 'order' and 'permission'. This category is defined here as 'obligation of attempt' and refers to all other expressions found in treaties which cannot be described as 'direct orders' as they are called in this study.

Direct orders obligate states to accomplish what is required in the norm-sentence, for instance, to establish an EIA procedure; [383] obligations of attempt only require that an attempt be made to do so. Direct orders are expressed normally through the modals such as 'shall'[384] and 'must' whereas obligations of attempt can be paraphrased as 'to try to' and are expressed through modals such as 'should' and 'ought to'.[385] Yet, importantly, the latter still establish a legal order for the contracting states. The parties are at least obligated to try to implement the prescribed pattern of behavior. If states provide that they 'should' establish, for instance, an EIA procedure for plans, programs and policies, they are under a legal obligation to try to do so. It is enough that some attempt is made, e.g., by designating a governmental organ to study an issue. If nothing is done, the obligation has clearly been breached. In this sense, obligations of attempt operate in a binary manner like all the norms of international law.

Especially in international environmental treaties, there is a wealth of expressions that make it clear that the states only opted for an obligation of attempt. One much

[383] Preisler defines 'will' and 'shall' as belonging to the category of 'commitment' and 'must' as part of 'insistence'. Preisler 1992, p. 121.

[384] In the Dictionary of Modern Legal Usage, Garner (1995, pp. 939-942) points out the difficulties related to the use of the 'words of authority', such as 'shall', in English-speaking jurisdictions and describes drafting guidelines pertaining to these words.

[385] The modals 'should' and 'ought to' have been categorized by Preisler as 'preference' as distinct from the categories of 'insistence' and 'commitment'. See Preisler 1992, pp. 121-122.

used group of expressions are additional verbs to the main verb which qualify the requirements of the norm-sentence in question such as 'shall' and the main verb, e.g., 'endeavour', 'promote' and 'seek' as in the sentence 'states shall seek to establish an EIA procedure'.[386] Another group of expressions used in international treaties consists of phrases such as 'as far as possible', 'to the extent appropriate', 'as far as practicable', etc. These phrases convey that states have not committed themselves to execute the order as such and thus they can most appropriately be treated as obligations of attempt. As will be shown below, the same applies to provisions which are very vague in scope or indeterminate in content even though they are formulated using a modal such as 'shall'.[387]

Finally, if the desirable course of behavior is expressed in such general terms as to approach a statement of social values rather than a prescription of norms, there may be problems in determining whether it is an independent order or something else. If the parties agree that 'we shall live as good neighbours', questions arise as to whether the parties have really reached an agreement at all. For example, this was the reasoning of the ICJ in the *Oil Platforms* case.[388] The Court refused to base its jurisdiction ratione materiae on the very generally worded obligation 'There shall be firm and enduring peace and sincere friendship between the United States...and Iran', despite the argument of Iran to the effect that this was a legal obligation. The ICJ saw the provision as 'fixing an objective, in the light of which the other Treaty provisions are to be interpreted and applied'[389]. Hence, if a provision is expressed in very

[386] A good example is the 1992 Convention on Biological Diversity, which contains a large number of expressions such as 'shall seek' or 'as far as practicable'.
[387] See *infra* chapter III (3.3.).
[388] *Oil Platforms (Islamic Republic of Iran v. United States of America), Preliminary Objection, Judgment, I.C.J. Reports 1996, p. 803.*
[389] Ibid. The relevant parts of the *Oil Platforms* judgment read: '24. Article I of the Treaty of 1955 provides that: "There shall be firm and enduring peace and sincere friendship between the United States...and Iran". 25. Iran contends that this provision "does not merely formulate a recommendation or desire...but imposes actual obligations on the Contracting Parties, obliging them to maintain long-lasting peaceful and friendly relations". This interpretation is said to be required by the context, and to be reinforced by the circumstances in which the Treaty was

general terms, it does not qualify as an agreement but can be used in interpreting the other provisions of the treaty, a practice that is studied in detail below.[390] There are also various kinds of expressions that parties use not to provide any legal normative guidance but to declare their general future goals as well as to describe the existence of states of affairs.[391] The instruments of the Arctic cooperation process contain direct orders and obligations of attempt, but mainly non-binding political statements.

3.3. Problem 3: The Status of Written Instruments Adopted by Inter-Governmental Organizations

A distinction must be drawn between two legal powers of states. As shown above, as soon as states enter into commitments with other states over future behavior, these commitments become binding in international law. However, states are, of course, entitled to decide that a new entity they create, e.g., an inter-governmental organization, does not have any binding decision-making power. In fact, Schermers and Blokker state in their leading textbook on international organizations: 'As a general rule of modern international institutional law, it has been accepted that international organizations cannot take binding external decisions unless their constitutions expressly so provide'[392]. This is an important rule from the viewpoint of this study since if the forms of Arctic cooperation, i.e., the AEPS cooperation and the

---

concluded. It is described as the only interpretation which would enable "effectiveness" to be imparted to Article I. That Article would, then, impose upon the Parties "the minimum requirement...to conduct themselves with regard to the other in accordance with the principles and rules of general international law in the domain of peaceful and friendly relations" 26. The United States considers, on the contrary, that Iran "reads far too much into Article I". That text, according to the Respondent, "contains no standards", but only constitutes a "statement of aspiration"...28. It follows that Article I must be regarded as fixing an objective, in the light of which the other Treaty provisions are to be interpreted and applied'.

[390] See *infra* chapter III (4.).

[391] For instance, the preface of the Nuuk Declaration (paragraph 4) states: 'Determined, individually and jointly, to conserve and protect the Arctic environment for the benefit of present and future generations, as well as for the global environment'. The preface of the Inuvik Declaration (paragraph 5) provides: 'Aware also that environmental changes in the Arctic regions have significant local, regional and global impact'.

Arctic Council, are deemed international organizations for the purposes of international law, it can also be claimed that all the decisions these inter-governmental organizations have taken are non-binding in international law with regard to the Arctic states because there exists no express provision according binding decision-making power to these forms of cooperation.

The first question must then be whether the forms of cooperation in the Arctic cooperation process can be considered inter-governmental organizations for the purposes of international law. Schermers and Blokker provide an important criterion[393] in this regard:

> [I]nternational organizations are defined as forms of cooperation founded on an international agreement creating at least one organ with a will of its own, established under international law.[394]

All the instruments concluded between the representatives of the Arctic states can be treated presumptively as binding in international law since they contain some commitments between the states.[395] Perhaps it could then be argued that both the AEPS cooperation and the Arctic Council are inter-governmental organizations for the purposes of international law. Yet, the second element required of the forms of Arctic cooperation if they are to qualify as inter-governmental organizations for the purposes of international law is that they should have at least one organ with a will of its own. This organ should be formed by delegates of two or more states and should not be dependent on any particular state.[396] More importantly, at least from the viewpoint of this study:

---

[392] Schermers and Blokker 1995, p. 813 (§1320).
[393] Ibid., pp. 22-33.
[394] Ibid., p. 23 (§33).
[395] See *supra* chapter III (1).
[396] Schermers and Blokker 1995, pp. 29-30 (§44).

> The requirement that the organ in question have a will of its own distinguishes organizations from bilateral or multilateral treaties, whereby parties lay down a common will, which remains their own, however, and is not entrusted to a newly created body.[397]

It would seem that from the beginning there has been no organ with a will of its own in the Arctic cooperation process. In establishing both the AEPS cooperation and the Arctic Council, the Arctic states have laid down their common will, which has remained their own and has not been entrusted to an organ representing the will of the Arctic Council.[398] It can therefore be concluded that no inter-governmental organization has as yet emerged.

But are these instruments adopted in the Arctic cooperation process binding in international law? Can clear evidence be found that the Arctic states intended not to agree with each other but only to record their political aspirations for the future? Since these instruments contain clear normative guidance for the Arctic states, the presumption is that these instruments are binding in international law. There does not exist any hard evidence to rebut the presumption of bindingness: the Arctic states have neither expressly provided that the instruments are non-binding in international law nor have they made public their intentions as to the legal status of these instruments.[399] Consequently, the AEPS document, the Nuuk Declaration, the Inuvik Declaration, the constitutive instrument of the Arctic Council, the Alta Declaration, the Iqaluit Declaration and the Barrow Declaration are binding in international law.

There are two kinds of rules that have been adopted within the Arctic cooperation process: rules regulating the cooperation process itself and rules regulating the protection of the Arctic environment. The problem with the latter group of rules is

---

[397] Ibid., p. 30 (§44).

[398] A similar approach is provided by Brownlie, who lays down three criteria for the personality of an inter-governmental organization. See Brownlie 1998, pp. 679-680.

that there exists no explicit definition of what can be regarded as the southernmost limit of the Arctic; none can be found in the instruments adopted in the Arctic cooperation process. The instruments only refer to the 'Arctic environment'[400] or 'Arctic ecosystems'[401]. Those rules that obligate the Arctic states to cooperate with each other evidently do not require an explicit definition of the Arctic to become operable.

The AEPS contains very far-reaching obligations for the parties. Section 2 (2) of the Strategy provides that:

> iii) Management, planning and development activities which may significantly affect the Arctic ecosystems shall:
> a) be based on informed assessments of their possible impacts on the Arctic environment, including cumulative impacts...
> e) take into account the results of scientific investigations and the traditional knowledge of indigenous peoples.

This provision is written very much as a direct order in that it uses the modal 'shall'. If it were taken to be a direct order, it would mean that the Arctic states had agreed on a wide variety of measures in respect of all activities that 'may significantly affect the Arctic ecosystems' irrespective of whether these activities are located in the Arctic or in the southern areas of the Arctic states. It would also mean that the Arctic states had agreed on assessing the impact of 'management, planning and development activities' on not only the ecosystems located in their territory but also the territory of other Arctic states and international areas. On balance, this provision can be viewed as being so general in substance and vague in scope that it merely establishes an obligation of attempt for the parties. Similar considerations apply to all the environmental protection norms adopted within the Arctic cooperation process.

---

[399] In fact, in the AEPS document, for instance, it was provided that the: 'implementation of the Strategy will be carried out through national legislation and in accordance with international law'.
[400] See, e.g., Section 2.1 (iii) of the Strategy of the AEPS document.
[401] Ibid. See, e.g., Section 2.1 (i).

Another problem is the instruments adopted by the environmental protection working groups of the AEPS cooperation, now the Arctic Council. Are these intended to be binding themselves or are they merely part of the negotiation phase in which final consent is given by the ministerial meetings of the Arctic states?[402] As was shown above, the conferences of the Arctic states have concentrated on providing guidelines for the work of SAOs and the working groups. In this respect, a procedure has been institutionalized whereby the normative activity of the working groups is mandated by the meetings of the Arctic states and is thus part of the negotiation procedure in which final approval is given by the ministerial conferences of the Arctic states, now the meetings of the Arctic Council. This view is supported by the fact that the normative instruments produced by the working groups lack all formalities, including signatures,[403] an indication that the results of their work are only negotiations leading to the Declarations.

The most important instruments in this thesis are the above-mentioned Guidelines for EIA in the Arctic and the Arctic Offshore Oil and Gas Guidelines, both of which the Arctic states agreed to apply in the Alta meeting:

> We receive with appreciation the "Guidelines for Environmental Impact Assessment (EIA) in the Arctic" and the "Arctic Offshore Oil and Gas Guidelines" developed under the AEPS, and **agree** that these guidelines be applied. (original emphasis)

---

[402] This problem is heightened by the fact that an instrument concluded by an environmental working group, the Arctic Offshore Oil and Gas Guidelines, uses the expression 'while recognizing the nonbinding nature of these guidelines'. However, in the Alta Declaration the Arctic states state in paragraph 3 that: 'We receive with appreciation the "Guidelines for Environmental Impact Assessment (EIA) in the Arctic" and the "Arctic Offshore Oil and Gas Guidelines" developed under the AEPS, and agree that these guidelines be applied'.

[403] This is not in itself a decisive criterion, as shown by the *Aegean Sea Continental Shelf* case. See *supra* chapter III (3.1.).

The wording suggests that the Arctic states agree to apply the normative substance of these instruments; i.e., direct orders are established for the parties. Yet, the same problems that pertain to the environmental protection norms of the ministerial Declarations also appear in the case of these instruments. These documents are written in very vague, indeterminate language, as evidenced by the use of the modals 'should' or 'may', and the recommendatory purpose of both instruments is made explicitly or implicitly clear in the documents themselves.[404] Yet, as was studied above, the working groups carry out the preparatory work, after which the final approval is given by the ministerial conferences. In Alta, the Arctic states agreed to apply these instruments. On balance, then, these two sets of guidelines can best be seen as conveying obligations of attempt.

It can thus be concluded that the instruments adopted in the ministerial meetings of the AEPS cooperation and the ministerial meetings of the Arctic Council are binding in international law. The presumption of bindingness applies also to the future outcomes of the ministerial meetings if the Arctic states do not explicitly provide that the instrument is non-binding in international law. In a similar vein, if the Arctic states agree through ministerial conferences to apply the normative instruments adopted in the working groups, these are also binding in international law. If the environmental protection norms adopted within the cooperation process are very vague in scope or if the provisions they contain are more recommendatory in tone, they are better qualified as obligations of attempt.

---

[404] The Guidelines for Environmental Impact Assessment in the Arctic provide in Section 1 'Introduction': 'These guidelines are not intended to replace existing procedures adopted by international, national or provincial laws, land claim agreements, regulations or guidelines...They aim at providing suggestions and examples of good practice'. The Arctic Offshore Oil and Gas Guidelines provides expressly in its Section 1.2. 'Goals': 'while recognizing the nonbinding nature of these guidelines'. Archer and Scrivener state (2000, p. 612): 'At the Inuvik Ministerial Meeting it

4. Conclusion

It will be useful to summarize the basic method by which the legal status of all the intentionally produced norms will be analyzed in chapters IV and V:

1. All instruments are presumptively binding in international law if it cannot be clearly shown that the states intended the instrument to be non-binding at the time of its conclusion. If it can be shown that the states did not intend the instrument to be binding, it is not binding.
2. The norm-texts of these instruments fall into three distinct categories: direct orders, obligations of attempt and permissions. If a norm is expressed in very general language, it is more appropriately seen as fixing the object and purpose of a treaty.
3. With respect to separate organs of inter-governmental organizations there exists a presumption that if the constitutive treaty of an organization does not contain any express rules on the decision-making power of its organs, their decisions are non-binding. This is evident when provided for explicitly in the constitutive treaty. However, when an inter-governmental institution adopts its decisions through meetings of its states parties and decisions are not taken by independent organs of the organization, such as in the case of the Arctic Council, the instruments and decisions of the organization are agreements and thus binding in international law.

Since the approach adopted in this thesis differs from that of the mainstream research in international environmental law, it is necessary to use different terminology to underline the differences in substance. There is an extensive number of normative instruments which are not legally binding but may nevertheless influence the

was decided that the outcome should be a set of practical guidelines for voluntary use by states, rather than a legally-binding code'.

development of international law. These range from decisions adopted by organs of inter-governmental organizations that do not have binding decision-making power and treaties which have not yet entered into force to instruments adopted by non-governmental expert bodies on international law. These will be referred to here as **'legally relevant documents'**. Although these legally relevant documents are not legally binding per se, they do have legal relevance in numerous ways.

As was argued above, if a certain instrument is found to be non-binding, it does not create any legal obligations for the parties to it, and all the norms contained in it are non-binding. The rules in such an instrument may nevertheless have legal consequences.[405] For instance, the ICJ provided in the *Nicaragua* case that certain non-binding instruments may be used as evidence of the *opinio juris* of states, especially when they are of an institutional character, such as UNGA resolutions.[406] Another route by which the decisions and instruments adopted by organs of inter-governmental organization can acquire legal force is through other rules of the constitutive document of the organization.[407]

Non-binding instruments may also be looked at from the viewpoint of the customary law of treaties. Article 31 (3) of the Vienna Convention, which sets out the generally accepted rules of treaty interpretation, provides that:

---

[405] For arguments in a similar vein, see Ebbesson 1996, pp. 20-22.

[406] *Military and Paramilitary Activities in and against Nicaragua (Nicaragua v. United States of America), Merits, Judgment, I.C.J. Reports 1986, p.14, paragraphs 202-209.* As clarified by Chinkin (1989, p. 858): 'There the Court held that the opinio juris of States with respect to the prohibition against the use of force could be deduced from their attitude towards the relevant resolutions of the General Assembly. Without any more evidence of the alleged *opinio juris* than mere acceptance of the resolutions, the Court determined that there was indeed an established principle of customary international law. This conclusion was reached despite many instances of inconsistent State practice which were not considered by the Court to be fatal to the claim, but were dismissed as illegal breaches of the rule'.

[407] According to Horn (1988, p. 72): 'Article 65 of the Helsinki Convention stipulates a duty to report on measures taken pursuant to recommendations. This duty to report in combination with the very aims of the Nordic co-operation presupposes a duty to deal with a recommendation'.

> There shall be taken into account, together with the context:
> (b) any subsequent practice in the application of the treaty which establishes the agreement of the parties regarding interpretation

The idea here is that although a non-binding instrument does not establish any legal rights and obligations for the parties, it is possible that the states behave in accordance with the norms of the instrument and thus provide 'subsequent practice in the application of the treaty'. This applies primarily to non-binding instruments which are preceded by, and connected with, an international convention. Again, it is not the instrument which produces legal effects, but the behavior prompted by the norms contained within this non-binding instrument. Finally, as was shown above, provisions which are expressed in terms so general as not to manifest agreement between the states parties are still legally relevant. They form part of the object and purpose of the treaty and thus can be used in several contexts in the law of treaties, for instance, in interpreting the more specific treaty provisions and in studying whether a material breach of treaty has taken place.[408]

As was shown above, the informal instruments adopted by the ministerial conferences of the Arctic states are binding in international law, although they regularly set out only a minimum of requirements for the parties to them. These informal instruments were previously called 'treaties in simplified form'[409], but many scholars now use the term 'soft law instruments'. As was argued above, the term 'soft law' captures the essence of these instruments because they normally provide only for 'soft' requirements, albeit with the understanding that these are binding in international law.[410]

---

[408] See Articles 31 (1) and 60 (3) of the Vienna Convention. For problems in determining the object and purpose of treaties, see Klabbers 1997, pp. 138-160.
[409] See Klabbers 1996 I, p. 5 and the notes thereat. See also Elias 1974, p. 13.
[410] In this vein, see Ebbesson 1996, pp. 19-23. See also Chinkin 1989, pp. 850-867.

The problem with the term 'soft law' is that many international lawyers understand it to mean non-binding instruments with some legal consequences.[411] Indeed, in light of the disparate meanings associated with the term, it seems ill-advised to continue to call these instruments 'soft law instruments'. It is proposed here that these instruments be designated **'informal arrangements'** instead, with the understanding that their status in international law can be equated with that of international treaties, but that their normative substance generally provides for more modest regulation than formal international treaties. The term 'arrangement' can be justified on the basis that it has been increasingly used in international treaties.[412]

[411] See, e.g., Birnie and Boyle 1992, pp. 26-30.
[412] See, e.g., the 1991 Espoo Convention on Environmental Impact Assessment in a Transboundary Context, Article 8: 'The Parties may continue existing or enter into new bilateral or multilateral agreements, or other arrangements'; the 1992 Convention on the Transboundary Effects of Industrial Accidents, Article 24: 'The Parties may, in order to implement their obligations under this Convention, continue existing or enter into new bilateral or multilateral agreements or other arrangements'. Most strikingly, Article 9 (1) of the 1992 Transboundary Watercourses Convention requires the riparian states to conclude 'bilateral or multilateral agreements or other

IV ENVIRONMENTAL IMPACT ASSESSMENT (EIA) NORMS IN THE ARCTIC

In this chapter, the main study aim of this thesis is taken up, namely the systematization of the so called EIA norms. The idea is to interpret and systematize all the norms that apply to the planning phase of environmentally harmful stationary activities in the Arctic. The chapter starts with an introduction to the concept of EIA in general and EIA in the Arctic conditions in particular. Four basic categories of EIA norms are then distinguished: EIA norms applicable in national, transboundary and international contexts as well as EIA norms applicable to transboundary watercourses. After establishing the present state of EIA norms in the Arctic, one case is taken up in which the EIA norms have been applied in the Arctic, the case of the planned construction of the Vuotos Reservoir.

As a work aiming to systematize the norms of international law applicable in the Arctic, this thesis does not pretend that it is the first of its nature. As discussed in chapter III, the strategy of the Arctic cooperation process has been to rely heavily on existing international treaties and informal arrangements. The AEPS document already identified the existing international environmental treaties that could be applied to combat the six prioritized pollution problems[413] and environmental protection working groups collected the available international instruments in certain specified fields.[414] Scholars of international law have taken a corresponding interest

---

arrangements...in order to define their mutual relations and conduct regarding the prevention, control and reduction of transboundary impact'.

413 See especially Section 4 of the Strategy. Already the prepatory process to the AEPS document produced a report on the applicable international environmental treaties in the Arctic, the 1990 Report on the Yellowknife Preparatory Meeting.

414 On the PAME home page (at http://www.grida.no/pame/FRprogramme_activities.htm [12.3.2001]), it is provided: 'an update of the current status of the international conventions and agreements as presented in the 1996 PAME report and continues to assess the adequacy of existing international agreements and arrangements related to the protection of the Arctic marine

in the instruments, which is hardly surprising since the Arctic states have been at the forefront of international environmental policy-making and there are many international environmental treaties that apply to their Arctic territories.[415] Still, much work remains to be done, as noted by the Barrow ministerial meeting of the Arctic Council:

> **Call on** the Arctic States to accede to, ratify and implement relevant existing agreements designed to protect and restore the Arctic environment, and to identify gaps where new agreements may be needed.[416] (original emphasis)

The present study is not, however, confined to existing international treaties and informal arrangements but takes account of customary international law and the goals of international society as well.

It will first be necessary to define the main terms used in this chapter. As the thesis does not go too deeply into the complexities of defining environmental damage, the terms 'pollution' and 'transboundary pollution' are not defined in detail.[417] 'Pollution' will refer to a physical phenomenon produced by humans that causes adverse effects on the environment understood in a broad sense; e.g., it also includes negative effects on natural ecosystems[418] and biodiversity;[419] 'transboundary

environment'. The EPPR Working Group Meeting which took place 13 to 15 June 2000 also dealt with the completion of the analysis of international agreements relevant in the field of EPPR, agenda item 5 (2). The report can be found on the home page of the Arctic Council at http://www.arctic-council.org/eppr.asp [12.3.2001]. CAFF has also provided a 'Global Overview of the Conservation of Migratory Arctic Breeding Birds outside the Arctic', which can be found on its home page at http://www.grida.no/caff/global.htm [28.3.2001].

415 Rothwell has done a general study of the international environmental treaties applicable in the Arctic. See Rothwell 1995 II and 1996. Brubaker has examined the international environmental treaties applicable in the Barents region in particular. See Brubaker 1993.

416 Paragraph 16 of the 2000 Barrow Declaration.

417 Several attempts have been made by inter-governmental and non-governmental organizations alike to define what is meant by the term 'environmental interference' or 'pollution'. For an analysis, see Springer 1983, pp. 63-89, and Lefeber 1996, pp. 8-17.

418 The concept of 'ecosystem' can be defined according to Gates (1985, p. 2): 'An ecosystem is an ecological unit, a subdivision of the landscape, a geographic area that is relatively homogenous and

pollution' is a sub-concept of 'pollution' and will refer only to pollution that crosses international boundaries of states.[420]

'EIA norms' are understood to include all those norms of international law, both principles and rules, and of whatever origin, which apply to the planning phase of environmentally harmful stationary activities in the Arctic. The concept of environmental impact assessment has been given various legislative forms in the national legal systems of states of the world as well as in international treaties. It is intended that the term 'EIA norms' should capture the core of the concept of EIA, namely that it should refer to governmentally controlled procedures by which at least some kind of evaluation is made of the potential environmental impacts of a proposed environmentally harmful stationary activity. The concept will thus cover, for example, treaties by which states agree to notify other states in case of proposed activities since notification already includes at least some kind of evaluation of the possible transboundary impacts. The term is defined to be as inclusive as possible and thus also covers the rights of the private legal subjects to participate in the planning phase as well as other norms that apply to the planning phase of environmentally harmful stationary activities. In order to highlight the qualitative difference between the different procedures, the following distinctions are made.

---

reasonably distinct from adjacent areas...An ecosystem is made up of three groups of components: organisms, environmental factors, and ecological processes. That is, it comprises organisms, species, populations, and communities; soil, climate, and other physical factors; and processes such as energy flow, nutrient cycling, water flow, freezing, and thawing'.

[419] For an analysis of the problems in interpreting the concept of 'environment', see Birnie and Boyle 1992, pp. 2-3. Birnie and Boyle state (p. 2): 'None of the major treaties, declarations, codes of conduct, guidelines etc. referred to throughout this work attempts directly to define the term. No doubt this is because it is difficult both to identify and to restrict the scope of such an ambiguous term, which could be used to encompass anything from the whole biosphere to the habitat of the smallest creature or organism'.

[420] It is necessary to define the term 'transboundary pollution' here since the term can also be used for pollution that crosses the administrative boundaries within one state or of sub-units of federal states.

One sub-category of these EIA norms comprises **'prior procedures'**, which require states to establish governmentally controlled procedures in their own territorial jurisdiction for the planning phase of environmentally harmful stationary activities. These are mostly obligations to establish procedures for evaluating the environmental impact of proposed activities and obligations to establish licencing procedures. The most developed forms of 'prior procedures' are **'EIA procedures'**, which are defined in the following chapter. **'Prior transboundary procedures'** mean procedures that require the states concerned, i.e., the origin state[421] and the affected state[422], to communicate on the potential transboundary impacts of a proposed activity. **'Transboundary EIA procedures'** are prior transboundary procedures that are based on the operation of national EIA procedures, as will be studied in chapter V (1).

1. Introduction to the EIA Procedure

It is generally understood that the first national EIA procedure established was the National Environmental Policy Act (NEPA) of the United States in 1969. Thereafter, the EIA procedure first spread to the Anglo-American countries of Canada and Australia, and then to Europe and also to some developing countries.[423] At the moment, 'around 200 systems for environmental assessment (EA) have been introduced in countries, states and international organisations around the world'[424]. This is no wonder since the EIA procedure is seen as one of the most important means of implementing the goal of sustainable development.[425]

---

421 The term 'origin state' is used of a state in whose jurisdictional area the activity potentially causing transboundary pollution is proposed.
422 The term 'affected state' is used of a state in whose jurisdictional area potential damage from transboundary pollution may occur.
423 For a review of the history of EIA procedures, see Gilpin 1995, pp. 2-4, Sheate 1996, pp. 16-22 and Harrop and Nixon 1999, pp. 2-7. In Finnish, see Saastamoinen 1996, pp. 211-223.
424 Harrop and Nixon 1999, p. 2.

There exist differences between the national EIA procedures since the idea of EIA was implanted into different national traditions of environmental protection.[426] Yet, certain basic elements of a typical EIA procedure can be outlined. An EIA procedure is frequently established either via a separate piece of legislation or by including EIA provisions in existing sectoral laws.[427] The main function of an EIA procedure is to produce scientific information for the decision-making on the proposed activity. In this sense, an EIA procedure is not a procedure leading to a decision, but one producing information for a separate procedure that deals with the actual licencing decision. Frequently, it is provided in national EIA laws and regulations that the licencing decision-maker is obligated to take into account the results of the EIA procedure.

Importantly, the EIA procedure provides comprehensive information on all potential environmental impacts of the proposed activity. It does not only produce information that studies the potential environmental impact of the proposed activity; it also requires that the potential environmental impact of alternatives to the main plan, including the so called zero-alternative, in which nothing is constructed, be assessed and possible mitigation measures outlined. The decision-maker is thus provided with information that facilitates its task of prescribing the permit conditions for the proposed activity, not only its task of comparing the costs and benefits of the activity.

In accordance with the precautionary principle, most EIA procedures require that the uncertainty confronted in making the impact studies be explicated.[428] Since the language of the impact studies can be hard to understand, based as it is on scientific terminology, it is normally required that the results of the impact studies be written

---

[425] See, e.g., Sands 1994, pp. 357-358 and Gilpin 1995, pp. 9-10.
[426] See Gilpin 1995, pp. 2-4.
[427] Ibid., p. 2.
[428] For an analysis of the precautionary approach in the EIA Directive, see Peters 1996, pp. 210-213 and Sifakis 1998, pp. 349-352.

and presented in such a way that other than experts are also able to understand and assess the potential impact of the proposed activity. Since most EIA procedures require that the public and relevant governmental authorities be entitled to participate in an EIA procedure one way or another, it is important that the results of the impact studies be written in an understandable manner.

A special terminology has developed to describe the different phases of this assessment procedure, called an EIA procedure here.[429] Because some differences in terminology exist, it is useful to define what is meant by each term used in this thesis. Not all proposed activities go through an EIA procedure. Since only those activities that are known to be environmentally harmful are to be assessed, a decision must be made as to which activities are harmful enough to require an assessment procedure. In some jurisdictions, the preferred system is a mandatory list of activities, normally included in the general legislative act on EIA. These activities have been evaluated as being environmentally harmful in all situations, and thus as requiring an EIA procedure. In some jurisdictions, no such list exists, and the responsible agency thus has discretion to decide whether to apply an EIA procedure or not. In most jurisdictions, elements from both of these extreme systems are included.[430] The

[429] In some jurisdictions, the term describing the whole procedure is environmental assessment (EA), e.g., in the United Kingdom. Sheate (1996, p. 14) gives a good explanation why the term EIA is preferable: 'This book refers throughout to EIA (environmental impact assessment) although EA (environmental assessment) is used more widely in the EC and the UK. However, EA in the United States of America is used to describe an early assessment of whether an EIA is required. The use of EIA therefore avoids confusion'.

[430] Frequently, a distinction is made between the screening systems of Europe and North America, the former upholding the mandatory list approach and the latter, the United States and Canada, operating on the basis of case-by-case analysis. Yet, the EIA Directive and the Amendment Directive combine both approaches in the sense that there is a list of activities which have to be assessed (Article 4 (1) and Annex I) but also a list of activities with regard to which the member states have discretion (Article 4 (2) and Annex II). In the United States, the federal agency in charge of making the assessment under the NEPA frequently possesses a list of activities that are excluded from the impact studies. If the agency is not certain whether significant impact is likely, it can make a preliminary impact study, the result of which can be that no significant impact is likely, and thus there is no need for full-blown impact studies; or that significant impacts are likely and a full-blown impact study must be organized. In Canada, the federal screening system under the Canadian Environmental Assessment Act (CEAA) falls between the screening systems described above

system by which the state determines those activities that go through the EIA procedure can be called a **screening** system, whereas a decision to apply the EIA procedure to a certain individual activity can be called the screening phase of an EIA procedure.

As was outlined above, an EIA procedure provides not only information on potential environmental impacts but also all kinds of other information, such as possible mitigation measures. The requirements as to the scope of the studies to be made are normally laid down in a general legislative act, but in some jurisdictions a special **scoping** procedure must be organized. A scoping procedure is the phase of an EIA procedure in which the public and relevant governmental authorities are entitled to make comments on which issues should be taken up in the assessment studies. Many jurisdictions provide both a list of issues that have to be addressed in the assessment studies and a separate scoping procedure.[431] The document that records the results of the impact studies which have been conducted is called the **'Environmental Impact Statement (EIS)'**[432].

---

because there is no threshold requirement such as 'significant impact' as in the system in the United States and no list of projects that require environmental assessment absolutely. Yet there is a list, called the 'comprehensive study list', based on the type of projects. If the project is of a type described in this list, a more thorough impact study should be made. For information on the EIA systems of Canada and the United States, see the home page of the NACEC at http://www.cec.org/pubs_info_resources/law_treat_agree/index.cfm?varlan=english [26.3.2001].

[431] A good example is the Finnish EIA procedure. The Finnish Act on EIA contains an Article which defines the minimum scope of the impact study (Section 2). The procedure starts with the screening decision, and if the procedure is deemed applicable, the project proponent is obligated to first prepare an assessment program, which is reviewed by a coordination authority. Thereafter, the public and other governmental ministries can state their views on the adequacy of the program and the issues that should be addressed in the report. For the Finnish EIA procedure in general, see Environmental Impact Assessment, Finnish Ministry of the Environment 1996.

2. EIA in the Arctic Conditions

As was briefly examined above[433] and will be studied in detail below,[434] the Arctic states have adopted Guidelines to adapt the EIA procedure and the transboundary EIA to Arctic conditions: the EIA Guidelines and the Offshore Guidelines adopted by the 1997 Alta Declaration. These two set of guidelines do not establish direct orders but only obligations of attempt for the eight Arctic states. But could it be argued that the norms of these guidelines are, in fact, direct orders because of the special conditions in the Arctic?

From the standpoint of environmental protection alone, the Arctic presents a number of problems in comparison to the temperate regions. First, as a rule, it is more risky to carry on environmentally harmful stationary activities in the Arctic. In particular, the long, dark and cold Arctic winter puts human technology to the test. It seems safe to presume that if a certain environmentally harmful human activity is practiced in the Arctic as well as in more temperate climatic areas, the Arctic poses the more challenging environment for the safe operation of these activities. A good example is offshore hydrocarbon exploitation, in which ice creates additional hazards not only in the actual drilling phase but also in the transportation phase. It also hampers efforts to contain the effects of an accident.[435]

---

432 In some jurisdictions the term 'environmental statement (ES)' is preferred. See Harrop and Nixon, p. 3.
433 See *supra* chapter III (3.3.).
434 See especially *infra* chapter IV (3.3.1.).
435 In Section 10.8., 'Conclusions and recommendations', the AMAP Assessment Report (1998, p. 693) states: '8. The effects of oil pollution may be more severe and persistent in the Arctic than for corresponding situations in other regions. This is due to the effects of the extreme Arctic climatic conditions on the distribution, composition, and physical state of petroleum hydrocarbons as well as on the rates of biological factors, such as growth rates, productive seasons lengths, generation turnover times, and ages at maturity, that determine the rate of ecosystem recovery from oil damage...10. Primary concerns associated with the major new Arctic oil and gas developments involve the increased risks of accidents and the increased difficulties of taking remedial measures in such severe environments'.

Second, the Arctic environment, and especially its ecosystems, seem to be more vulnerable to anthropogenic pollution than the ecosystems of more temperate regions. Osherenko and Young have listed some possible factors which can make the Arctic ecosystems more fragile:

1. Low temperatures retard the decomposition of natural and man-made substances and the breakdown of pollutants;
2. Regeneration is a protracted process because of the short growing season;
3. Large concentrations of animal populations heighten vulnerability to catastrophes;
4. Marine areas are particularly important in the Arctic in comparison with other regions of the globe;
5. Climatic conditions are likely to produce a more pronounced CO2-induced warming trend; and
6. Severe weather and ice dynamics make environmental protection and clean-up extremely difficult.[436]

The authors conclude that 'combinations of these Arctic conditions intensified by several phenomena peculiar to the region make Arctic ecosystems extremely vulnerable to physical and chemical disturbance'[437]. The view of Arctic ecosystems as especially sensitive can be presumed here, and one must ask whether the special sensitivity of the Arctic ecosystems and the fact that the Arctic conditions are more risky from the viewpoint of operating industrial activities should or must have an effect on the implementation and application of EIA and transboundary EIA procedures that apply to the whole territory of the eight Arctic states, both below and

[436] Osherenko and Young 1989, pp. 111-117. Samson (1997, pp. 34-39) argues that this kind of perspective is too simplistic. Also, the AMAP Assessment Report (1998, p. 118) states: 'Arctic species are generally not considered 'sensitive', since environmental tolerances of most species are broad. However, the physical environment of the Arctic is sensitive'.

[437] Osherenko and Young 1989, p. 116.

above the Arctic Circle.[438] It is important to note that since the Arctic as defined here consists of varying ecological systems, the factors enumerated above figure more prominently in some places than in others.[439]

The principle of due diligence applies between all eight Arctic states and international EIA treaties apply between those Arctic states that are parties to them. The national EIA laws of the Arctic states do not, as a rule, make any distinctions between the Arctic and southern areas of the states.[440] Yet, it can be asked:

1. Do special characteristics of the Arctic require that EIA and transboundary EIA be implemented and applied differently, according to the circumstances?
2. Does the fact that the Arctic presents a more risky environment mean that the EIA and transboundary EIA must be applied in a stricter manner in the Arctic areas of the Arctic states than in their sub-Arctic areas?

Many times the EIA procedures must be applied differently by the Arctic states in their Arctic areas than in their southern areas because the environmental conditions of the Arctic differ so much from the conditions in more temperate regions. A good example is the definition of the term 'environment' which provides the scope of issues that must be, at a minimum, addressed in the EIS. Since the environment of the Arctic is different from that of southern latitudes, the scope of the impact studies

438 The similar question can be posed whether international areas, both of the Arctic and of other regions of the globe, require a different kind of application of EIA procedures. Yet, because this question is of such marginal importance in the Arctic, and since the basic ideas developed in this chapter apply also to these cases, this issue is not explicitly treated here.

439 The AMAP Assessment Report (1998, p. 119) makes a distinction between the high Arctic, low Arctic and subarctic.

440 Some EIA procedures apply only in the areas above the Arctic Circle, such as the EIA procedure of the Nunavut settlement area. The Nunavut Impact Review Board (NIRB), established by the Nunavut Land Claims Agreement, is in charge of these procedures, which specifically concentrate on integrating the scientific studies and traditional knowledge of indigenous peoples. A general presentation of the NIRB and its activities can be found at http://www.wkss.nt.ca/html/nirb.html [23.4.2001] pending the construction of NIRB's own home page.

covers different issues than the EIA procedures organized in the southern areas of the Arctic states. Another example is the organization of public participation, which surely must be carried out differently in an area where human population is sparse and indigenous peoples are the main inhabitants. The EIA Guidelines and the Offshore Guidelines provide guidance in this respect as to how to proceed with the implementation of EIA in the Arctic conditions, which must in any event be conducted in a different manner than in the sub-Arctic parts of the Arctic states.

Although the Arctic areas of the Arctic states may be considered a more risky environment than the more temperate areas of these states, international law obligations apply equally to the whole of the territory of the states. The basic principle of freedom of contracting entitles states to a freedom of choice in concluding treaties; i.e., it is their option to not participate in treaty-making, to choose to take unilateral action or to accept only obligations of attempt. Since they have chosen to adopt only obligations of attempt, they are only obligated to try to observe these two sets of guidelines. As a consequence, although it is prescribed that the threshold levels for the application of the EIA procedure should be lower in the Arctic than in the southern areas of the Arctic states[441] and that the determination of whether the potential impact of the project is likely to cause a significant impact should use different thresholds in the Arctic,[442] these stipulations constitute no more than obligations of attempt, meaning that the Arctic states need only try to observe them.

[441] Paragraph 1 of Section 2 of the EIA Guidelines provides: 'The decision to conduct an EIA for a project in the Arctic should take into account the special conditions in the Arctic. Arctic-specific thresholds and sensitivity criteria are strongly recommended'.

[442] Ibid. Section 5 (4) contains the following guidance: 'The sensitivity of the arctic environment demands special attention, possibly in the form of special arctic thresholds for significance'.

3. EIA Obligations on the National Environmental Protection Systems of the Arctic States

International law contains an increasing number of principles and rules which guide states to improve their national environmental protection system.[443] A good example is the extensive set of such principles in the non-binding Rio Declaration on Environment and Development (hereinafter 'the Rio Declaration').[444] As will be shown below,[445] many international treaties have incorporated corresponding rules and principles. These harmonization norms are of great importance in the planning stage of stationary activities, for they require states to modify their national environmental protection system to include prior procedures for studying the potential environmental impacts of activities and for organizing the participation of the public of the origin and the affected states. In many cases they also require that prior licencing systems be established for environmentally harmful activities.

3.1. Territorial Scope of Obligations

As was shown in chapter II, the Arctic states enjoy full sovereignty over much of the Arctic's physical space. Large areas of the Arctic are also under their jurisdictional competence. The states possess different rights in regard to their areas of territorial sovereignty and jurisdictional competence. With respect to areas of territorial sovereignty, the states are in principle free to decide the content of all policies which affect the area. In the case of stationary activities in the Arctic, this means that the content of all policies which cover these activities, for instance, defense policy,

443 Here the term 'national environmental protection system' is used in its broadest sense to refer not only to national laws and institutions that protect the environment but also to laws and institutions that deal with issues of sustainable development in general.

444 The Declaration can be found from 31 I.L.M. 874 (1992).

445 See *infra* chapter IV (3.3.).

industrial policy and environmental policy, are freely decided by the Arctic states. This applies, of course, to the planning phase of these activities as well: states are free to decide the level of requirements in the planning phase within the limits of international law.

The situation is different where maritime areas of jurisdictional competence are concerned. What is common to all of the different areas of jurisdictional competence, the Exclusive Economic Zone (EEZ) and the continental shelf, is that the states' power of decision is limited to those areas of policy which are expressly permitted by the law of the sea for each different zone. In these specific areas of policy, the Arctic states have rights similar to those they enjoy with respect to territorial sovereignty. An EEZ accords a state exclusive rights mainly in regard to natural resource exploitation. A continental shelf area provides governance of the sea-bed. The main question here is whether the obligations of international environmental law/ the law of sustainable development apply in the areas of jurisdictional competence. The main principle of international environmental law, the due diligence principle, obligates the states to ensure that 'activities within their jurisdiction or control do not cause damage to the environment of other States or of areas beyond national jurisdiction'[446]. Since the term 'jurisdiction' is used, the areas of jurisdictional competence are covered.

There are some international treaties studied below that do not provide exact parameters as to the physical areas to which they apply. The 1969 Vienna Convention on the Law of Treaties (hereinafter 'the Vienna Convention') and the customary law of treaties[447] provide: 'Unless a different intention appears from the treaty or is

[446] Principle 21 of the 1972 Stockholm Declaration and principle 2 of the 1992 Rio Declaration. The Stockholm Declaration can be found in 11 I.L.M. 1416 (1972) and the Rio Declaration in 31 I.L.M. 874 (1992).

[447] In effect, a majority of the rules of the 1969 Vienna Convention can be regarded as part of customary international law; see, e.g., Villiger 1985. As Brownlie states (1998, p. 608), if not yet clearly part of customary international law, they 'constitute presumptive evidence of emergent rules

otherwise established, a treaty is binding upon each Party in respect of its entire territory'[448]. The use of the word 'territory' means that the treaty applies only to states' sovereign territory, i.e., the states' land area, internal waters, territorial sea and the airspace above these. If a treaty does not contain an explicit clause to the contrary, or if it refers merely to 'territory', it only applies to a state's area of sovereignty.[449] Some of the treaties studied use the term 'jurisdiction' to refer to areas of jurisdictional competence.[450]

3.2. Obligations in General International Law

Especially after the Rio Declaration, a number of potential principles of general application have been identified, e.g., the principle of precaution and the polluter-pays principle.[451] These principles require measures from states irrespective of whether the activities within their territory cause damage to the territory of other states and thus '[have broadened] the idea of what is international about environment'[452] from the traditional perspective of transboundary pollution. Some of these principles also apply to the situations studied here, for instance, the principles of precaution and sustainable development.

A considerable number of interpretations of the content and scope of each of these principles can be found, but much of the confusion these entailed was clarified by the ICJ in the *Gabcíkovo-Nagymaros* case of 1997.[453] First of all, the Court opted to

---

of general international law'. Aust (2000, p. 11) goes even further: 'For most practical purposes treaty questions are resolved by applying the rules of the Convention. To attempt to determine whether a particular provision of the Convention represents customary international law is now usually a rather futile task'.

448 Article 29 of the Vienna Convention.

449 See, e.g., Sinclair 1984, pp. 89-90 and Aust 2000, p. 162.

450 See e.g., Article 1 (ii) of the 1991 Convention on Environmental Impact Assessment in a Transboundary Context.

451 See, e.g., Birnie and Boyle 1992, pp. 89-112, Sands 1994, pp. 303-381, and Hohmann 1994.

452 Boyle 1997, p. 19.

453 *The Gabcíkovo-Nagymaros Project (Hungary/Slovakia), Judgment, I.C.J. Reports 1997, p. 7.*

express these new principles in terms of the concept of sustainable development rather than the principle of precaution.[454] Second, the ICJ noted that it did not have to study the issue of whether any norms of jus cogens had arisen in the field of international environmental protection, because neither party invoked the issue.[455] The Court opined:

---

[454] Ibid. The Court did refer to the precautionary approach but only in passing. In paragraph 113 it states: 'The Court recognizes that both Parties agree on the need to take environmental concerns seriously and to take the required precautionary measures'. Interestingly, the International Tribunal of the Law of the Sea (ITLOS) resorted to the principle of precaution in its judgment on provisional measures in the Southern Bluefin Tuna Case: '79. Considering that there is scientific uncertainty regarding measures to be taken to conserve the stock of southern bluefin tuna and that there is no agreement among the parties as to whether the conservation measures taken so far have led to the improvement in the stock of southern bluefin tuna; 80. Considering that, although the Tribunal cannot conclusively assess the scientific evidence presented by the parties, it finds that measures should be taken as a matter of urgency to preserve the rights of the parties and to avert further deterioration of the southern bluefin tuna stock'. See also *the Request for an Examination of the Situation in Accordance with Paragraph 63 of the Court's Judgment of 20 December 1974 in the Nuclear Tests (New Zealand v. France) Case, I.C.J. Reports 1995, p. 288, paragraph 64.*

[455] Ibid. In paragraph 112, the Court states: 'Neither of the Parties contended that new peremptory norms of environmental law had emerged since the conclusion of the 1977 Treaty, and the Court will consequently not be required to examine the scope of Article 64 of the Vienna Convention on the Law of Treaties'. Indeed, there is no evidence that norms which permit no derogation have come to prevail in a field of international law characterized by what might be termed balancing norms. Some norms of international law, the so-called *jus cogens* norms, are undoubtedly treated as being of higher status than 'normal' rules of international law. Norms of *jus cogens* are norms that are accepted by the whole international community and are absolute in character in the sense that they permit no derogation. International morality and international law clearly overlap in *jus cogens* norms; these norms regulate behavior which is morally condemnable, such as genocide, and thus permit no derogation. Yet even these norms must meet the criteria of general law formation to become accepted in international law. There exists no equivalent to the national constitutional law-making procedures for the creation of *jus cogens* norms in international law; the norms are a product of the operation of normal law-making processes which produce general norms of international law. However, jus cogens norms override all other 'normal' rules of international law in their field of application and can only be changed by norms of similar status. Do there exist norms of *jus cogens* applicable to the situations studied in this thesis? In the field of international environmental protection, there evidently exists at least one norm of *erga omnes*, i.e., a norm which establishes a legal relation between a state and all other states of the world, or between a state party to a treaty and the whole treaty community of states. This is the principle of due diligence, which will be studied in detail in chapter IV (2). This principle requires states to prevent pollution from the activities within their jurisdiction and control that affect the environment of international areas. The concepts of *erga omnes* and *jus cogens* do not coincide in scope, however; only some *erga omnes* norms are *jus cogens* norms. It seems clear that the principle of due diligence is not a norm of *jus cogens*, a norm that would permit no derogation, as it does not contain any prohibition of an absolute character. For an analysis of the relationship between the concepts of *jus cogens* and *erga omnes*, see Byers 1997, pp. 211-239. Another distinction that is relevant here, one introduced by the state responsibility project of the ILC, is that between international crimes and international delicts.

> Owing to new scientific insights and to a growing awareness of the risks for mankind – for present and future generations – of pursuit of such interventions at an unconsidered and unabated pace, new norms and standards have been developed, set forth in a great number of instruments during the last two decades. Such new norms have to be taken into consideration, and such new standards given proper weight, not only when States contemplate new activities but also when continuing with activities begun in the past. This need to reconcile economic development with protection of the environment is aptly expressed in the concept of sustainable development.[456]

The concept of sustainable development was used by the Court as the most general normative expression, one requiring that the two important values of international society - economic development and environmental protection - be reconciled; i.e., neither can override the other. Is the concept of sustainable development, then, part of *lex lata*, or is it only an expression of salient societal values?

Sustainable development seems best described as a goal of the international community rather than a legal principle.[457] According to Schachter, a goal of the

---

Evidently, international crimes are normally covered by *jus cogens* norms because both of these concepts are based on the same idea: the norms derived from these concepts are seen as more important for the community of states than normal rules and are thus elevated to a higher status within primary norms (*jus cogens*) and their breach is regarded as requiring a stronger reaction from the community of states (international crimes). Interestingly, the ILC introduced examples of international crimes in its draft articles of which Article 19 (3) is relevant here: 'A serious breach of an international obligation of essential importance for the safeguarding and preservation of the human environment, such as those prohibiting massive pollution of the atmosphere or of the seas.' Could this also be seen as a norm of *jus cogens*? The problem here lies primarily in the fact that, as will be shown below, such a norm has not become accepted in general international law, which clearly is a condition for the norm to become a norm of *jus cogens*. For a thorough analysis of the concept of *jus cogens*, see Hannikainen 1988.

[456] Ibid., paragraph 140 of the judgment in the *Gabcíkovo-Nagymaros* case. The Court identified the following as one important part of sustainable development: 'The awareness of the vulnerability of the environment and the recognition that environmental risks have to be assessed on a continuous basis have become much stronger in the years since the Treaty's conclusion' (paragraph 112).

[457] For arguments in a similar vein, see Birnie and Boyle 1992, pp. 3-6, Malanczuk 1995, p. 51, and Fuentes 1998, pp. 119-200. Boyle (1997, p. 18), however, argues: '*Gabcíkovo-Nagymaros* nevertheless suggests that we are dealing with a legal principle, however confined, rather than a mere policy or moral invocation'. Handl (1991, pp. 79-83) also takes a positive stance on the matter

international community can be defined as 'a statement of the policy or objective underlying a legal rule or principle'[458]. Schachter asserts that these goals 'differ from principles in that they do not in themselves give rise to claims of entitlements'[459] but if 'linked to principles and rules, they become relevant to choices among opposing principles and rules'[460]. In a similar vein, Lowe finds that the concept of sustainable development is not a norm of customary law but that it nevertheless has a normative status in international law.[461] He asserts that the concept 'is a meta-principle, acting upon other legal rules and principles – a legal concept exercising a kind of interstitial normativity, pushing and pulling the boundaries of true primary norms when they threaten to overlap or conflict with each other'[462]. Both scholars conclude that such

---

(p. 80): 'it [sustainable development] has thus turned out to be a notion around which legally significant expectations regarding environmental conduct have begun to crystallize. Indeed, before long the concept might turn into a mandatory standard of international law evaluation, a peremptory norm of international law'. Brownlie (1998, p. 287) does not share this enthusiasm, stating '[f]or the present, the concept remains problematic and nebulous, appearing more as a statement of the issues than as a resolution of the basic problems'.

458 Schachter 1991, p. 21. Schachter himself treats the goals of the international community as part of international law, which itself can be divided into rules, principles and goals on the basis of level of abstraction. Although Schachter does not say so, he seems to have been inspired by Dworkin (1986), who introduced the idea of having an approach to legal decision-making between conventionalism (pp. 114-150; Schachter's 'positivism', in which rules are validated on the basis of criteria that are internal to the system) and pragmatism (pp. 151-175; Schachter's 'instrumentalism', in which decision-making based on perceived preferable goals from the viewpoint of good of the whole society) which he called 'the law as integrity' (pp. 176-275) and combines the most worthwhile aspects of both approaches. In a similar vein, Schachter (1991, pp. 18-33) perceives that if rules validated through the sources of law doctrine exist in international law, they take primacy, but if such cannot be found, goals and objectives also become important in legal decision-making. There are scholars who do not accept the position that the international community is an association that generates its own shared goals and objectives. For instance, Nardin (1983) distinguishes two basic modes of international associations: purposive and practical associations. In the purposive association (p. 10): 'the authority of the rules governing the relations among the associates is derived from the shared purposes that the association exists to promote'. According to Nardin (p. 12), 'international society could not be regarded as a kind of purposive association'.

459 Schachter 1991, p. 21.

460 Ibid.

461 Lowe (1999, p. 30) argues: 'I find it hard to see that the concept of sustainable development can be regarded as having sufficient identifiable normative meaning to be capable of generating a self-contained norm of customary international law, no matter what its utility as a description of policy goals in international treaties might be'.

462 Ibid., p. 31.

goals or concepts are not part of *lex lata* from the strict legal point of view but nevertheless fulfill several tasks in the international legal system.[463]

The ICJ's decision shed light on how the social goal of sustainable development should be understood. According to Boyle:

> the Court's use of sustainable development has implications primarily for the process of decisionmaking by the parties and not for the decision itself. It is in this sense that requiring states to evaluate and assess environmental impacts and apply new environmental norms and standards becomes part of the process for giving effect to the objectives of sustainable development. From this perspective, sustainable development may also entail commitments to public participation in decisionmaking, environmental impact assessment, and the application of the precautionary principle.[464]

Boyle also states: 'it [sustainable development] dictates no particular outcome or result and leaves the parties free to give effect to this fundamental value in almost any way they choose'[465]. In a similar vein, Lowe considers that the concept 'seem[s] to be more of a procedural than of a substantive character'[466].

Boyle's proposal will be seen as providing a relevant starting point here for defining the parameters of the goal of sustainable development. His proposal can be refined by interpreting it in light of the Rio Declaration, which clearly is the most authoritative declaration of the content and scope of the goal.[467] Because the principle is very

---

[463] For an analysis, see Schachter 1991, pp. 21-31 and Lowe 1999, pp. 31-37.
[464] Boyle 1997, p. 18.
[465] Ibid.
[466] Lowe 1999, p. 36.
[467] Unlike most other legally relevant documents, which have been adopted by inter-governmental organizations, a regional grouping of states only, or by non-governmental actors, the Rio Declaration differs in that it was signed by representatives of almost all states of the world. Interesting recent suggestions regarding the content and scope of the goal of sustainable development in international law have been made by Ebbesson 1996, pp. 233-256, and Boyle and Freestone 1999, pp. 1-19. For a theoretical analysis of the concept in Finnish, see Karvo 1994, pp. 279-300.

general in nature, different elements of it are relevant in different contexts. Its main function is to provide guidance to the national environmental protection systems of the states of the world in general, including the Arctic states. Each chapter to follow begins by examining what the goal of sustainable development requires in the context described.

3.3. Treaty Obligations to Harmonize Prior Procedures

In the situations studied in this thesis, the most important aspect of the goal of sustainable development is that states have been prompted to introduce some form of EIA procedure. An EIA procedure can be defined as 'the official appraisal of the likely effects of a proposed...project on the environment; alternatives to the proposal; and measures to be adopted to protect the environment'[468] before a licence is given to undertake the activity. According to the Rio Declaration:

> Environmental impact assessment, as a national instrument, shall be undertaken for proposed activities that are likely to have a significant adverse impact on the environment and are subject to a decision of a competent national authority.[469]

This principle provides only general outlines of the requirement of the goal of sustainable development. At the very least, a national EIA procedure, which is applicable to the planning stage of stationary activities, should contain a procedure whereby the potential environmental impacts of an activity are evaluated and the public is given some kind of right to participate in the procedure. The goal of sustainable development also implies that a licencing system should be introduced for those activities that are known to be environmentally harmful. As will be shown below,[470] the customary law principle of due diligence can also be viewed as

---

[468] Gilpin 1995, pp. 4-5.
[469] Principle 17 of the Rio Declaration.
[470] See *infra* chapter IV (4.).

requiring the establishment of national EIA procedures and national prior licencing procedures for environmentally harmful activities.

3.3.1. Treaties Providing for Prior Procedures

There exist prior procedures that are designed for certain stationary activities only, such as the management of the use of nuclear power for peaceful purposes.[471] Two treaties are relevant here that are applicable in the Arctic and have been negotiated under the framework of the International Atomic Energy Agency (IAEA). *The 1994 Convention on Nuclear Safety*[472] concentrates generally on 'the application of fundamental safety principles, rather than of detailed safety standards' in relation to all nuclear installations. *The 1997 Joint Convention on the Safety of Spent Fuel Management and on the Safety of Radioactive Waste Management (hereinafter 'the 1997 Joint Convention')*[473] concentrates only on the management of radioactive waste.

The Convention on Nuclear Safety contains three provisions on prior evaluation. The corresponding provisions of the 1997 Joint Convention are quite similar in content, and the two conventions are usefully considered together. The general provision on EIA in the Convention on Nuclear Safety is contained in Article 14, where it is required that 'comprehensive and systematic safety assessments are carried out

[471] Neither the Convention on Nuclear Safety nor the Joint Convention on the Safety of Spent Fuel Management and on the Safety of Radioactive Waste Management applies to military uses of nuclear power. Yet, paragraph (viii) of the Joint Convention provides a non-binding recommendation that: 'spent fuel and radioactive waste excluded from the present Convention because they are within military or defence programmes should be managed in accordance with the objectives stated in this Convention'. Article 3 (3) of the Joint Convention also provides for the possibility of a declaration to include 'spent fuel or radioactive waste within military or defence programmes'. If the spent fuel and radioactive waste are managed by 'exclusively civilian programmes', the Convention applies even though the origin of waste is in a military or defence program.
[472] For further information on the Convention, see *infra* chapter IV (4.1.3.).
[473] For further information on the Convention, see *infra* chapter IV (4.1.3.).

before the construction and commissioning of a nuclear installation and throughout its life'[474]. These assessments must also be 'well documented...and reviewed under the authority of the regulatory body'[475]. This provision does not require that the prior evaluation contain an evaluation of the dangers posed throughout the life cycle of the nuclear installation, but only that safety assessments be made after the commissioning of the facility.[476] The 1997 Joint Convention takes a somewhat different approach in terms of both spent fuel and radioactive waste. Articles 8(i) and 15(i) require in identical terms that 'a systematic safety assessment and an environmental assessment appropriate to the hazard presented by the facility and covering its operating lifetime shall be carried out'. Hence, the safety assessment must also cover the assessment of the life cycle of the activity.

Prior evaluation of the life cycle of the nuclear installation is not totally lacking in the Convention on Nuclear Safety; it is required in Article 17 as part of siting requirements. Appropriate procedures must also be established and implemented for evaluating all site-related factors 'likely to affect the safety of a nuclear installation for its projected lifetime'[477] and 'the likely safety impact of a proposed nuclear installation on individuals, society and the environment'[478]. The 1997 Joint Convention contains corresponding requirements in relation to both spent fuel and radioactive waste but adds that this safety information must be made available to the members of the public.[479] Finally, Article 19 of the Convention on Nuclear Safety requires that all states parties ensure that safety analysis is a precondition for authorization.[480]

---

[474] Article 14 (i) of the Convention on Nuclear Safety.
[475] Ibid.
[476] Ibid., Article 14 (ii).
[477] Ibid., Article 17 (i).
[478] Ibid., Article 17 (ii).
[479] Article 6 (iii) and Article 13 (iii) of the 1997 Joint Convention.
[480] Ibid. Articles 9 and 16 make the same requirement in terms of spent fuel and radioactive waste.

Some treaties provide for prior procedures as part of the management of accidents. Most of the international treaties that deal with accidents concentrate on issues other than requiring the contracting states to establish preventive procedures such as EIA and licencing. Frequently, these treaties obligate the states parties to notify each other if an accident takes place and require the parties to combat the early damage of the accident together.[481] Also common are strict liability conventions which try to ensure that compensation is paid to those who suffer from the accident.[482] In recent years, some treaties have gone further and required the establishment of prior procedures as part of the measures to prevent accidents from happening in the first place. These procedures have a distinct function in these treaties; i.e., environmental information is collected mainly for the purposes of accident prevention. There exist two treaties applicable in the Arctic that are aimed at combatting accidents in general: *the ILO 1993 Convention on Prevention of Major Industrial Accidents* and *the 1992 Convention on the Transboundary Effects of Industrial Accidents (hereinafter 'the 1992 Industrial Accidents Convention')*.

The 1992 Industrial Accidents Convention[483] contains many obligations requiring the contracting states to establish prior evaluation procedures for industrial accidents; an accident is defined as 'uncontrolled development in the course of any activity involving hazardous substances'[484]. Many categories of activities have been excluded from the scope of the Convention, however, e.g., nuclear accidents and accidents at military installations.[485] First of all, the contracting states are obligated to identify all hazardous activities, including proposed ones.[486] Article 6 requires the contracting states to ensure that operators of hazardous activities, including proposed ones,

---

[481] Good examples are the IAEA Conventions of the 1986 Convention on Early Notification of a Nuclear Accident and the 1986 Convention on Assistance in the Case of a Nuclear Accident or Radiological Emergency.
[482] For an overview of these conventions, see Wetterstein 1991, pp. 49-71.
[483] For further information on the Convention, see *infra* chapter IV (4.1.3.).
[484] Ibid., Article 1 (a).
[485] Ibid., Article 2 (2).

provide important safety information on the proposed hazardous activity to the competent authorities; this information is specified in Annex V. Finally, in Article 7, all the contracting states are required to establish siting policies for new hazardous activities.

In the International Labour Organization's (ILO) 1993 Convention on the Prevention of Major Industrial Accidents[487], the scope of the Convention is defined using the term 'major hazard installation'[488]. Such an installation is one which 'produces, processes, handles, uses, disposes of or stores...one or more hazardous substances in quantities which exceed the threshold quantity'[489]. Nuclear and military installations are excluded from the scope of the Convention, as is 'transport outside the site of an installation other than by pipeline'[490]. Article 10 of the Convention requires operators of the major hazard installations to prepare a safety report. This safety report must be prepared before any new major hazard installation is put into operation.[491] The requirements for the content of the safety report are defined in Article 9. Essentially, the operator must specify the measures aimed at ensuring that no accident will follow from operation of the installation. These include 'the identification and analysis of hazards and the assessment of risks including consideration of possible interactions between substances'[492]. Article 12 obligates operators to deliver these safety reports to competent authorities.

In some treaties on economic cooperation, states have included environmental protection measures in general and an EIA procedure in particular as a supplement to

---

[486] Ibid., Article 4 (1).
[487] ILO Convention C 174, Prevention of Major Industrial Accidents, was adopted on 22 June 1993, and entered into force on 3 January 1997. It has been ratified by only five states, of which one, Sweden (21.12.1994), is an Arctic state. Information on the status obtained from the home page of ILO at http://ilolex.ilo.ch:1567/public/english/docs/convdisp.htm [28.4.2001].
[488] Ibid., Article 3 (c).
[489] Ibid.
[490] Ibid., Article 1 (3).
[491] Ibid., Article 10 (2b).

the treaty. In *the 1994 Energy Charter Protocol on Energy Efficiency and Related Environmental Aspects*[493], the EIA procedure is in fact promoted: Article 8, 'Domestic Programmes', provides that domestic programs on energy efficiency 'may include activities such as...assessment of the energy, environmental and economic impact of actions taken'; the Illustrative and Non-Exhaustive List of Possible Areas of Co-Operation Pursuant to Article 9 includes 'environmental impact assessment, including major energy projects' as one possible area of cooperation. Clearly, however, the two provisions only declare the possible areas of policy, not mandatory ones. *The 1994 North American Agreement on Environmental Cooperation (hereinafter 'the 1994 Environmental Cooperation Agreement')*[494], in Article 2 (e), also contains a general duty to 'assess, as appropriate, environmental impacts'. Since the duty is qualified by the term 'as appropriate', it is best treated as an obligation of attempt.

Certain prior procedures focus on evaluating effects on certain parts of the environment. *The 1982 United Nations Convention on the Law of the Sea ('the LOSC')*[495] contains a duty of environmental impact assessment for the contracting states.[496] Article 206 requires that if there are reasonable grounds to believe that planned activities under the jurisdiction or control of contracting states 'may cause

---

[492] Ibid., Article 9 (a).

[493] The 1994 Energy Charter Protocol on Energy Efficiency and Related Environmental Aspects is an integral part of the European Energy Charter Treaty. Nevertheless, the contracting parties to the Treaty must give their consent to the Protocol before it becomes binding on them. The Treaty and the Protocol entered into force on 16 April 1998. Of the Arctic states, the Protocol has entered into force for Denmark (16.4.1998), Finland (16.4.1998), Sweden (16.4.1998) as well as the EC (16.4.1998). The Arctic states that have ratified the Protocol but not the Treaty, and thus are not yet parties to the former, are Iceland, Norway and the Russian Federation. Information obtained on the home page of the European Energy Charter at http://www.encharter.org/index.jsp [29.4.2001]. Last update: 7 December 2000.

[494] For further information on the Agreement, see *infra* chapter IV (4.1.2.).

[495] The 1982 United Nations Convention on the Law of the Sea. For further information on the Convention, see *infra* chapter IV (4.1.3.).

[496] Ibid. Another reference to the EIA procedure can be found in Article 202 (c): 'States shall, directly or through competent international organizations: c) provide appropriate assistance, especially to developing States, concerning the preparation of environmental assessments'.

substantial pollution of or significant and harmful changes to the marine environment' the potential effects of such activities must be assessed 'as far as practicable'. The obligation here can again be best regarded as an obligation of attempt because of the wording used. It should be noted here that the object of assessment in the LOSC is the whole marine environment, not merely the area of jurisdictional competence of the coastal states.

The 1992 Convention on Biological Diversity[497] concentrates on preserving biological diversity through varying measures, one of these being an EIA procedure. In Article 14 (1a), each contracting party is required 'as far as possible and as appropriate' to establish EIA procedures for activities 'that are likely to have significant adverse effects on biological diversity'. Since the provision is expressed in such vague terms, e.g., the use of 'appropriate', it is best considered an obligation of attempt. 'Biological diversity' has been defined in very broad terms in Article 2 to mean 'variability among living organisms from all sources...this includes diversity within species, between species and of ecosystems'[498]. It is also stipulated that the EIA procedure 'where appropriate, allow for public participation in such procedures'[499]. The contracting states are obligated, in Article 7 (c), to 'identify processes and categories of activities which have or are likely to have significant adverse impacts on the conservation and sustainable use of biological diversity'.

According to *the 1992 Convention on the Protection and Use of Transboundary Watercourses and International Lakes (hereinafter 'the 1992 Transboundary Watercourses Convention')*[500], the parties are required to develop and harmonize their

[497] For further information on the 1992 Convention on Biological Diversity, see *infra* chapter IV (4.1.3.).
[498] Ibid., Article 2, paragraph 1.
[499] Ibid., Article 14 (1a).
[500] For further information on the 1992 Transboundary Watercourses Convention, see *infra* chapter IV (5.1.).

measures of environmental policy 'as far as possible'[501] in order to ensure that EIA procedures and other means of assessment are applied and that waste-water discharges are authorized.[502] *The 1999 Protocol on Water and Health*[503], negotiated to supplement the Convention also contains requirements for prior procedures. In Article 4 (6), it provides that the parties 'shall require public authorities which are considering taking action, or approving the taking by others of action, that may have a significant impact on the environment of any waters within the scope of this Protocol to take due account of any potential impact of that action on public health'[504].

In contrast to the prior procedures studied above, which concentrate only on certain sectors, general prior procedures establish domestic procedures for all kinds of environmentally harmful activities. *The 1991 Convention on Environmental Impact Assessment in a Transboundary Context (hereinafter 'the 1991 Espoo Convention')*[505] focuses on preventing potential transboundary pollution from stationary industrial activities. Since only a national EIA procedure can provide scientific information on potential effects in general and transboundary impacts in particular, the Convention harmonizes – or, as the case may be, requires the establishment of - EIA procedures for the contracting states. The Convention does not require much harmonization of the national EIA procedures per se, only that an EIA procedure be established for the categories of activities listed in its Appendix I and

---

[501] Ibid., Article 3.
[502] Ibid. Article 3 provides that the parties must 'develop, adopt, implement and, as far as possible, render compatible relevant legal, administrative, economic, financial and technical measures, in order to ensure that: (b) Transboundary waters are protected against pollution from point sources through the prior licencing of waste-water discharges by the competent national authorities, and that the authorized discharges are monitored and controlled...(h) Environmental impact assessment and other means of assessment are applied'.
[503] The 1999 Protocol on Water and Health to the 1992 Convention on the Protection and Use of Transboundary Water courses and International Lakes. For further information on the Protocol, see *infra* chapter IV (5.1.).
[504] Ibid., Article 4 (6).
[505] For further information on the 1991 Espoo Convention, see *infra* chapter V (2.).

which are likely to cause significant adverse transboundary impact.[506] The EIA procedure must also permit public participation,[507] 'public' being defined as 'one or more natural or legal persons'[508]. Unfortunately, the Convention does not specify when and how the public is to participate in the EIA procedure.

A national EIA procedure must also permit the 'preparation of the environmental impact assessment documentation described in Appendix II' of the Convention.[509] The appendix sets out rather strict requirements for the national EIA procedure of the contracting states. An environmental impact statement (EIS) must describe not only the main plan of the activity but also reasonable alternatives to this main plan, including a no-action alternative. The environment likely to be significantly affected as well as the potential environmental impact must be evaluated from the viewpoint of the main plan as well as the reasonable alternatives and the no-action alternative. In addition, an estimation must be given of the significance of the potential environmental impact. Crucially, the EIS produced by the national EIA procedure must also contain a 'description of mitigation measures to keep adverse environmental impact to a minimum'. The precautionary principle is manifested in that the procedure must provide 'an identification of gaps in knowledge and uncertainties encountered in compiling the required information'[510]. The Convention also contains an implicit requirement, in Article 2 (3), that there be a prior licencing procedure in place for all listed categories of activities.[511]

The EC has also been active in promoting the EIA procedures to be adopted in its member countries. The Council of Ministers adopted *the 1985 Directive on the Assessment of the Effects of Certain Public and Private Projects on the Environment*

[506] Ibid., Article 2 (2).
[507] Ibid.
[508] Ibid., Article 1 (x).
[509] Ibid., Article 2 (2).
[510] Ibid., Appendix II (g).
[511] Koivurova 1997, pp. 171-174.

(hereinafter 'the 1985 EIA Directive')[512], which was amended through *the 1997 Amending Directive on the Assessment of the Effects of Certain Public and Private Projects on the Environment (hereinafter 'the 1997 Amendment Directive')*[513]. The EIA Directive requires the member states as well as states parties to the EEA Agreement[514] to adopt strict measures with regard to establishing a comprehensive national EIA procedure. Although the amendment Directive only amends the EIA Directive, it is considered useful to treat them as separate pieces of legislation as this permits a comparison of the two.

The screening system of the Directive is dual. First, certain classes of activities, i.e., those listed in Annex I, must always go through the entire EIA procedure.[515] The main change effected by the Amendment Directive is that the number of categories of activities which must be assessed has expanded considerably.[516] Second, the member states have broad discretion in relation to the classes of activities specified in Annex II of the EIA Directive.[517] This discretion was very extensive in the EIA Directive since the activities go through the whole EIA procedure 'where Member States consider that their characteristics so require'[518]. The discretion of the member states is reduced by the Amendment Directive in that they must establish a national system which evaluates for every case of an activity in Annex II whether it should go through the entire EIA procedure.[519] The criteria informing this determination are provided in Annex III.

---

[512] Council Directive of 27 June 1985 on the Assessment of the Effects of Certain Public and Private Projects on the Environment. For further information on the Directive, see *infra* chapter IV (4.1.1.).

[513] Council Directive 97/11/EC of 3 March 1997 Amending Directive 85/337/EEC on the Assessment of the Effects of Certain Public and Private Projects on the Environment. For further information on the directive, see *infra* IV (4.1.1.). For a comparison of the EIA Directive and its amendment, see Sheate 1997, pp. 235-243.

[514] See *infra* chapter IV (4.1.1.).

[515] Article 4 (1) of the EIA Directive.

[516] Annex I of the Amendment Directive.

[517] Article 4 (2) of the EIA Directive.

[518] Ibid.

[519] Article 4 (2) of the Amendment Directive.

According to the 1985 EIA Directive, the scope of the assessment is determined through the minimum requirements set for the impact study to be made by the project proponent. The project proponent must submit to the governmental authority at least the minimum information specified in Article 5 (2) on the potential environmental impacts of the proposed activities. This includes only basic information on the project, 'the data required to identify and assess the main effects'[520] of the project and a description of the mitigation measures available. Neither alternatives to the main plan for the proposed activity nor any scientific study of the potential environmental effects are required of the project proponent; a collection of data relevant for the licencing authority suffices. However, the project proponent must submit more extensive information, detailed in Annex III[521] of the Directive, if the exception provided in Article 5, paragraph 1 (a and b) does not apply.[522] The Amendment Directive requires somewhat more extensive requirements of the EIS. The main difference is that the minimum information, required in Article 5 (3), also entails 'an outline of the main alternatives studied'. Another important difference is that an optional formal scoping procedure[523] is established although the member states may introduce a mandatory formal scoping procedure if they wish.[524] The Amendment Directive obligates the member states to provide an opportunity for the project

[520] Article 5 (2) of the EIA Directive.

[521] Ibid. Annex III requires a description of the project, an outline of alternatives, a description of the aspects of the environment likely to be significantly affected, a description of the likely effects of the proposed project, a description of possible mitigation measures as well as an account of any uncertainties in obtaining the information.

[522] Ibid. Certain discretion is thus left for the member states, as they are allowed to decide whether the Annex III requirements are applicable on the basis of the following criteria in Article 5 (1) of the Directive: '(a) the Member States consider that the information is relevant to a given stage of the consent procedure and to the specific characteristics of a particular project or type of project and of the environmental features likely to be affected; (b) the Member States consider that a developer may reasonably be required to compile this information having regard inter alia to current knowledge and methods of assessment'.

[523] The scoping procedure is defined by Gilpin (1995, p. 19): 'a process that starts early, involves all affected parties, and enables agencies and the writers of EISs to pinpoint significant issues warranting study and analysis'.

[524] Article 5 (2) of the Amendment Directive.

proponent to ask for the advice of the competent authority on the content of the impact study.[525] In addition, the member states are given an opportunity to introduce a mandatory formal scoping procedure.

The comments to be submitted on the completed impact study are stipulated in Article 6. The EIA Directive requires that 'authorities likely to be concerned by the project by reason of their specific environmental responsibilities' be accorded the right to express their opinion 'on the request for development consent'. The Amendment Directive introduces an important change here in that these authorities not only are consulted for development consent but also have a right to express their opinion 'on the information supplied by the developer and on the request for development consent'[526]. With regard to the rights of the public, the EIA Directive requires only that the member states organize their national EIA procedure in such a way that the public is informed of 'any request for development consent and any information gathered pursuant to Article 5'[527]. The members of the public also have a right to express their opinion 'before the project is initiated'[528].

The Amendment Directive extends the right of the public by requiring that it be informed in good time so that it can 'express an opinion before the development consent is granted'[529] and not merely before a project is initiated. The information produced by a national EIA procedure 'must be taken into consideration in the development consent procedure'[530]. According to the EIA Directive, the final licencing decision, as well as the content of the decision and any conditions attached,

---

525 Ibid.
526 Ibid., Article 6 (1).
527 Article 2 of the EIA Directive.
528 Ibid.
529 Article 6 (2) of the Amendment Directive.
530 Ibid. Article 8 was amended only to ensure that the licencing decision would take into account the consultations between the states in a transboundary EIA procedure.

must be delivered to the public.[531] The Amendment Directive provides an additional requirement that the 'main reasons and considerations on which the decision is based'[532] must also be made available.

There are some treaties that require the states parties to adapt their national prior procedures to the special conditions in the Arctic. *The 1987 Agreement Between the Government of Canada and the Government of the United States of America on the Conservation of Porcupine Caribou Herd (hereinafter 'the 1987 Porcupine Caribou Agreement')*[533] requires the parties to conduct an impact assessment of activities that may have a potential impact on the conservation of the porcupine caribou herd or its habitat.[534] An additional precondition for imposing such an impact assessment duty is that the category of activity in question is subject to licencing according to the national laws of that state.[535] Consideration must be given to cumulative impacts of the activity on the herd and its habitat in conducting such an assessment.[536]

The 1973 Agreement on the Conservation of Polar Bears[537] also contains a provision requiring prior measures from the five Arctic states parties to it. Article II provides:

> Each contracting Party shall take appropriate action to protect the ecosystems of which polar bears are a part, with special attention to habitat components such as denning and feeding sites and migration patterns, and shall manage polar bear populations in accordance with sound conservation practices based on the best available scientific data.

[531] Article 9 of the EIA Directive.
[532] Article 9 (1) of the Amendment Directive.
[533] For further information on the 1987 Porcupine Caribou Agreement, see *infra* chapter IV (4.1.2.).
[534] Ibid., Article 3 (a and b).
[535] Ibid., Article 3 (c).
[536] Ibid., Article 3 (g).
[537] This Agreement was signed by the five Arctic states (Canada, Denmark (Greenland), Norway, the United States and the Soviet Union/Russian Federation) in Oslo on 15 November 1973. It was not until 25 January 1978 that all the signatory states had ratified the agreement, which entered into force on 26 May 1976.

This provision requires that unless prior procedures are otherwise available, the states parties must establish some kind of prior evaluation procedure for proposed activities that may have injurious consequences to, for instance, the denning and feeding sites of polar bears; if a prior procedure is available, these factors must be taken into account in its application in the Arctic areas. An agreement has recently been concluded between the United States and the Russian Federation on this matter.[538]

The Arctic cooperation process has produced many norms in adjusting the requirements of general prior procedures to Arctic conditions. As explained above, these obligations are most properly considered as obligations of attempt.[539] Both of the instruments comprising the AEPS document contain norm-texts pertaining to the focal issue of this thesis. The Declaration commits the Arctic states, for instance, to '[a]ssessment of potential environmental impacts of development activities'. This provision in the Declaration is specified in the AEPS:

> ii) Use and management of natural resources shall be based on an approach which considers the value and interdependent nature of ecosystem components;
> iii) Management, planning and development activities which may significantly affect the Arctic ecosystems shall:
> a) be based on informed assessments of their possible impacts on the Arctic environment, including cumulative impacts...
> e) take into account the results of scientific investigations and the traditional knowledge of indigenous peoples[540]

[538] The 2000 Agreement Between the United States of America and the Government of the Russian Federation on the Conservation and Management of the Alaska-Chukotka Polar Bear Population was signed on 16 October 2000 but is not yet in force. The Agreement also contains articles on the conservation of ecosystems and important habitats of polar bears. One noteworthy point is that the Joint Commission established to implement the Agreement includes representatives of Alaskan and Chukotka native peoples. Information on the Agreement can be found in WWF Arctic Bulletin, No. 4, p. 17, and on the home page of the United States Fish and Wildlife Service at http://news.fws.gov/newsreleases/display.cfm?NewsID=A11C3DA2-AC20-11D4-A179009027B6B5D3&Keywords=polar,bear [9.3.2001].

[539] See *supra* Chapter III (3.3.).

[540] Sections 2.2 (ii) and (iii) of the Strategy of the AEPS document.

Although there are also other provisions that are important for prior procedures in the Arctic,[541] this is the most significant. It prescribes that the Arctic states apply their national EIA procedures and prior licencing systems in a way that takes into account the special circumstances in the Arctic. First of all, a national EIA procedure is to take into account cumulative impacts if possible. Second, a procedure applied in the Arctic should take into account the special position of indigenous peoples as the main inhabitants of the region. The Arctic states should also ensure that traditional knowledge of indigenous peoples, as well as their 'social, economic...cultural needs and values'[542] are incorporated in EIA procedures. The general provisions of the AEPS document were later elaborated in *the 1997 Guidelines for Environmental Impact Assessment in the Arctic (hereinafter 'the 1997 EIA Guidelines')*[543] and *the*

[541] There are also other provisions pertaining to application of EIA procedures in the Arctic. Section 5 (4) of the Strategy reads: 'i) The effects of noise associated with Arctic marine and terrestrial projects should be evaluated as part of the project planning and approval processes, and if significant adverse noise effects on the specific components of Arctic ecosystems are predicted, then measures should be implemented to avoid or mitigate the impact. ii) Efforts should be made to improve the knowledge on marine mammal auditory function, communication and behavior and the current noise exposure assessment techniques. For specific project evaluations, site-specific data should be addressed before and during the evaluation. This includes determining how much exposure migrating stocks are encountering throughout the year'. These provisions are written as obligations of attempt and thus require the Arctic states to at least try to include the effects of noise as part of the prior procedures if such an impact is possible. The next potential legal obligation pertaining to the present study is found in Section 8 (1): 'They will also take steps to convene a meeting of experts to consider and recommend the necessary system of cooperation, which could include, inter alia, the following elements: iv) Assessment of the risks for significant accidental pollution and of the adverse effects in such cases so as to enable the parties to take the necessary preventive, preparedness and response measures'. Clearly, the legal obligation here only extends to the obligation to convene a meeting of experts, not to any assessment duty. The Barrow Ministerial Declaration the Arctic states also adopted the Arctic Council Action Plan to Eliminate Pollution in the Arctic (ACAP), which contains one program of interest here. Annex B of the ACAP contains potential future cooperative activities, one of which is 'Guidelines for Performing Environmental Impact Assessments of Handling and Storage of Radioactive Waste in Russia'. The task is to 'develop guidelines for performing Environmental Impact Assessments of Handling and Storage of Radioactivity, based on Russian regulations'. Information on the ACAP program can be found on the home page of the Arctic Council at http://www.arctic-council.org/acap.asp [20.4.2001].

[542] Section 2 (2iii, vii) of the Strategy.

[543] The document can be found on the home page of the Arctic Environmental Impact Assessment (ARIA) at http://finnbarents.urova.fi/aria/ [18.2.2001]. For an analysis of EIA in the Arctic, see Furman and Hildén 1998.

1997 Arctic offshore Oil and Gas Guidelines (hereinafter 'the 1997 Offshore Guidelines')[544], both of which provide obligations of attempt.

The 1997 EIA Guidelines contain important screening criteria for Arctic activities:

> In the Arctic, EIA should be applied to activities associated with the exploitation of both renewable and non-renewable natural resources, public use, **military activities** and the development of infrastructure for different purposes that may cause significant environmental impacts.[545] (emphasis added)

Of principal importance here is Section 2 of the EIA Guidelines that national EIA procedures in the Arctic should include military activities, since these are normally excluded from the scope of the procedures. Another important part of Section 2 is that the Arctic states should provide for specific Arctic thresholds.[546] This kind of obligation of attempt is particularly important in those national EIA procedures whose applicability is determined by a mandatory list of activities, since 'national laws do not always take into account the sensitivity of the Arctic areas'[547]. If the EIA system operates on a case-by-case basis, i.e., the decision to apply the national EIA procedure is made on the basis of contextual factors, sensitivity criteria should be taken into account. In the Arctic context this means a consideration of the possibility of cumulative impacts as well as proximity to sensitive areas since 'almost every activity in the Arctic will be close to one or several sensitive areas'[548].

The EIA Guidelines require that the Arctic states include 'all potential environmental, socio-cultural and economic impacts, especially impacts on the traditional uses of

---

[544] This document can also be found on the ARIA site at http://finnbarents.urova.fi/aria/ind.asp. [20.4.2001].

[545] Section 2 (2) of the EIA Guidelines.

[546] Ibid. Section 2 (1) provides: 'The decision to conduct an EIA for a project in the Arctic should take into account the special conditions in the Arctic'.

[547] Ibid., Section 2 (4).

[548] Ibid., Section 2 (5).

resources and livelihoods of indigenous people and also the consideration of alternatives'[549] in the scope of their impact assessment. An Arctic impact study should specify the project and its alternatives in sufficient detail as well as set realistic temporal, spatial and jurisdictional boundaries for the assessment.[550] In addition, it should specify the key environmental criteria and methods to be used.[551] Impact prediction and the evaluation phase of an Arctic EIA should contain three main elements.[552] The first is the identification of the types of impact, especially cumulative impacts, which refer to the 'accumulation of changes to the environment caused by human activities (e.g. past, existing and proposed activities, including activities associated with the project under assessment)'[553]. The second is a prediction of the magnitude, the probability of occurrence, and the extent of the impact. The third is a determination of the significance of the impact.[554] Decisions on the significance of the impact should be presented in a clear manner, and the different points of view on this significance should be taken up.[555] The determination of the significance of the impact is to be based on 'existing standards, discussions, judgement and agreement'[556].

Especially in Arctic EIA, the interpretation of the assessment findings should take into account that many impacts cannot be identified due to the lack of a common base for comparing the significance of different types of impacts, the differences in world-views of the Arctic stakeholders, and the different structure of scientific and indigenous knowledge.[557] The Arctic states are encouraged to include an extensive

549 Ibid., Section 3 (1).
550 Ibid., Section 3 (2).
551 Ibid.
552 Ibid., Section 5.
553 Ibid., Section 5 (1).
554 Ibid., Section 5.
555 Ibid., Section 5 (4).
556 Ibid., Section 5 (4).
557 Ibid.

amount of information in their Arctic impact assessment document.[558] Sections 9, 10 and 11 of the Guidelines, which deal with public participation, the use of traditional knowledge and transboundary impacts, respectively, are studied below in the appropriate chapters.[559]

The 1997 Offshore Guidelines also contain obligations of attempt for the Arctic states in regard to EIA procedures. Section 2 concentrates on providing for preliminary EIA and regular EIA in terms of 'impacts of offshore oil and gas exploration, development, transportation and infrastructure'[560]. It is stipulated first of all that before commencing the two stages of impact assessment procedures 'the assessment must be carefully planned'[561]. The scope of an EIA should be clearly defined immediately after the decision to initiate an EIA.[562] Public participation in scoping is urged 'for efficient and full use of traditional knowledge'[563]. Sufficient time and resources for public participation are also encouraged.[564] Rules for preliminary EIA (PEIA) are laid down in Section 2 (3). PEIA 'should contain sufficient detail to

[558] Ibid., Section 8 (1). The following kinds of information should be provided: '1. A description of the proposed project and its alternatives, including information about the location and the design and size or scale of the project. 2. A description of the environment that could be affected by the proposed project or alternatives. 3. The data and other information that have been used to identify and assess the main effects which the project is likely to have on the environment, including a description of the traditional knowledge incorporated into the EIA. 4. The estimated type and quantity of expected impact factors resulting from the proposed project when in operation. 5. The methods used in the assessment. 6. An identification of the impact area. 7. The likely significant impacts. 8. Where significant adverse effects are identified, a description of the measures proposed to avoid, reduce or rectify these effects taking into consideration the slow recovery and regeneration factors in the Arctic. 9. An evaluation of the different alternatives, including the alternative of no action. 10. A description of the integration of EIA, public participation and public consultation into planning and decision-making throughout the process. 11. A summary in non-technical language. If need be, other means of displaying this information should be based on the cultural heritage of the local and indigenous people'.

[559] See *infra* chapter IV (3.3.2.), (3.3.3.) and chapter V (2.7.).

[560] Section 2 of the Offshore Guidelines.

[561] Ibid., Section 2 (5).

[562] Ibid.

[563] Ibid.

[564] Ibid.

permit assessment of whether a proposed activity may have a significant impact'[565] and should already include quite an extensive amount of information.[566] The regular impact assessment, described in Section 2 (4) of the Guidelines, 'should be based on the best available information'[567] and must include more extensive information than the PEIA.[568] Both PEIAs and EIAs should consider many kinds of factors which are characteristic of the Arctic.[569]

---

565 Ibid., Section 2 (3).

566 Ibid. The PEIA should contain a description of the proposed activity that includes 'its purpose, location, duration and intensity', consideration of alternatives to the proposed activity, and 'any impacts that the activity and its alternatives may have, including consideration of cumulative impacts'. The PEIA should also contain information on all potential impacts on the environment, of which the following are specified: 'a. humans; b. flora and fauna; c. soil and permafrost; d. air and water; e. climate; f. cultural heritage; g. the overall landscape, and h. the interaction among any of the above'. 'A determination whether significant impacts, that would require further assessment, are likely to occur'.

567 Ibid., Section 2 (4).

568 Ibid. The regular impact assessment should include the following information: A description of the proposed activity. This requirement is much extended from the PEIA requirement in that the description 'shall also include land use requirements during construction and operation phases' as well as 'main characteristics of the development process proposed, including type and quantity of materials to be use'. The estimated type and quantity of expected residues and emission. This is seen to include 'air, water, soil, vibration, light, heat and radiation pollution'. A risk analysis 'including oil spill trajectory modeling'. The exposure of the methods used in the assessment as well as any lacking data. An identification of the area of impact. The likely significant effects on the environment of the proposed activity. This includes 'direct and indirect effects...secondary, cumulative, short, medium and long term, permanent, temporary, positive and negative'. Interestingly, the possibility of transboundary impacts must also be considered. A description of the likely significant effects, which are the same as in the PEIA study. A description of the measures 'proposed to avoid, reduce or rectify these effects'. It is particularly important that the Arctic states should take into consideration 'the slow recovery and regenerative capacity of the Arctic'. Other alternatives, including that of no action. A summary in non-technical language.

569 Ibid. These are enumerated in Section 2 (7): 'marine mammals; Arctic human communities; subsistence lifestyles (e.g. harvest practices and availability of food supply); oil spill preparedness and response in sea ice conditions; the permafrost and transition zones; sustainability of renewable resources; flora and other fauna; other human activities (e.g. tourism, scientific research); air and water quality; and ports and shore reception facilities; and the effects of ice scour and movement'.

3.3.2. Obligations of the Arctic States to Guarantee the Participation of Private Legal Subjects

The Rio Declaration can be seen as setting out the requirement of sustainable development with regard to the issue of public participation:

> Environmental issues are best handled with the participation of all concerned citizens, at the relevant level. At the national level, each individual shall have appropriate access to information concerning the environment that is held by public authorities, including information on hazardous materials and activities in their communities, and the opportunity to participate in decision-making processes. States shall facilitate and encourage public awareness and participation by making information widely available. Effective access to judicial and administrative proceedings, including redress and remedy, shall be provided.[570]

This requirement of sustainable development contains two dimensions in the situations studied here.[571] First, the national environmental protection system must be able to provide access to information concerning the environment. Second, the public must be guaranteed the right to participate in the decision-making processes concerning the environment. In the situations studied here, the planning stage of stationary activities, these tasks are frequently handled as part of the EIA and prior licencing procedures. It is the international conventions to be studied below that try to

---

[570] Principle 10 of the Rio Declaration. For a theoretical analysis of the concept of public participation in international law, see Ebbesson 1997, pp. 51-97.

[571] It is possible to distinguish three dimensions of principle 10: access to information concerning the environment, access to decision-making processes concerning the environment and access to judicial and administrative proceedings. Given to the context of this thesis, the access to decision-making processes is considered to include access to judicial and administrative proceedings. For a general overview of the requirements of sustainable development in this field, see Sands and Werksman 1995, pp. 178-204.

ensure that the national EIA procedures and prior licencing procedures include these two elements of public participation. Another important task of these conventions is to guarantee the rights of the public of a potentially affected state to participate in the national EIA procedures and prior licencing procedures of the origin state.

Certain international conventions, as well as some legally relevant documents[572] that are applicable in the Arctic, aim at guaranteeing that the public of the contracting states have access to information concerning the environment. *The 1974 Paris Convention for the Prevention of Marine Pollution from Land-Based Sources (hereinafter 'the 1974 Paris I Convention')*[573] did not contain any rules on access to information. *The 1992 Convention for the Protection of the Marine Environment of the North-East Atlantic (hereinafter 'the 1992 Paris II Convention')*[574] changed this and provides for access to information by private legal subjects in its Article 9. According to the Article, any natural or legal person has a right to obtain information 'on activities or measures adversely affecting or likely to affect [the state of the maritime area]'[575]. The request must be reasonable, and no unreasonable charges may

[572] One such document is the 1993 Convention on Civil Liability for Damage Resulting from Activities Dangerous to the Environment, the so-called Lugano Convention, which applies mainly to issues of transnational liability of the project operator but also contains rules providing rights for private legal subjects. The Convention was adopted in 1993 under the auspices of the Council of Europe and does not seem to enter into force at all. Of the Arctic states, Iceland and Finland have signed the Convention. Information obtained on the home page of the Council of Europe at http://conventions.coe.int/treaty/EN/cadreprincipal.htm [18.2.2001]. The basic principle governing access to information is Article 14 (1): 'Any person shall, at his request and without his having to prove an interest, have access to information relating to the environment held by public authorities. The Parties shall define the practical arrangements under which such information is effectively made available'. Significantly, the concept of person is defined very broadly to mean not only natural and legal persons but also a 'State or any of its constituent subdivisions'. The concept of public authority is also defined broadly (Article 13). The same access to environmental information is granted to the bodies with public responsibilities for the environment as stipulated in Article 15. Interestingly, the Convention also provides for a right of request for the environmental organizations (Articles 18 (1a) and (1b)). Although access to environmental information is conceived very broadly in principle, there exist numerous exceptions to this main rule.

[573] For further information on the 1974 Paris I Convention, see *infra* chapter IV (4.1.1.).

[574] For further information on the 1992 Paris II Convention, see *infra* chapter IV (4.1.1.).

[575] Ibid., Article 9 (2).

be levied for the information provided.[576] The private legal subject does not have to prove an interest and is entitled to obtain the information 'as soon as possible and at the latest within two months'[577]. Paragraph 3 sets out many reasons which can be used by the origin state to refuse the information,[578] although the 'reasons for a refusal to provide the information requested must be given'[579].

Certain conventions provide a more ambitious approach, not only requiring the governmental authorities to provide information on request but in fact obligating them to take active measures in ensuring that the public is informed. The 1992 Transboundary Watercourses Convention requires certain measures of the riparian contracting states in relation to planned stationary activities potentially affecting transboundary waters.[580] According to Article 16, 'the Riparian Parties shall ensure that information on...measures planned to be taken to prevent, control and reduce transboundary impact...is made available to the public'. This information includes, for instance, water-quality objectives, and 'permits issued and the conditions to be met'[581]. This information must be made available free of charge, and 'on payment of reasonable charges, copies of such information'[582] must also be provided.

Article 9 of the 1992 Industrial Accidents Convention contains a general clause on the participation rights of the public. The parties are obligated to 'ensure that

---

576 Ibid., Article 9 (1).
577 Ibid.
578 Ibid., Article 9 (3).
579 Ibid., Article 9 (4).
580 An access-to-information provision is also to be found in Article 10 of the 1999 Water Protocol to this Convention. The Protocol has not yet entered into force. In Article 10 (1), all states parties are required to make available varying kinds of information to the public. This information must be provided to the public for inspection and copies of it made available 'on payment on reasonable charges' (Article 10 (3)). On the other hand, Articles 4 and 5 set out rather extensive grounds on which an authority can deny access to information, albeit, as provided in Article 10 (5), these grounds must be interpreted in a restrictive way. For further information on the Protocol, see *infra* chapter IV (5.1.).
581 Article 16 of the 1992 Transboundary Watercourses Convention.
582 Ibid., Article 16 (2).

adequate information is given to the public in the areas capable of being affected by an industrial accident arising out of a hazardous activity'[583]. The information to be provided to the public is defined in very extensive terms in Annex VIII of the Convention. It includes, for instance, a description of the hazardous substances used 'with an indication of their principal characteristics'[584], the nature of a possible accident occurring as a result of the activity, 'adequate information on how the affected population will be warned'[585], and guidelines on how the members of the public should conduct themselves in case of an accident. The information supplied should also take into account the matters outlined in Annex V.[586]

In one treaty, only the second element of the public participation, the right to participate in decision-making processes, is stipulated.[587] *The 1974 Nordic Environment Protection Convention (hereinafter 'the 1974 NEPC')*[588] provides the private legal subjects of the contracting states with many procedural rights in the state of origin, including, for example, the right to participate in the prior licencing procedure. The private legal subjects of the affected state are entitled to 'bring before the appropriate Court of Administrative Authority of that State the question of the permissibility of [environmentally harmful] activities'[589]. In accordance with the non-discrimination principle, the private legal subjects are entitled to avail themselves of these rights only if the private legal subjects of the origin state possess such rights.

Both elements of public participation - access to information and the right to participate in decision-making processes - are contained in the 1994 Environmental

---

[583] Article 9 (1) of the 1992 Industrial Accidents Convention.
[584] Ibid., Annex VIII (3).
[585] Ibid., Annex VIII (6).
[586] Ibid., Article 9 (1).
[587] It should be borne in mind that the 1991 Espoo Convention and the 1992 Industrial Accidents Convention both contain participation rights as well. These Conventions are studied *infra*, in chapters V (2.) and IV (4.1.3.).
[588] For further information on the NEPC, see *infra* chapter IV (4.1.1.).
[589] Ibid., Article 3 (1).

Cooperation Agreement. Article 4 requires the contracting states to ensure that their 'laws, regulations, procedures and administrative rulings of general application' are published in order for interested persons and the other contracting states to 'become acquainted with them'. In Paragraph 2, this obligation is specified by stipulating that '[t]o the extent possible', all the contracting states must ensure that this publication must be done before the planned measure. Interested persons as well as other contracting states are also provided with a reasonable opportunity to 'comment on such proposed measures'[590]. In addition, in Article 6 (2), the contracting states are required to provide the private legal subjects of each contracting state with 'appropriate access to administrative...proceedings for the enforcement of the Party's environmental laws and regulations'. Moreover, it is guaranteed that the competent authorities will be requested 'to take appropriate action to enforce that Party's environmental laws and regulations in order to...avoid environmental harm'[591].

The Convention on Access to Information, Public Participation in Decision-Making and Access to Justice in Environmental Matters (hereinafter 'the 1998 Aarhus Convention')[592] merits study here although it has not entered into force, as it has implications for the Arctic and will in all likelihood come into force in the near future. Significantly, it is the first convention to provide comprehensively and exclusively for both of the elements of public participation of the contracting states as directed by principle 10 of the Rio Declaration.[593] Of the Arctic states, Denmark has

---

[590] Article 4 (2b) of the 1994 Environmental Cooperation Agreement.

[591] Ibid., Article 6 (3c).

[592] The Convention enters into force on the ninetieth day after the deposit of the sixteenth instrument of ratification, acceptance, approval or accession (Article 20 (1)). Since 40 states have signed the Convention, and 10 have accepted the Convention as binding on them, the Convention will in all likelihood come into force in the near future.

[593] A new kind of method to provide information to the public is to be negotiated under the auspices of the UN ECE, as provided in Press Release (ECE/ENV/01/01, of 28 February 2001): 'Negotiations on new information law on pollution start today. The United Nations Economic Commission for Europe (UNECE) is launching negotiations in Geneva today on a new legally binding instrument requiring companies to report to the public on their polluting emissions to the environment. Under the new law countries will have to set up pollution inventories known as pollutant release and transfer registers (PRTRs). PRTR systems require corporate polluters to report

ratified the Convention, and Finland, Iceland, Norway, Sweden and the EC have signed it.[594]

The Aarhus Convention contains two sets of norms that are important for the rights of private legal subjects in relation to proposed stationary activities. First, Article 4, 'Access to Information', lays down general rules concerning the right of 'the public'[595] of all the contracting states to obtain environmental information from the authorities of all the contracting states.[596] The contracting states are also granted the right to refuse to give the requested information on certain grounds.[597] Article 5 obligates the states parties to collect and update the environmental information, and, importantly, to disseminate this information actively.[598]

Particularly important for the focus of the present study are the rules in Article 6 providing for public participation in decisions on specific activities. These 'specific activities' are enumerated in Annex I of the Convention and are all environmentally harmful stationary activities. The system set out in Article 6 is based on three different sets of activities. First, the contracting states are obligated to guarantee the rights specified in Article 6 in relation to all proposed activities which qualify under Annex I.[599] Second, the parties are also obligated to apply Article 6 if the proposed

---

periodically on their emissions of certain polluting substances to air, water and land, as well as on their off-site transfers of such substances for treatment or disposal. All this information is reported electronically and made accessible to the public, including through the Internet, subject to limited exemptions'.

594 Denmark has declared, however, that 'signing by Denmark of the Convention, therefore, not necessarily means that Danish ratification will in due course include the Faroe Islands and Greenland' Information obtained from the United Nations Treaty Series (UNTS) on-line service. The last change of status of all the information related to the status of treaties studied in this thesis and obtained from the UNTS is 31 January 2001.

595 The concept of public is defined very broadly in the Aarhus Convention. According to Article 2 (4): '"The public" means one or more natural or legal persons, and, in accordance with national legislation or practice, their associations, organizations or groups'.

596 Ibid. The concept of public authority is defined in Article 2 (2).

597 Ibid., Article 4 (3 and 4).

598 Ibid., Article 5.

599 Ibid., Article 6 (1a).

activity 'may have a significant effect on the environment'[600]. Third, the contracting states have a right not to apply Article 6 in terms of certain proposed activities if a party deems that this 'would have an adverse effect'[601] on national defense purposes.

The general obligation to ensure public participation in decisions on specific activities is laid down in Article 6 (2): 'The public concerned shall be informed, either by public notice or individually as appropriate, early in an environmental decision-making procedure, and in an adequate, timely and effective manner'. Article 6 (2) specifies the kind of information that must be provided to the public, e.g., basic information on the proposed activity, the phases of the decision-making procedure, and, especially, how exactly the public is to participate in this procedure. In addition, the notification must contain information on whether 'the activity is subject to a national or transboundary environmental impact assessment procedure'[602]. Notification of the public must also be carried out using a reasonable timetable in order for 'the public to prepare and participate effectively during the environmental decision-making'[603]. Early public participation is also required in Article 6 (4), but no particulars are given. Similarly, no specification is provided of how public participation is to be organized; a guidance is given to the operators of the proposed activities that they should identify the public concerned, enter into discussion with the public and provide information regarding the objectives of their application.[604]

Article 6 (6) is designed to guarantee that the public has access to information in the case of specific activities as well. The competent authorities are obligated to provide information to any private legal subjects who request the environmental information related to the planning phase of these activities. This information must be provided

[600] Ibid., Article 6 (1b).
[601] Ibid., Article 6 (1c).
[602] Ibid., Article 6 (2e).
[603] Ibid., Article 6 (3).
[604] Ibid., Article 6 (5).

free of charge and as soon as it becomes available.[605] Significantly, Article 6 (7) requires the contracting states to organize not only the notification of the public but also its active participation in the procedure. The public must be given some way of expressing its opinion on the proposed activity 'in writing or, as appropriate, at a public hearing or inquiry with the applicant'[606]. The contracting states are also obligated to ensure that due account is taken of comments by the public in the licencing decision.[607] Moreover, in accordance with Article 6 (9), the public must also be informed of the licencing decision. In fact, the text of the decision must be made accessible to the public along with the 'reasons and considerations on which the decision is based'[608].

The 1997 EIA Guidelines contain important adaptation rules for Arctic EIA in relation to public participation. Significantly, the Guidelines define the concept of public very broadly to include 'indigenous people, groups, organizations or communities that have an interest in or could be affected by the proposed action'[609]. The Guidelines stipulate that the Arctic states ensure that the public is informed when a project is first proposed.[610] In addition, 'unique features such as culture, socio-economic and remoteness factors should be considered in planning and carrying out public consultations in the Arctic'[611]. The public should also be given the opportunity to review and comment on each phase of the EIA[612] as well as be entitled to review the final analysis and recommendations and submit comments on these to the licencing authority.[613]

---

605 Ibid., Article 6 (6).
606 Ibid., Article 6 (7).
607 Ibid., Article 6 (8).
608 Ibid., Article 6 (9)
609 Section 9 (2) of the 1997 EIA Guidelines.
610 Ibid., Section 9 (7).
611 Ibid., Section 9 (1).
612 Ibid.
613 Ibid.

The Guidelines also require that the Arctic states try to establish clear rules and procedures for organizing meetings and consultations.[614] The methods of communication used by government agencies, project proponents and researchers 'should not be intrusive and should be sensitive to indigenous peoples' customs and life-styles'[615]. Indigenous peoples should be entitled to provide their knowledge throughout the Arctic EIA process. An early plan should be made for using traditional knowledge[616], and contacts should be established early so that indigenous people who are experts in traditional knowledge can be identified.[617]

3.3.3. Obligations of the Arctic States in Regard to Indigenous Peoples Within Their Territory

One salient characteristic of the Arctic is its small population. Many of people in the region can be considered members of indigenous peoples, and of the eight Arctic states only Iceland does not have an indigenous people living in its territory.[618] The indigenous peoples in the Arctic have frequently settled in the territory of more than one state, and, increasingly, have established international organizations of their own. Many of these international organizations have been accepted as permanent participants in the Arctic Council.[619]

Sustainable development requires the Arctic states to provide efficient means for indigenous peoples to participate in EIA procedures related to planned projects.[620]

---

614 Ibid., Section 9 (8).
615 Ibid., Section 9 (10).
616 Ibid.
617 Ibid., Section 9 (11).
618 For a review of peoples living in the Arctic, see the AMAP Assessment Report 1998, pp. 141-182. For a brief overview of the problems related to the definition of indigenous peoples and minorities, see Thornberry 1995, pp. 57-65.
619 See *supra* chapter III (1.2.). For an overview of the status of Arctic indigenous peoples in the Arctic cooperation process, see Nuttall 2000, pp. 621-637.
620 The Arctic cooperation process has provided obligations of attempt relevant to the status of indigenous peoples in planning processes in the Arctic. The 1991 AEPS contains an important

Principle 22 of the Rio Declaration outlines the requirement of the goal of sustainable development in respect of indigenous peoples:

> Indigenous people and their communities, and other local communities, have a vital role in environmental management and development because of their knowledge and traditional practices. States should recognize and duly support their identity, culture and interests and enable their effective participation in the achievement of sustainable development.

The reason why indigenous peoples have a role to play in environmental management is their 'knowledge and traditional practices'[621]. This wording clearly refers to knowledge which is different from the knowledge obtained through the methods of modern science. It is knowledge which has accumulated in the course of history because of the close relationship of indigenous peoples to their environment.[622] In Arctic EIA, the goal of sustainable development requires that if a proposed activity might cause environmental effects in those areas which are used by indigenous peoples, the evaluation of these effects should take into account not only the scientific findings but also the traditional knowledge and practices of indigenous

---

provision from the viewpoint of the situations studied here. Section 2.2 (iii) provides that the traditional knowledge of indigenous peoples should be taken into account in prior procedures and recommends that the scope of these EIA studies be enlarged to cover the health, social, economic and cultural needs and values of indigenous peoples. These general obligations of attempt have been specified in the 1997 EIA Guidelines. Arctic EIA should contain several elements that take into account the interests of indigenous peoples. Since many sections of the 1997 EIA Guidelines contain obligations of attempt pertaining to the indigenous people, the ARIA provides a compilation of the parts of the Guidelines in which reference is made to indigenous peoples. See the ARIA home page at http://finnbarents.urova.fi/aria/trakno.asp [20.4.2001].

[621] Principle 22 of the Rio Declaration. See also Agenda 21, especially Chapter 26, which contains the following passages: '26.3. (a) Establishment of a process to empower indigenous people and their communities through measures that include...Recognition of their values, traditional knowledge and resource management practices with a view to promoting environmentally sound and sustainable development' and '26.6. (a) Develop or strengthen national arrangements to consult with indigenous people and their communities with a view to reflecting their needs and incorporating their values and traditional and other knowledge and practices in national policies and programmes in the field of natural resource management and conservation and other development programmes affecting them'.

[622] See Section 10 of the 1997 EIA Guidelines.

peoples. The concept of traditional knowledge has been included in some international agreements as well.[623]

The difference in status in Arctic EIA between Arctic indigenous peoples and other private legal subjects, i.e., associations, companies and natural persons, is clear: other private legal subjects are only entitled to receive information and comment on the proposed activity in accordance with the national EIA procedure, whereas the indigenous peoples are also partly responsible for making the impact studies. In light of the goal of sustainable development, it can be suggested that EIA procedure in the Arctic requires the Arctic states to ensure that traditional knowledge forms part of the impact studies. Traditional knowledge is defined in the EIA Guidelines as 'accumulated knowledge held by indigenous people on the arctic environment'[624].

The 1997 EIA Guidelines stipulate that the methods of communication used by the government agencies, project proponents and researchers 'should not be intrusive and should be sensitive to indigenous peoples' customs and life-styles'[625]. Indigenous peoples should be entitled to offer their knowledge throughout the process. A plan for incorporating traditional knowledge ought to be established early, as should contacts with indigenous persons who are expert in such knowledge.[626] The Guidelines also require that when meetings are planned, 'the remoteness of many Arctic communities should be taken into consideration'[627]. The EIA document produced is to include a

---

[623] The concept of traditional knowledge has a prominent place in the Convention on Biological Diversity (Articles 8 (j) and 17). The Convention is important in the Arctic context since seven of the Arctic states are parties to it and the United States has signed it. Another important agreement here is the 2000 Agreement Between the United States of America and the Government of the Russian Federation on the Conservation and Management of the Alaska-Chukotka Polar Bear Population, Article 8 (7 b) of which defines the tasks of the Commission as follows: 'determining on the basis of reliable scientific data, including traditional knowledge of the native people, the polar bear population's annual sustainable harvest level'.

[624] Section 10 (2) of the 1997 EIA Guidelines.

[625] Ibid., Section 9 (9).

[626] Ibid., Section 9 (10).

[627] Ibid., Section 9 (12).

description of the traditional knowledge that has been used in the impact assessment.[628]

In strict legal terms, neither general international law nor the treaties providing for the participation rights of the public contain norms entitling the Arctic indigenous peoples to participate in the prior procedures as a distinct group.[629] In other words, their position in these procedures, e.g., EIA procedures, is determined by the criteria applied to other private legal subjects. Indigenous persons and groups can thus participate in EIA procedures as private landowners, associations, etc. There are, however, a few international treaties[630] and legally relevant documents[631] that provide

---

[628] Ibid., Section 8 (3).

[629] Some scholars are of the opinion that customary law has come to include principles which provide participation rights for indigenous peoples. For instance, Anaya (1996, especially pp. 49-58) treats certain norms of ILO Convention No. 169 that are applicable to the situations studied in this thesis as part of customary law or at least as emerging customary law. However, it seems premature to argue that customary law has come to include participation rights for indigenous peoples. A more cogent approach may be to point to the injustice of the present arrangements regarding the indigenous peoples living within the established nation-states. See, e.g., Brölmann and Zieck 1993, pp. 187-220, and Spiry 1995, pp. 129-152.

[630] Article 27 of the International Covenant on Civil and Political Rights has been interpreted by the Human Rights Committee, the organ monitoring observance of the Covenant, as applicable to the planning stage of environmentally harmful activities. Article 27 of the Covenant reads: 'In those States in which ethnic, religious or linguistic minorities exist, persons belonging to such minorities shall not be denied the right, in community with the other members of their group, to enjoy their own culture, to profess and practice their own religion, or to use their own language'. The concept of culture has been given a broad meaning in the case law of the Human Rights Committee, which is reflected in General Comment No. 23 adopted in 1994: 'With regard to the exercise of the cultural rights protected under Article 27, the Committee observes that culture manifests itself in many forms, including a particular way of life associated with the use of land resources, specially in the case of indigenous peoples. That right may include such traditional activities as fishing and hunting and the right to live in reserves protected by law. The enjoyment of those rights may require positive legal measures of protection and measures to ensure the effective participation of members of minority communities in decisions which affect them'. Official Records of the General Assembly, Forty-Ninth Session, Supplement No. 40 (A/49/40), Annex V.

[631] Four legally relevant documents are worth mentioning here. The first is the Draft United Nations Declaration on the Rights of Indigenous Peoples as agreed upon by the members of the UN Sub-commission on Prevention of Discrimination and Protection of Minorities by its Resolution 1994/45, especially Article 30 (August 26, 1994. U.N. Doc. E/CN.4/1995/2, E/CN.4/Sub.2/1994/56, at 105, 1994 (34 I.L.M. 541 (1995)). For an in-depth analysis of the draft Declaration, see Gayim 1994. The second is the Draft Inter-American Declaration on the Rights of Indigenous Peoples approved by the Inter-American Commission on Human Rights on February 26, 1997, at its 1333rd session, 95th regular session, especially Article XIII (2). The third document worthy of note is the

for the rights of indigenous peoples to participate in EIA procedures as a distinct group.[632] The principal instruments are two international treaties negotiated under the auspices of the International Labour Organization (ILO), *the Convention Concerning the Protection and Integration of Indigenous and Other Tribal and Semi-Tribal Populations in Independent Countries*[633], and *the 1989 Convention Concerning Indigenous and Tribal Peoples in Independent Countries (hereinafter 'the 1989 ILO Convention')*[634]. These treaties have rather few parties, however; for example, only two Arctic states, Norway and Denmark, have become parties to the latter, although Finland is expected to ratify it as well in the near future.[635] Interestingly, there are some prospects for a convention between Finland, Norway and Sweden on the protection of the Sami people.[636]

1992 Declaration on the Rights of Persons Belonging to National or Ethnic, Religious and Linguistic Minorities adopted by the UNGA in Resolution 47/135 (The Raoul Wallenberg Compilation of Human Rights Instruments 1997, pp. 313-316), which recognizes the rights of participation belonging to minorities, thus including the indigenous peoples. Significantly, these rights may be exercised not only individually but also as a group. The fourth instrument of interest is the declaration made by a Meeting of Experts convened by the UN on the rights of indigenous peoples in Nuuk, Greenland, 'The Nuuk Conclusions and Recommendations on Indigenous Autonomy and Self-Government' (UN Doc. E/CN.4/1992/42).

632 For an overview of these international instruments, see Barsh 1994, pp. 33-86.

633 Convention No. 107 Concerning the Protection and Integration of Indigenous and Other Tribal and Semi-Tribal Populations in Independent Countries. The Convention was adopted on 26 June 1957 and entered into force on 2 June 1959. As of 5 February 2001, there were 27 ratifications although none by an Arctic state. Information obtained on the home page of ILO at http://ilolex.ilo.ch:1567/public/english/docs/convdisp.htm [28.4.2001].

634 Convention No. 169 Concerning Indigenous and Tribal Peoples in Independent Countries. The Convention was adopted on 27 June 1989 and entered into force on 5 September 1991. As of 5 February 2001, there were 14 ratifications, of which two had been made by Arctic states, i.e., Denmark (22.2.1996) and Norway (19.6.1990). Information obtained on the home page of ILO at http://ilolex.ilo.ch:1567/public/english/docs/convdisp.htm [28.4.2001].

635 The latest development has been examined by Tirronen (2001): 'The Ministry of Justice appointed in November 2000 a new Rapporteur...to make a legal assessment from the perspective of real estate law on the significance of the reports so far made to the land ownership question in the Sami Homeland. At the same time, the Ministry also appointed a new Working Group with the task to assess whether, or to what extent, it would be advisable to realise the proposals of Dr. Vihervuori report, or whether they should be modified. The Working Group's proposals should meet the minimum criteria of the ILO Convention'. See also in Finnish, Oikeusministeriön säädöshankkeet 2001(1.2.2001), pp. 24-25.

The most significant instrument relating to the planning stage of stationary activities is the 1989 ILO Convention, Articles 15 and 16 of which accord special participation rights to the indigenous peoples in the territories they traditionally occupy or otherwise use.[637] Article 15 (1) stipulates: 'The rights of the peoples concerned to the natural resources pertaining to their lands shall be specifically safeguarded. These rights include the right of these peoples to participate in the use, management and conservation of these resources'[638]. This general right applies to any planning procedures, including the EIA and prior licencing procedures, and means that the indigenous peoples are entitled to participate in these procedures as a group with special status. This is made even clearer in Article 15 (2):

> In cases in which the State retains the ownership of mineral or sub-surface resources or rights to other resources pertaining to lands, governments shall establish or maintain procedures through which they shall consult these peoples, with a view to ascertaining whether and to what degree their interests would be prejudiced, before undertaking or permitting any programmes for the exploration or exploitation of such resources pertaining to their lands.

These principles go further than the goal of sustainable development in that if the proposed activity is to be located on lands traditionally occupied by indigenous peoples, or the activity may infringe their subsistence and traditional activities, indigenous peoples must be accorded a special position in the prior procedures. They must be treated as a distinct group, for instance, in the EIA procedure in the sense

---

[636] See Alfredsson 1999, pp. 397- 411. For a general analysis of the rights of Scandinavian Sami in light of international law, see Hannikainen 1996, pp. 315-326, and Fitzmaurice 1997, pp. 200-243.

[637] Article 13 (2) of Convention No. 169 specifies what is meant by the term 'land' in Articles 15 and 16: '[land] shall include the concept of territories, which covers the total environment of the areas which the peoples concerned occupy or otherwise use'. It should be noted that Article 6 (a) provides that governments' shall consult the indigenous peoples when consideration is being given to legislative or administrative measures which may affect them directly'. This does not mean that the government must obtain consent from the indigenous peoples, but only that it must strive to achieve agreement or consent in good faith. See Brölmann and Zieck 1993, p. 209. For a detailed study on the requirements of the Convention in Swedish, see Myntti 1998, pp. 152-166.

[638] Article 15 (1) of the 1989 ILO Convention.

that they are to be consulted 'with a view to ascertaining whether and to what degree their interests would be prejudiced'[639].

3.4. Conclusion

The main goal of international society at present is sustainable development. This goal reflects the present era of human history, in which economic activities are evidently considered very important for development but their adverse environmental effects are seen as requiring protective measures. Neither value can be given priority and thus both find expression in the goal of sustainable development, which primarily urges procedural reforms in states' policies and laws in general and national environmental protection systems in particular. Although not part of *lex lata* in strict legal terms, the goal can be seen as requiring a wide variety of procedural reforms, for instance, the increasing use of EIA procedures and the adoption of public participation procedures that include opportunities for indigenous peoples to participate in prior procedures as a distinct group.

The means by which states may implement the requirements of the goal of sustainable development at the planning stage of stationary activities are either international or national: they can conclude international treaties in which they harmonize their national environmental protection systems and/or implement the goal through national measures only. As discussed above, there exist international treaty requirements on the scope and content of the prior procedures to be followed by the Arctic states. There are also treaties which require the Arctic states to establish prior procedures, EIA procedures and licencing systems, either as part of other regimes, or, in the case of the Espoo Convention, as the main requirement. Some treaty obligations require the Arctic states to adapt their prior procedures to the circumstances of a certain issue, for instance, polar bears. Most importantly,

[639] Ibid.

however, the Arctic cooperation process has yielded obligations of attempt on how to conduct EIA in the Arctic and even elaborated EIA for offshore hydrocarbon activities.

A number of the treaties on record guarantee the participation rights of the private legal subjects. Some guarantee access to information, whereas others give the public the right to participate in the prior procedures. Of particular importance here is the 1998 Aarhus Convention, which, after its entry into force, will offer guarantees enabling the private legal subjects of both the origin and the affected state to enforce their right of participation. Still other treaties in force in the Arctic require that indigenous peoples be given special participation rights but apply only in a very limited manner at the moment. Almost all the treaties that regulate the content of the national environmental protection systems of the Arctic states are of recent origin. This reflects the comparatively recent 'coming into force' of the goal of sustainable development, which has effectively eroded the idea that these issues belong exclusively to the domestic jurisdiction of states.

The most problematic sets of activities are environmentally harmful military activities, particularly acute concern in the Arctic, for these are excluded from the scope of the prior procedures. Significantly, however, the 1997 EIA Guidelines for the Arctic requite an attempt to be made to include military activities in national EIA procedures. Other international measures that have been taken to prevent environmental pollution from these activities include financial assistance and technological assistance agreements, especially between the Russian Federation and the other Arctic states.[640]

[640] For further information, see *infra* chapter IV (7.).

4. EIA Norms in Transboundary Context

The principle which governs conflicts over transboundary pollution between states was first elaborated in the *Trail Smelter* case of 1941.[641] Many formulations have been subsequently advanced by the scholarly world for a principle that would govern the responsibility for transboundary pollution between state territories.[642] A consensus on the formulation of the expanded[643] version of this principle started to emerge when the 1972 Stockholm Declaration was adopted, principle 21 of the Declaration in particular was increasingly cited as the authoritative formulation:

> States have, in accordance with the Charter of the United Nations and the principles of international law, the sovereign right to exploit their own resources pursuant to their own environmental (and developmental) policies and the responsibility to ensure that activities within their jurisdiction or control do not cause damage to the environment of other States or of areas beyond national jurisdiction.[644]

---

[641] *Trail Smelter Case, Award of Arbitral Tribunal, 11 March 1941, part three*. The Tribunal in this case stated: 'under the principles of international law, as well as the law of the United States, no state has the right to use or permit the use of its territory in such a manner as to cause injury by fumes in or to the territory of another or the properties of persons therein, when the case is of serious consequence and the injury is established by clear and convincing evidence'. This formulation came to be known as the principle of *sic tuo utere* in accordance with an old maxim derived from the Roman legal system. For a detailed analysis in Finnish, see Kuusiniemi 1994, pp. 118-136.

[642] For an overview, see Koskenniemi 1984, pp. 106-128 and Pissillo-Mazzeschi 1991, pp. 15-35.

[643] As will be shown below, the formulation contained in principle 21 of the Stockholm Declaration was much larger in scope than the norm formulations made before the Stockholm Conference. For a comparison, see Handl 1980, pp. 527-535.

[644] Article 21 of the Stockholm Declaration and Article 2 of the Rio Declaration of 1992. The Stockholm Declaration can be found from 11 I.L.M. 1416 (1972) and the Rio Declaration from 31 I.L.M. 874 (1992).

This formulation of the principle has been included in an extensive number of multilateral treaties and declarations[645] and has been designated 'the principle of due diligence' because, as will be shown below, the consensus on the content of the principle has grown. The ICJ confirmed the principle as *lex lata* in two recent cases:

> The existence of the general obligation of States to ensure that activities within their jurisdiction and control respect the environment of other States or of areas beyond national control is now part of the corpus of international law relating to the environment.[646]

Since the consensus on the status of the principle of due diligence as *lex lata* is so widespread, the scholarly debate on the principle has moved on to study what exactly it requires of states.[647] Above all, the scope of the principle has broadened from that seen in its earlier formulations as a norm governing transboundary pollution.[648]

---

645 See e.g., the Preamble of the 1979 ECE Convention on Long-Range Transboundary Air Pollution, paragraph 21 (d) of the 1982 World Charter for Nature, Article 193 of the LOSC, the Preamble of the 1985 Vienna Convention and the Preamble of the 1992 Climate Change Convention and Article 3 of the 1992 Biodiversity Convention.

646 The ICJ first stated this view in its *Advisory Opinion in the Legality of the Threat or Use of Nuclear Weapons, Advisory Opinion, I.C.J. Reports 1996, p. 226, paragraph 29.* The Court also repeated this passage in the case concerning the *Gabcíkovo-Nagymaros Project (Hungary/Slovakia), Judgment, I.C.J. Reports 1997, p. 7, paragraph 53.*

647 For evidence that the principle is part of *lex lata*, see Lefeber 1996, pp. 19-25. There are still a few scholars who refuse to accept that the principle of due diligence is a valid principle of international law. See, e.g., Zemanek 1991, p. 188, and Malanczuk 1997, pp. 245-247. For instance, Schachter (1991, pp. 364-365) argues: ' Principle 21 of the Stockholm declaration is, at best, a starting point. On its own terms, it has not become state practice: States generally do not "ensure that the activities within their jurisdiction do not cause damage" to the environments of others. Nor have governments given any significant indication that they regard this far-reaching principle as binding customary law. Environmental treaties, though numerous, are limited in scope and in participation. On the whole, they are not accepted as expressions of customary law and are regarded as binding for the parties alone. These facts suggest that the legal principles expressed in the above-mentioned texts are de lege ferenda and still await the imprimatur of state practice and *opinio juris* communis to endow them with the authority of customary international law'. Okowa (2000) replies to Schachter's statement as follows (p. 76): 'this argument confuses the existence of a right and its observance in actual practice. While it is true that there are still numerous cases of transboundary harm, these should in principle be treated as breaches of the rule rather than the affirmation of a contrary principle'.

648 The best known formulations of principle governing transboundary pollution are the above-mentioned principle of *sic tuo utere* and the so-called principle of non-harmful use of territory provided by the ICJ in the Corfu Channel case: 'every State's obligation not to allow knowingly

The present due diligence principle refers not only to the area of territorial sovereignty of states, as the old formulations do, but to states' own resources. It thus refers to areas where states have exclusive territorial jurisdiction to exploit natural resources, i.e., areas of jurisdictional competence. In addition, the principle affirms states' 'responsibility to ensure that activities within their jurisdiction or control do not cause damage to the environment of other States or of areas beyond the limits of national jurisdiction'. Here the differences to the earlier statements of the principle are clear.

Crucially, what is at issue is not merely the physical pollution that occurs within a state's sovereign areas but pollution that originates from any activity which is under the jurisdiction or control of a state. The concept of 'jurisdiction' refers not only to the areas of sovereignty but also to the areas of jurisdictional competence, i.e., the areas of the EEZ and the continental shelf. The concept of 'control' extends the scope of due diligence considerably *vis-à-vis* older formulations of the principle in that a state's responsibility now encompasses its areas of jurisdictional competence as well as all activities over which it has control. The principle thus now includes all mobile activities whose home state the origin state is, e.g., the state of registration of an oil tanker. In addition, damage to the environment of other states now covers damage to the areas of maritime jurisdiction of states and damage to international areas.[649] Whereas the older formulations of the principle were restricted to *inter se* relations, its present elaboration protects the environment of international areas as well, making it a norm of *erga omnes*.[650]

[their territories] to be used for acts contrary to the rights of other States' (*Corfu Channel case, Judgment of April 9th, 1949: I.C.J. Reports 1949, p. 22).*

649 See *infra* chapter IV (6.).

650 For an analysis, see Handl 1980, pp. 527-535 and Birnie and Boyle 1992, pp. 91-92.

The due diligence principle applies to all possible stationary activities, military or civilian, except for those that are regulated by the special law relating to transboundary watercourses[651] or are prohibited. There are some stationary activities that are prohibited *per se* in international law.[652] One problematic question is whether nuclear weapons testing falls within the scope of the principle. Some scholars argue that all nuclear weapons testing is presently prohibited by customary law, or at least testing in areas regulated by the 1963 Nuclear Test Ban Treaty, i.e., the atmosphere, outer space and underwater;[653] others assert that no conclusive evidence exists outside of the treaty regimes that customary law has come to prohibit nuclear weapons testing.[654] The practical importance of this question is greatly lessened by the fact that all the major nuclear powers, and all the Arctic states as well, are parties to the 1963 Nuclear Test Ban Treaty.[655] In addition, the Arctic states have committed themselves commendably in the 1996 Comprehensive Test Ban Treaty.[656]

---

[651] See *infra* chapter IV (5.).

[652] If a stationary activity is planned for the production of a certain item or substance which itself is prohibited, the entire activity is clearly illegal in international law. A good example can be found in the 1992 Convention on the Prohibition of the Development, Stockpiling and Use of Chemical Weapons. A recent convention that prohibits the use of certain chemicals is the 2001 Stockholm Convention on Persistent Organic Pollutants. Another example is where states have prohibited certain stationary activities in a certain area, for instance, mining in the Antarctic Treaty Area, prohibited on the basis of Article 7 of the 1991 Protocol on Environmental Protection to the 1959 Antarctic Treaty.

[653] See, e.g., Birnie and Boyle 1992, pp. 359-360. Goldblat (1995, p. 51) is more cautious on the matter, providing that the normative development 'may lead to the conclusion that the ban on nuclear testing in the atmosphere, outer space and under water is on the way of becoming customary law binding on all states'.

[654] Okowa 2000, pp. 101-110.

[655] Where the Arctic states are concerned, the 1963 Nuclear Test Ban Treaty has entered into force for Canada (28.1.1964), Denmark (15.1.1964), Finland (9.1.1964), Iceland (29.1.1964), Norway (21.11.1963), the Russian Federation (10.10.1973), Sweden (9.12.1963) and the United States (10.10.1963). Information obtained on the home page of the International Earth Science Information Network (CIESIN) at http://sedac.ciesin.org/prod/charlotte [14.3.2001]. However, according to Article IV of the 1963 Treaty, if supreme interests of the country so require, a party may withdraw from the Treaty, thus at least making it possible that an Arctic state could conduct nuclear testing in these areas in the future.

[656] The Treaty has been signed by 160 states, of whom 74 have also ratified it. The following dates of ratification are relevant here: Canada (18.12.1998), Denmark (21.12.1998), Finland (15.1.1999), Iceland (26.6.2000), Norway (15.7.1999), the Russian Federation (30.6.2000) and Sweden (2.12.1998). The United States has signed the Treaty (24.9.1996). Information obtained from the UNTS.

From the substantive viewpoint, although there is an emerging consensus that states are required to exercise due diligence, there exist numerous and diverse proposals as to what this means in practice.[657] The majority opinion seems to be that the principle is an international minimum standard providing a test whereby a state's conduct is compared to what a 'reasonable government' would do in a specific situation of transboundary pollution.[658] The standard does not allow the poor condition of the origin state's national environmental protection system to be used as a pretext for transboundary pollution nor is it an objective standard that is breached merely by a state's causing material damage to the environment of other states.[659] The international minimum standard is objectified when states conclude an international treaty in which the degree of care is agreed upon. The standard consists of objective and subjective aspects, of which the former will be examined first. Two elements are important in defining the objective criteria of the due diligence principle:

1. Threshold of transboundary pollution. Broad consensus exists that the principle of due diligence allows minor transboundary pollution. As pointed out by Fitzmaurice, the threshold seems to have lowered a bit in recent years from 'considerable' to 'significant'.[660]

---

[657] For an overview, see Pisillo-Mazzeschi 1991, pp. 15-35.

[658] According to Okowa (2000, p. 82): 'International tribunals that have considered the question of due diligence have generally reached the conclusion that the correctness of a state's application of the law to the facts is subject to international supervision. In each case the conduct of the state must be subjected to an international standard'. Lefeber (1996, p. 65) states: ' the degree of care which a state has actually exercised must be compared with the degree of care which a..."good government" would have exercised in similar circumstances'. See also, e.g., Handl 1980, pp. 535-540; Corten and Schaus 1989, pp. 254-256; Doeker and Gehring 1990, p. 2; Pisillo-Mazzeschi 1991, pp. 34-35; and Birnie and Boyle 1992, pp. 92-93.

[659] In a literal reading, infringement of the due diligence principle does not seem to require anything other than the causing of damage. However, it was clear already at the Stockholm Conference that the additional requirement of diligence was an integral part of the concept. See Handl 1980, pp. 535-540.

[660] Fitzmaurice 1996, p. 58.

2. The impact of the transboundary pollution. The majority opinion seems to be that the due diligence is breached only if material damage is caused.[661]

A consensus is also forming on the subjective criteria for deciding whether the origin state has acted diligently or not in a specific case. Since the principle of due diligence is an international minimum standard, the requirements it sets out are always context-specific, requiring different measures in different circumstances. Accordingly, only some of the basic criteria for comparing states' conduct to the standard of a 'reasonable government' can be enumerated here. Lefeber has provided useful criteria to this end:[662]

Degree of control. The starting point here is the fact that 'the source state can only be required to exercise due diligence with respect to activities within its jurisdiction or control'[663]. This degree of control varies in two important aspects. The first is whether a certain location is densely or thinly populated, the degree of control being relaxed in respect of the latter. In the Arctic, the degree of control is particularly loose since the region is very thinly populated in general. The second aspect is whether the activity in question is governmental or non-governmental and whether the state exercises control, territorial jurisdiction or territorial sovereignty over it. It can be presumed that the degree of control required of states is less strict over non-governmental activities and for activities which operate outside a state's territorial jurisdiction.

Degree of diligence (object of control). The main distinction here is between continuous pollution activities and ultra-hazardous activities. Due diligence requires

661 See, e.g., Handl 1975, pp. 50-76. Fitzmaurice (1996, p. 59) is more cautious on the issue: 'There are very divergent views on the subject'. A broader approach can be found in Lefeber 1996, pp. 81-86.
662 Lefeber 1996, pp. 65-69. Lefeber mentions 'interests of states' as a fourth criterion.
663 Lefeber 1996, p. 67.

measures from states to prevent damage to the environment of other states or areas beyond national jurisdiction and applies most naturally to stationary activities of which pollution is a normal by-product. Where the pollution is a normal by-product, the origin state can take measures to reduce it to a level below the threshold of significant adverse transboundary material damage. However, due diligence does not apply easily to stationary activities which in normal circumstances do not pollute at all or produce only minor amounts of pollution but contain an inherent potential for major accident, e.g., nuclear power plants. While environmental protection measures must probably be stricter in regard to such activities, the lack of diligence on the part of a state rarely contributes to the occurrence of accidents; rather, accidents can be attributed to the risks inherent in the kinds of activities.[664]

The technical and economic abilities of the state concerned. This factor is predicated on the distinction between developing and developed states. It can be presumed that both the degree of control and the degree of diligence required will be mitigated for states whose economic and technical development warrants less stringent requirements. For instance, the degree of control may be lessened if a state simply does not have the financial resources to establish basic governance structures in some parts of its jurisdiction. The degree of diligence required may also be mitigated if a state does not possess the technology needed to reduce the effects of pollution or does not have the capital to obtain such technology. This criterion may be applicable to the Russian Arctic, for example.

What then does the principle of due diligence require in the planning stage of stationary activities? Some scholars argue that there already exist customary law obligations independent of the requirements of the due diligence principle that require states at least to notify other states in the planning stage of activities that may cause

[664] Some scholars go further and argue that with respect to ultra-hazardous activities, the due diligence principle comes close to an obligation of result. See, e.g., Pisillo-Mazzeschi 1991, p. 35.

significant adverse transboundary damage and/or to conduct transboundary impact assessments and to organize prior consultations in the planning stage of an activity.[665]

The first Special Rapporteur of the ILC project 'International Liability for Injurious Consequences Arising Out of Acts Not Prohibited By International Law'[666], Robert Q. Quentin-Baxter, proposed that states' preventive duties could be captured by what he called the 'compound primary obligation'. According to the schematic outline of articles,[667] Quentin-Baxter proposed that the origin state is obligated to inform the affected state of potential transboundary pollution, which in turn implies an obligation to conduct a transboundary impact assessment.[668] The states concerned are also obligated to negotiate on the measures to be taken in view of the potential transboundary impact.[669]

---

665 See, e.g., Francioni 1991, pp. 203-221 and Ebbesson 1996, p. 185.

666 For a short historical account of the project's earlier phases, see Magraw 1986, pp. 306-309. An analysis of the later phases of the project is provided by Boyle 1999, pp. 61-85.

667 Quentin-Baxter provided a schematic outline of articles in his Third Report on International Liability for Injurious Consequences Arising Out of Acts Not Prohibited by International Law (Document A/CN.4/360). His approach was criticized as being detrimental to the systematic structure of international law and this is the position of the present author as well. The existence of a special category of secondary norms for lawful acts that cause material damage is not considered tenable. There are only primary norms, of which some contain compensation as a requirement of conduct; if no compensation is made, then the normal secondary rules are invoked. As put by Brownlie (1983, p. 50): 'it is the content of the relevant rules which is critical and a global distinction between lawful and unlawful activities is useless'. Hence, only the breach of the whole 'compound primary obligation' results in a breach of the due diligence principle. It should be noted that the terminology may be confusing since the concepts of 'primary' and 'secondary' rules are also used by Hart, see *supra* chapter III (3.1.). In the ILC, primary rules mean the rules of conduct and secondary rules the rules which are triggered when primary rules are breached. Hart perceives primary rules in much the same way, but sees secondary rules as rules which govern the creation, change and termination of primary rules as well as provide rules of competence for agencies overseeing the primary rules.

668 Section 2 (1) of the schematic outline reads: 'When an activity taking place within its territory or control gives or may give rise to loss or injury to persons or things within the territory or control of another State, the acting State has a duty to provide the affected State with all relevant and available information'. Section 2 (2) provides for the right of the potentially affected state to initiate the procedure itself. Section 2 (3) provides for an exception to information exchange on the basis of national or industrial security.

669 Ibid., Section 3. Finally, a duty of reparation to be made by the origin to the potentially or actually affected state is to be provided for in the schematic outline (Section 4).

One relevant legal question is whether these preventive duties are independent legal norms which the potentially affected states can invoke already in the planning phase of activities? According to some authors[670] and some expert bodies of international law, such as the International Law Association (ILA)[671], these preventive duties are independent duties of customary law, but the majority opinion seems to perceive them as part of the principle of due diligence.[672] Although not part of international law as independent obligations, these preventive duties perform important function in that they determine whether due diligence has been observed. As Okowa argues:

---

[670] See Boyle 1989, pp. 278-281 and Handl 1991, p. 76.

[671] The ILA has a taken positive stance on the existence of these duties in the work of two committees: Committee on the Legal Aspects of the Conservation of the Environment and the Committee on Long-distance Air Pollution. In the Report of the former (pp. 171-176), it was opined that duties of prior notification and prior consultations are independent norms of international law and the duty of transboundary impact assessment a consequence of procedural and substantive duties. See the Report of the 60th Conference 1982, pp. 157-182. These draft rules later became the 1982 ILA Montreal Rules of International Law Applicable to Transfrontier Pollution. The Montreal Rules are reproduced in Basic Documents of International Environmental Law, pp. 249-250. The Preamble to the Draft Articles of long-distance air pollution state that the ILA is to restate the existing rules of international law applicable to transfrontier pollution (paragraph 6, p. 378). The modal 'should' is used for those norms that have not become part of general international law (paragraph 7, p. 379). The Draft Articles provide that prior notification and assessment of planned activities (pp. 405-406) as well as prior consultation (pp. 409-412) are, under these criteria, obligations in general international law. See the First Preliminary Report of the Committee on Legal Aspects of Long-Distance Air Pollution (1984 ILA Report of the 61st Conference, pp. 377-413).

[672] Okowa (1996, p. 324) argues that the ILA conclusions are based too much on international resolutions and international treaties and that they fail to take account of *opinio juris* of states, concluding on these grounds that these preventive duties are not independent obligations in customary international law. See also Okowa 2000, pp. 131-170. In a similar vein, Sands (1988, p. 35) argues that it would be desirable to have the duties of prior information and consultation relating to potentially harmful activities as part of customary international law but 'it is not yet supported by the requisite State practice or by *opinio juris*'. See also Lefeber 1996, pp. 35-41.

> [i]n the absence of mandatory standards the procedural rules are useful criteria for determining whether or not a state has complied with its due diligence obligations under customary law. A state that has failed to undertake environmental impact assessment, or enter into consultations with affected states may be precluded from asserting that the harm did not occur for want of diligence. Moreover, it has been noted that where a state fails to respond to a notification, it too may be taken to have acquiesced in the conduct of the activity and may be estopped from asserting that the activities as carried out are not compatible with its interests.[673]

In other words, the origin state does not breach its preventive duties by failing to perform them, but neglect of these duties may have an effect on the evaluation of whether the state has acted diligently or not.[674] In the terminology used in this thesis, these duties can be described as guidelines for the observance of the principle of due diligence. By way of conclusion, it is suggested that in the situations studied here, i.e., the planning stage of environmentally harmful stationary activities, two separate obligations can be identified:

The duty of notification. The origin state is obligated to notify the potentially affected states if significant adverse transboundary damage is likely to ensue from the proposed activity. This duty can be properly implemented only if the origin state has a national EIA procedure which is able to evaluate transboundary effects as well, for otherwise the content of the notification will remain too speculative; i.e., it will not be based on systematic scientific studies.[675] Notification should thus contain the results

---

[673] Okowa 2000, p. 169-170. See also Okowa 1996, p. 336.

[674] The only independent customary law duty is the obligation to notify in case of a threatening transboundary environmental accident. See Nordström 1996, and Okowa 2000, pp. 167-168.

[675] Interestingly, the ICJ noted in *the Request for an Examination of the Situation in Accordance with Paragraph 63 of the Court's Judgment of 20 December 1974 in the Nuclear Tests (New Zealand v. France) Case, I.C.J. Reports 1995, p. 288*, that: '64. Whereas moreover the present Order is without prejudice to the obligations of States to respect and protect the natural environment, obligations to which both New Zealand and France have in the present instance reaffirmed their commitment'. The ICJ referred to the principles of precaution and to the duty to evaluate environmental impacts as the duties to which France and New Zealand had reaffirmed their commitment.

of the impact study, including potential transboundary impacts.[676] For notification to proceed efficiently, the potentially affected state should be requested information about the environment within its jurisdiction. Here it should be noted that the affected state is obligated to respond to the notification within a reasonable time after receipt of it. If it does not, this may be taken to mean that it has acquiesced to the proposed activity and is estopped from opposing it.[677]

The duty of consultation. Upon request by the potentially affected state or on its own initiative, the origin state must consult the potentially affected states before a licence is granted for the activity. This duty entails the establishment of prior licencing systems for environmentally harmful activities since without such a system of control there would be few possibilities to include the results of prior consultations in the implementation of the proposed activity. Both states should negotiate in good faith. If the affected state refuses to participate in the consultations, this can be taken to mean that it has acquiesced to the activity and is thus also estopped from opposing its construction.

The duties outlined above are not in themselves legal obligations but serve as guidelines for deciding whether or not a state has observed due diligence under

---

[676] A similar approach was adopted in the commentary to the draft 'Rules of International Law Applicable to Transfrontier Pollution' (paragraph 12): 'the necessity for an environmental assessment is a consequence of the substantive and procedural duties'. See the Report of the Committee on the Legal Aspect of the Conservation of the Environment, pp. 157-182 (Report of the 60th Conference 1982). These draft rules later became the 1982 ILA Montreal Rules of International Law Applicable to Transfrontier Pollution. The Montreal Rules are reproduced in Basic Documents of International Environmental Law, pp. 249-250.

[677] An interesting case here is *the Passage Through the Great Belt (Finland v. Denmark), Order of 10 September 1992, I.C.J. Reports 1992, p. 348.* In this case Denmark claimed that Finland had failed to respond to its notification of a proposed bridge across the Great Belt. Finland argued that since construction of the bridge had been postponed many times by Denmark, it was under the impression that no rapid action was called for, and thus no presumption of estoppel or acquiescence was justified. Although the case was discontinued, it is important that both parties accepted the basic idea that failure to respond to a notification had legal consequences. See also the *Lake Lanoux* case in which an arbitral tribunal made interesting general notes on the preventive duties of states in

customary international law. Although it is possible to present these preventive obligations as means for implementing the due diligence principle in the planning stage of environmentally harmful stationary activities, it must be acknowledged that the requirements of the principle are context-dependent at this stage in the life cycle of the activity as well and thus cannot provide exact criteria for ascertaining whether the principle has been observed.

4.1. Applicable Prior Transboundary Procedures in the Arctic

Prior transboundary procedures are normally executed and agreed between states located close to one another, as physical pollution can rarely be traced to a stationary source if the distance between source and impact is great. Accordingly, two main geographical regions of the Arctic have been distinguished in order to study the applicable prior transboundary procedures: the Eurasian and North American regions. Within these regions, the state borders are quite close to each other, making it more likely that prior transboundary procedures will be invoked between the Arctic states.

Many of the treaties which contain prior transboundary procedures also contain a provision limiting the information exchange between the states concerned. These provisions contain certain reasons, normally expressed in very general terms which allow the origin state to withhold from the potentially affected state the information required by prior transboundary procedures. Since the prior transboundary procedures pertain above all to information exchange, it is important to examine the scope of these provisions.[678]

---

case of potential transboundary pollution from a hydro-electric development scheme. *Lake Lanoux case (France/Spain), Award of Arbitral Tribunal, 16 November 1957.*

[678] This will be done in the footnote attached to the name of each treaty examined. The starting point here is that the parties are perfectly free to limit their information exchange. In the treaties studied below, two principal kinds of provisions can be found: provisions which rely on the national law of the origin state to determine which kind of data are not to be delivered to the potentially affected state and provisions that lay down certain reasons which can be used to justify the non-

4.1.1. The Eurasian Region

The Eurasian region has been defined here on the basis of the proximity of the international borders of states and thus covers only the Nordic Arctic and the north-eastern part of the Russian Federation and the maritime zones these continental areas and islands generate. It comprises the mainland areas of Finland, Sweden and Norway and the Russian Kola Peninsula located above the Arctic Circle and the islands of Jan Mayen Land, Svalbard Islands, Franz Josef Land, Denmark (Greenland), and Iceland.

Inter-parliamentary cooperation between the Nordic states started in 1952 with the establishment of the Nordic Council.[679] The formal basis for this cooperation was created when the 1962 Helsinki Agreement was signed on 23 March 1962.[680] The areas of cooperation enumerated in Article 1 of the Agreement did not originally include environmental protection, but it was added later through an amendment.[681] Although a reference is made in the Agreement to inter-governmental cooperation,[682] the form, functions and competence of the Nordic Council of Ministers were not specified until the 1971 amendment to the instrument.[683]

---

delivery of information. Both groups of provisions provide only an exception to the main rule in that the information specified in the treaty must be delivered to the potentially affected state unless reasons can be found for not doing so. As exceptions, these provisions must be interpreted in a restricted manner.

679 The Nordic Council was not established through an international treaty, but, as reported by Berg (1983, p. 261): 'each parliament adopted the Council's Statute in the national language either through legislation...or by parliamentary resolution'. For an overview of the history of the Nordic Council, see Wendt 1981, pp. 33-49.

680 Agreement Concerning Co-Operation Between Iceland, Norway, Sweden, Finland and Denmark. The Agreement entered into force on 1 July 1962.

681 The amendment entered into force on 7 November 1993. For the Finnish version of the amendment, see SopS 77/1993.

682 Article 35 of the original Helsinki Agreement.

683 Amendment to the Helsinki Agreement. For the Finnish version, see SopS 21/1971.

The Nordic states concluded a number of international treaties on the basis of the Helsinki Agreement, several of which are relevant from the viewpoint of environmental protection.[684] One of the agreements is the 1974 NEPC[685] and the later environmental protection conventions[686] which were negotiated under the umbrella of the Nordic Council.[687] The NEPC contains two prior transboundary procedures, the choice of which depends on which governmental authorities of the origin state are in charge of the licencing decision on the environmentally harmful activity. The scope of both procedures relies on the concept of 'environmentally harmful activities', which is defined in detail in Article 1 (1); the definition only includes stationary activities.

The environmental effects of these stationary activities are not defined exhaustively since the Article only enumerates certain pollution effects and provides that other pollution effects are also covered.[688] Although both of the prior transboundary procedures of the NEPC are based on the concept of environmentally harmful activities, they have a somewhat different scope in that the procedure involving the examining authority and the supervisory authority of the concerned states is triggered

---

[684] See, e.g., the 1971 Agreement Concerning Co-Operation in Measures to Deal with Pollution of the Sea by Oil, which entered into force on 16 October 1971.

[685] The 1974 Nordic Environmental Protection Convention. The NEPC was signed on 19 February 1974 by Finland, Sweden, Norway and Denmark and came into force on 5 October 1976. It is still in force. The NEPC contains a Protocol, paragraph 1 of which provides: 'In connection with the signing today of the Nordic Environmental Protection Convention the duly authorized signatory agreed that the following comments on its application shall be appended to the Convention'. For an overview of the Convention, see Kiss 1974, pp. 808-814 and Broms 1986, pp. 141-152. The NEPC does not contain any provision limiting the information exchange, but the Protocol (paragraph 5) appended to the treaty requires that the supervisory authority of the potentially affected state observe 'professional secrecy as regard trade secrets, operational devices or business condition'.

[686] See, e.g., the 1992 Agreement Between Denmark, Finland, Norway and Sweden on Cross-Border Cooperation in Preventing or Limiting the Damage in Emergency Situations Having Effect on Humans, Property or the Environment (author's own translation). The Agreement entered into force on 1 March 1992.

[687] A brief review of the negotiation process of the NEPC can be found in Broms 1986, pp. 141-143. For an examination of the consequences of the entry into force of the Espoo Convention to the system of the NEPC, see Koivurova 1997 I, pp. 505-525. For a study on cases in which the NEPC has been applied, see Phillips 1986, pp. 153-180. See also Wendt 1981, pp. 237-240.

[688] Article 1 of the NEPC ends its enumeration of pollution effects with 'etc.'.

when a 'nuisance of significance'[689] takes place whereas that involving the central governments requires that 'considerable nuisance'[690] is caused. Hence, the threshold at which transboundary pollution is deemed to occur is lower in the former procedure.

If the licencing decision is made by a 'Court or the Administrative Authority'[691], the procedure is carried out by this examining authority and the supervisory authority[692] of the potentially affected state. All environmentally harmful activities which 'may entail nuisance of significance in another Contracting State'[693] are covered by a procedure carried out by the licencing authority of the origin state and the supervisory authority of the affected state. The origin state's licencing authority is obligated to 'send as soon as possible a copy of the documents of the case to the supervisory authority of the other State'[694]. The licencing authority must also allow the supervisory authority to give its opinion on the matter.

If the supervisory authority so requests, the permit authority must require from the applicant 'such additional particulars, drawings and technical specifications as the examining authority deems necessary for evaluating the effects in the other State'[695]. Hence, the origin state is obligated to conduct a transboundary impact assessment. In addition, the supervisory authority is entitled to conduct its own investigations on the potential transboundary impacts 'as it deems necessary'[696], which means that the affected state also has a right to have such information taken into account in the

[689] Ibid., Article 5.
[690] Ibid., Article 11.
[691] Ibid., Article 5.
[692] Ibid. Article 4 defines the concept of 'supervisory authority': 'Each State shall appoint a special authority (supervisory authority) to be entrusted with the task of safeguarding general environmental interests insofar as regards nuisances arising out of environmentally harmful activities in another Contracting State'.
[693] Ibid., Article 5.
[694] Ibid.
[695] Ibid., Article 6.
[696] Ibid., Article 7.

licencing procedure of the origin state. Interestingly, 'if necessary for determining the damage caused in another State'[697] the supervisory authority of the potentially affected state is obligated to organize an on-site inspection upon request by the examining authority. In Broms' estimation, 'this provision is intended to enable the examining authority to gather all possible evidence as to the damage caused before it reaches a decision in the matter'[698].

Environmentally harmful activities 'which may entail considerable nuisance in another Contracting State'[699] and which are controlled by the central government of the origin state are governed by a procedure that is carried out by central governments of the concerned states. If the potentially affected state so requests, the central governments must arrange consultations on the matter. In these cases, both of the concerned states are also entitled to demand that an advisory opinion be given by an inquiry commission.[700] When such a commission is appointed, the dispute cannot be decided until the commission has given its opinion.[701]

The 1976 Guidelines for Communication Between Finland, Norway, Sweden and Denmark on Security Issues Related To the Nuclear Installations Constructed Near the Border (hereinafter 'the 1976 Nordic Guidelines Treaty')[702] apply to situations in which a proposed nuclear installation[703] poses greater than a minor threat to the

---

[697] Ibid., Article 10.
[698] See Broms 1986, p. 148.
[699] Article 11 of the NEPC.
[700] Ibid., Article 12.
[701] Ibid.
[702] The Treaty was accepted as an international treaty through an exchange of notes between the four Nordic states (Finland, Sweden, Norway and Denmark) on 15 November 1976. There also exists a provision (Article 5) by which the contracting states accept that even if the national law of a party prohibits the transfer of information, the state in question is entitled to do so. (translation of the name of the document by the author).
[703] This is defined in a footnote to the Guidelines Treaty and covers the following categories of activities: 1) nuclear power plant with the exception of nuclear-powered vessels; 2) plant for the production and handling of nuclear substances; 3) plant for the extraction of isotopes from the nuclear fuels; 4) plant for the reprocessing of irradiated nuclear fuels; 5) plant for the storage of

security of the neighboring states. The prior transboundary procedure in this treaty begins with the requirement that the origin state is obligated to provide certain information on the proposed installation.[704] This includes information on the location of the proposed activity, the licence the activity must obtain, and the conditions under which the licence conditions may be revoked.[705] If information from the potentially affected state is necessary, which is almost always the case, the origin state must specifically request that information; the affected state is then obligated to supply it.[706] This information includes an account of the conditions in the potentially affected state. If one of the states provides a reasoned request on the basis of this treaty, the concerned states must consult one another on the matter.[707]

The Nordic cooperation which has taken place under the umbrella of the Nordic Council and the Nordic Council of Ministers has been challenged as the Nordic states have gradually established their relationship with the EC. Denmark joined the EC in 1973 but Greenland opted out in 1982. Norway has negotiated with the EC already several times, most recently in 1994, but has not joined and remains a European Free Trade Association (EFTA) state. At the moment, only the Arctic areas of Finland and Sweden are part of the area of the EC, both states having joined the EC in 1995.[708] Yet, because of the European Economic Area (EEA) Agreement, both Iceland and

---

nuclear substances or radioactive waste with the exception of temporary storage in transit; 6) other installations which according to official ordinances contain nuclear fuels or radioactive substances. (author's translation).

[704] Article 1 of the 1976 Nordic Guidelines Treaty.

[705] Ibid.

[706] Ibid., Article 3.

[707] Ibid., Article 4. Finland applied the Treaty in 1991 when the Foreign Ministry of Sweden and the governments of Norway and Denmark were informed of the construction of the nuclear power plant in Loviisa (author's personal copy of the documents, obtained from the Finnish Ministry of the Environment). The Finnish parliament later rejected the plan, but on 15 November 2000, Teollisuuden Voima Ltd. submitted an application for a Decision-in-Principle on Construction of a New Nuclear Power Plant Unit. Information obtained from the home page of the Finnish Ministry of Trade and Industry at http://www.vn.fi/ktm/eng/tiedote/tied2.htm [6.3.2001].

[708] See the Europa home page, which presents the major phases of the history of the European Union, at http://europa.eu.int/abc/history/1995/1995_en.htm [7.2.2001]. See also Shaw 2000, pp. 40-104.

Norway are required to implement the EIA Directive, as amended by the Amendment Directive in 1997. It is unfortunate from the viewpoint of Arctic environmental protection that Greenland withdrew from the EC[709] and Svalbard was excluded from the EEA Agreement,[710] for these decisions mean that the EIA Directive is not binding with regard to these areas.

The 1985 EIA Directive[711] is part of the EC legal system and in principle obligates only the member states of the system; however, as the Directive was included as part of *the 1992 Agreement on European Economic Area (hereinafter 'the 1992 EEA Agreement')*,[712] the members of EFTA are obligated by it in international law. The Agreement applies not only between the EC and the EFTA states but also between the EFTA states.[713] Since the EC is one of the signatories to the Espoo Convention

---

709 See *supra* chapter II (1.1.).

710 A special Protocol was adopted as part of the EEA Agreement to the effect that Norway may decide whether to apply the EEA Agreement to Svalbard or not (Protocol 40). Norway decided to exclude the Islands. See Ulfstein 1995, p. 299.

711 Council Directive of 27 June 1985 on the Assessment of the Effects of Certain Public and Private Projects on the Environment. The EIA Directive contains an extensive provision on exceptions to the requirement of free information exchange, Article 10 (2), which was not amended by the Amendment Directive: 'Where Article 7 applies, the transmission of information to another Member State and the reception of information by another Member State shall be subject to the limitations in force in the Member State in which the project is proposed'.

712 The Agreement was signed on 2 May 1992 and entered into force on 1 January 1994. For further information on the complex arrangement, see the home page of the EFTA Secretariat at http://secretariat.efta.int/euroeco/ [21.5.2001].

The EIA Directive is included in the EEA Agreement as Annex XX, article 74 (I (1)). It should be noted here that the Agreement includes certain other directives relevant to the situations considered in the present study but that since the EIA Directive is the most important instrument, and covers all the relevant situations, it is the only one examined. Other directives included in the EEA Agreement and of relevance here are, e.g., Council Directive 96/61/EC of 24 September 1996 Concerning Integrated Pollution Prevention and Control and Council Directive 96/82/EC of 9 December 1996 on the Control of Major-Accident Hazards Involving Dangerous Substances.

713 For an analysis, see Dedichen et al. 1993, pp. 73-75. According to Dedichen et al. (1993, p. 73) the Agreement 'is an agreement under public international law. However, due to its close relationship to the Community legal system, as well as its dynamic character, implying a development of the Agreement in step with the development of the Community legal system, it is of a unique character'.

and, for reasons that will be presented below,[714] the 1985 EIA Directive did not fulfill the obligations of the Espoo Convention, it became necessary to amend the Directive.

The Council of Ministers adopted the necessary Amendment Directive[715] in 1997, which the member states were required to implement by 14 March 1999.[716] The Amendment Directive also required certain measures to allow its incorporation into the EEA Agreement, and these were completed on 26 February 1999.[717] The EEA Agreement (EIA Directive) obligates the states parties to implement the provisions providing for a prior transboundary procedure. The parties can choose the 'choice of form and method of implementation'[718] With respect to prior transboundary

---

[714] See *infra* chapter V (1.).

[715] Council Directive 97/11/EC of 3 March 1997 Amending Directive 85/337/EEC on the Assessment of the Effects of Certain Public and Private Projects on the Environment. For the legislative history of the Amendment Directive, see Sheate 1995, pp. 77-81, 1996, pp. 36-43 and Sheate 1997, pp. 235-238. The member states were to have implemented the amendment by 14 March 1999. The latest development is an initiative to have a directive on strategic environmental assessment adopted by the Council of Ministers. The various stages of decision-making this initiative has gone through and the final Commission initiative to be presented to the Council of Ministers (COM (2000) 636) can be found on the Europa home page at http://europa.eu.int/comm/environment/eia/sea-legalcontext.htm [7.3.2001].

[716] As required by Article 249 of the 1992 Treaty Establishing the European Community: 'A Directive shall be binding, as to the result to be achieved, upon each Member State to which it is addressed, but shall leave to the national authorities the choice of form and methods'.

[717] Article 98 of the 1992 EEA Agreement provides: 'The annexes to this Agreement and Protocols 1 to 7, 9 to 11, 19 to 27, 30 to 32, 37, 39, 41 and 47, as appropriate, may be amended by a decision of the EEA Joint Committee in accordance with Articles 93 (2), 99, 100, 102 and 103'. The EEA Joint Committee of the EEA decided in 26 February 1999 to incorporate the Amendment Directive into the EEA agreement (OJ L 148, 22.6.2000, p. 45), decision No. 20/1999. According to Article 3 of the decision: 'This Decision shall enter into force on 27 February 1999, provided that all the notifications under Article 103 (1) of the Agreement have been made to the EEA Joint Committee'. According to the Acquis Implementation Database (AIDA) on the home page of the EFTA Surveillance Authority (http://www.efta.int/structure/SURV/efta-srv.asp [21.5.2001] both Iceland and Norway have implemented the Amendment Directive. Since the data on implementation status does not rely on what the states themselves have reported but only on what has been accepted as implemented by the Surveillance Authority, the information in this database can be considered reliable. The Surveillance Authority oversees the implementation of the EEA Agreement together with the Commission of the EC.

[718] As laid down in Part I, Article 7 of the EEA Agreement: 'Acts referred to or contained in the Annexes to this Agreement or in decisions of the EEA Joint Committee shall be binding upon the Contracting Parties and be, or be made, part of their internal legal order as follows...(b) an act corresponding to an EEC directive shall leave to the authorities of the Contracting Parties the choice of form and method of implementation'.

procedures, this can only be done through international treaties, such as the 1991 Espoo Convention. The two directives are studied here together since the Amendment Directive only changed certain parts of the EIA Directive.

The EIA Directive unequivocally excludes certain categories of activities from its scope. These include 'projects serving national defense purposes'[719] as well as projects 'the details of which are adopted by a specific act of national legislation'[720]. The Amendment Directive brought no changes in this regard. The EIA Directive provided that member states could exempt a project from the provisions of the Directive in exceptional circumstances, meaning that even if the potentially affected state requested the initiation of the transboundary EIA procedure, the origin state could exempt the activity from the operation of the national EIA procedure.[721] This provision was changed in the Amendment Directive in that the exemption cannot be invoked in cases to which Article 7 on the prior transboundary procedure applies.[722]

The prior transboundary procedure of the EIA Directive is initiated on the basis of a single fundamental criterion: if a project is likely to have significant effects on the environment in another member state, or in another EEA state, the procedure must be started.[723] The determination of whether this criterion is fulfilled in given circumstances takes two forms. First, if the origin state is aware that significant effects on the environment of other member states are likely, it must forward the information 'gathered pursuant to Article 5' to the potentially affected state 'at the same time as it makes it available to its own nationals'.[724] However, if the origin state

[719] Article 1 (4) of the EIA Directive.
[720] Ibid., Article 1 (5).
[721] Ibid. Article 2 (3) provides: 'Member States may, in exceptional cases, exempt a specific project in whole or in part from the provisions laid down in this Directive'.
[722] Article 2 (3) of the Amendment Directive provides: 'Without prejudice to Article 7, Member States may, in exceptional cases, exempt a specific project in whole or in part from the provisions laid down in this Directive'.
[723] Article 5 of the EIA Directive.
[724] Ibid.

does not initiate the procedure, the potentially affected state is entitled to request the origin state to forward such information, but only if the latter is likely to be significantly affected.[725] There exists no procedure for clarifying whether the criterion of applicability of the prior transboundary procedure has been fulfilled in a given case. The Amendment Directive does not provide for any inquiry commission procedure either to settle disputes as to the applicability of the prior transboundary procedure. The EIA Directive does not elaborate any rules on the confirmation of participation of the potentially affected state where the origin state has initiated the transboundary EIA procedure. The Amendment Directive does provide such rules, however, in accordance with the Espoo Convention.[726]

The EIA Directive only requires that 'information gathered pursuant to Article 5' be delivered to the potentially affected state at the same time as it is made available to its own nationals.[727] In terms of the EIA Directive this means application of the information requirements laid down in either Article 5 (2) or Annex III. Hence, the potentially affected state has no possibility to influence the content of the EIS, and, more importantly, the developer is not even required to study any transboundary impacts of the planned project. Moreover, the information required of the developer need not even contain any alternatives to the proposed project.[728] Here the Amendment Directive provides only marginal improvement, for while it extends the requirements of the EIS carried out by the developer to include alternatives to the project, it does not require that the study cover transboundary impacts.[729] In addition,

[725] Ibid.
[726] Article 7 (1) of the Amendment Directive.
[727] Article 7 of the EIA Directive.
[728] Ibid. Not even the more extensive requirements of Annex III absolutely require the studying of alternatives (Annex III (2)): 'Where appropriate, an outline of the main alternatives studied by the developer and an indication of the main reasons for his choice, taking into account the environmental effects'. The use of the phrase 'where appropriate' allows for considerable discretion.
[729] Both Article 5 (3) and Annex IV(2) of the Amendment Directive require the developer to outline the main alternatives, but neither obligates the developer to study transboundary impacts.

it does not accord the potentially affected state any possibility to contribute to the content of the EIS.

The EIA Directive does not absolutely require the origin state to initiate consultations with the potentially affected state; these take place only if 'necessary'.[730] This can be interpreted as meaning that if the potentially affected state considers consultations necessary, they must be arranged by the origin state. The Amendment Directive implements the obligations of the Espoo Convention in this respect by requiring that consultations take place.[731] The greatest improvement introduced by the Amendment Directive is that it guarantees the participation rights of the public of the potentially affected state; the EIA Directive does not provide for such rights at all. The states concerned must now arrange for the information provided by the developer to be supplied to the environmental authorities and the public of the potentially affected state.[732] The public also have a right to have their comments delivered to the licencing authority before consent for development is granted.[733]

The EIA Directive does not contain any rules on how the results of consultations between states are to be taken into account in the licencing decision. The Amendment Directive provides a clear change here by implementing the obligations of the Espoo Convention: it requires that the results of consultations and the comments of the public of the affected state 'must be taken into consideration in the development consent procedure'[734] and that this licencing decision must then be forwarded to the potentially affected state.[735] In fact, the Amendment Directive provides somewhat more detailed rules in this respect than the Espoo Convention, which only requires that the licencing decision be transmitted to the potentially affected state along with

[730] Article 7 of the EIA Directive.
[731] Article 7 (4) of the Amendment Directive.
[732] Ibid., Article 7 (3a).
[733] Ibid.
[734] Ibid., Article 8.
[735] Ibid., Article 9 (2).

the reasons and considerations on which the decision is based. The Amendment Directive requires that the potentially affected state be provided with 'the content of the decision and any conditions attached thereto', 'the main reasons and considerations on which it was based', and 'a description, where necessary, of the main measures to avoid, reduce and, if possible, offset the major adverse effects'[736].

The efforts to protect the marine environment of the Northeast Atlantic also apply in Arctic waters as defined in this study. The first relevant convention in this respect was the 1972 Convention for the Prevention of Marine Pollution by Dumping from Ships and Aircraft. This was followed by the 1974 Paris I Convention, which dealt with pollution of the marine environment from land-based sources in general and contained a prior transboundary procedure.[737] These two conventions have been replaced by the 1992 Paris II Convention between the parties to both.[738]

The Paris I Convention was signed on 4 June 1974 and entered into force on 6 May 1978. The geographical scope of the Convention is limited to parts of the North Atlantic, the Arctic Ocean and the North Sea.[739] Among the Arctic states, it has entered into force for Norway, Denmark, Iceland and Sweden.[740] The Convention provides for a consultation right for the potentially affected state if pollution from land-based sources...by substances not listed in Part 1 of Annex A'[741] is likely to

[736] Ibid., Article 9 (1).

[737] The 1974 Convention for the Prevention of Marine Pollution from Land-Based Sources was concluded at Paris on 4 June 1974 and came into force on 6 May 1978. The Convention does not contain any provisions on limiting information exchange.

[738] As provided in Article 31 (1) of the Paris II Convention: 'Upon its entry into force, the Convention shall replace the Oslo and Paris Conventions as between the Contracting Parties'.

[739] Article 2 (a) of the Paris I Convention reads as follows: 'those parts of the Atlantic and Arctic oceans and the dependent seas which lie North of 36' north latitude and between 42' west longitude and 51 east longitude, but excluding...(b) that part of the Atlantic Ocean north of 59' latitude and between 44' west longitude and 42 west longitude'.

[740] The Convention entered into force as follows for the Arctic states: Denmark (06.05.1978), EC (06.05.1978), Iceland (19.07.1981), Norway (06.05.1978), Sweden (06.05.1978). Information obtained from the UNTS.

[741] Article 9 (1) of the Paris I Convention .

prejudice its interests. Hence, substances which are listed in Part I of Annex A are not governed by the procedure, because they are prohibited altogether. This provision can be interpreted as covering potential pollution as well due to its use of the word 'likely'. The aim of the procedure is to negotiate a co-operation agreement under the guidelines provided in paragraph 3.[742] If any one of the concerned states so requests, the Commission - a joint organ established by the Convention - 'shall consider the question' and 'may make recommendations'[743].

The 1992 Paris II Convention[744] was signed on 22 September 1992 and entered into force on 25 March 1998. The geographical scope of this Convention is larger than that of the Paris I Convention in that it includes certain parts of the Baltic and Mediterranean Seas in addition to the parts of the North Atlantic Ocean, the North Sea and the Arctic Ocean covered by its predecessor. The Arctic states parties are the countries of the Nordic Arctic, i.e., Iceland, Norway, Denmark, Finland and Sweden. Denmark excluded Greenland and Faroe Islands from the application of the Convention when it signed the instrument but withdrew this reservation upon ratifying it.[745] Hence, the 1992 Paris II Convention has replaced the 1974 Paris I Convention altogether in the relations between the Nordic Arctic states. Interestingly, the Paris II Convention provides for a possibility to adjust the standards it prescribes to certain regions on the basis of differences in ecological conditions.[746]

---

742 Ibid. Article 9 (3) reads: 'The special agreements specified in paragraph 1 of this Article may, among other things, define the areas to which they shall apply, the quality objectives to be achieved, and the methods for achieving these objectives including methods for the application of appropriate standards and the scientific and technical information to be collected'.

743 Ibid., Article 9 (2).

744 The 1992 Convention for the Protection of the Marine Environment of the North-East Atlantic. The Convention does not contain any provisions on limiting information exchange.

745 Denmark made the following declaration upon signature of the Convention: 'The present Convention is subject to ratification and with reservation for application to the Faroe Islands and Greenland'. Yet, after it had ratified the Convention, Denmark notified the depositary state, France, that it had withdrawn its reservation. Information obtained on the Ospar Commission home page at http://www.ospar.org/eng/html/welcome.html [17.2.2001].

746 Article 24 of the Paris II Convention reads: 'The Commission may decide that any decision or recommendation adopted by it shall apply to all, or a specified part, of the maritime area and may

The Paris II Convention provides for a prior transboundary procedure that is similar to that in the Paris I Convention, albeit with a different material scope. The procedure is triggered if 'pollution originating from a Contracting Party is likely to prejudice the interests of one or more of the other Contracting Parties'[747]. The material scope has been widened very much from the Paris I Convention in that the initiation of the procedure is triggered merely by the prejudicing of the interests of other states parties. However, since the Paris II Convention focuses only on protecting the marine environment, the interests of the affected state can be prejudiced only when the maritime area under its sovereignty or competence is threatened. If these criteria are fulfilled, the potentially affected state or states have a right to consultation with the origin state.[748] The aim of the procedure is to negotiate a cooperation agreement under the guidelines provided in paragraph 3, which are identical to those in the Paris I Convention. If any one of the concerned states so requests, the Commission 'shall consider the question' and 'may make recommendations'[749].

4.1.2. The North American Region

The North American region is also defined here from the perspective of the topic to include the areas of the Arctic where the international borders of the Arctic states are close to one other. Thus, agreements between the Russian Federation, Canada and the United States are relevant since the international borders of these states are quite close in the Canadian Basin. Another area belonging to the North American region is the international border between Canada and Denmark (Greenland).

provide for different timetables to be applied, having regard to the differences between ecological and economic conditions in the various regions and sub-regions covered by the Convention'.

747 Ibid., Article 21 (1).

748 Ibid.

749 Ibid., Article 21 (2).

The North American region includes only one tradition of environmental cooperation between states, that between Canada and the United States. Although there are some environmental cooperation agreements between Canada and the United States with the Soviet Union/the Russian Federation, no applicable prior transboundary procedure is contained in these treaties. The neighboring states of Canada and Denmark (Greenland) have only one treaty providing for a prior transboundary procedure, *the 1983 Agreement Between Denmark and Canada for Cooperation Relating to the Marine Environment (hereinafter 'the 1983 Marine Agreement')*[750].

The territorial scope of the 1983 Marine Agreement is defined in Article 1 and covers in general the marine areas located in Nares Strait, Baffin Bay and Davis Strait. The Agreement applies to plans of 'works or undertakings' in the area of responsibility of the parties which 'may create a significant risk of pollution'[751]. The prior transboundary procedure is initiated by the origin state 'on its own initiative or at the request of the other Party'[752]. If a dispute arises over whether the Marine Agreement is applicable, the dispute-resolution procedure in Article XIII of the Agreement is invoked. If the parties cannot find a solution to the dispute through negotiations, either is entitled to initiate an ad hoc tribunal procedure within six months of the end of negotiations.[753] The origin state is obligated to first provide 'all relevant information and data' and invite comments on this information from the potentially affected state.[754] Consultations are arranged if either party requests them and must be

[750] The 1983 Agreement Between the Government of the Kingdom of Denmark and the Government of Canada for Cooperation Relating to the Marine Environment. Article IV contains certain exceptions to the information exchange rules. If the information transfer is prohibited by the national laws of the states parties, or the information is understood to be confidential, these may provide the party with an exemption from the information exchange rules of the transboundary procedure. The Agreement was signed on 26 August 1983, and entered into force on the same day by virtue of Article XIV. The Marine Agreement is still in force.
[751] Article IV (1) of the 1983 Marine Agreement.
[752] Ibid.
[753] Ibid., Article XIII (2).
[754] Ibid., Article IV (1).

conducted 'over a reasonable period of time'[755]. It is also cautioned that consultations must not be used by a party to postpone unreasonably or to impede the works or undertakings.[756]

Some general bilateral environmental cooperation agreements have been concluded by the Russian Federation with Canada and the United States. *The 1993 Agreement Between the Government of Canada and the Government of the Russian Federation Concerning Environmental Cooperation*[757] provides that the parties agree that cooperation 'may include the following: environmental impact assessment and post-project analysis...environmental protection laws, regulations and policies; and national environmental plans'[758]. According to Article III, the parties may exchange information on the items in Article II 'including their management, policy and regulatory elements, socio-economic implications and major environmental studies'. The provisions only outline possible areas of cooperation, as indicated by their use of the word 'may', and thus do not legally obligate parties to cooperate in issues pertaining to EIA, for example.

The 1994 Agreement on Cooperation in the field of protection of the Environment and Natural Resources[759] between the United States and the Russian Federation provides that 'parties shall develop cooperation in the field of protection of the environment'[760]. These include 'arctic and subarctic areas and resources', 'environmental impact assessment' as well as 'participation of the public, including

---

[755] Ibid., Article IV (2).
[756] Ibid.
[757] The Agreement was signed on 5 August 1993 and entered into force on the same day.
[758] Ibid., Article II (1).
[759] The 1994 Agreement on Cooperation in the field of protection of the Environment and Natural Resources, with Annex, was signed on 23 June 1994 and entered into force on the same date. As provided in Article VIII of the Agreement, the Agreement supersedes the earlier treaty on environmental cooperation between the United States and the Soviet Union which entered into force on 23 May 1972, namely, the Agreement on Cooperation in the Field of Environmental Protection.
[760] Ibid., Article I (3).

non-governmental organizations, in environmental decision-making'[761]. Yet, it is provided that these forms of cooperation 'may be implemented'[762], and thus the cooperation in EIA remains optional for the parties.

The main tradition of environmental cooperation between Arctic states has evidently been that between Canada and the United States, which dates from the 1909 Boundary Waters Treaty. These states have concluded agreements providing for prior procedures that are applicable also in the Arctic. *The 1975 Agreement Between the United States of America and Canada Relating to the Exchange of Information on Weather Modification Activities (hereinafter 'the 1975 Weather Agreement')*[763] applies to weather modifications carried out by private actors and governmental agencies. Weather modification activities have been defined broadly as 'activities performed with the intention of producing artificial changes in the composition, behavior, or dynamics of the atmosphere'[764]. Such activities are covered only if they can be considered as 'weather modification activities of mutual interest'[765]. Two criteria apply here: if the weather modification activity is carried out in or over the territory of a party which is within 200 miles of the international boundary, the procedure is always applied; if the activity is outside this limit, the procedure can be applied if a party deems that it may 'significantly affect the composition, behavior, or dynamics of the atmosphere over the territory of the other Party'[766].

---

[761] Ibid., Article II (5, 6 and 12).
[762] Ibid., Article II.
[763] There are two provisions providing for limitations on the exchange of information. In regard to weather modification activities carried out by private legal subjects, Article II (3) stipulates that the responsible agency is obligated not to disclose information if prohibited by law. In addition, the responsible agency has a certain level of discretion in regard to other kinds of information. With respect to weather modification activities conducted by the responsible agency itself, no such limitation exists, since Article II clearly applies to private activities only. The Agreement was signed on 26 March 1975 and entered into force on the same day by virtue of Article IX.
[764] Article I (a) of the 1975 Weather Agreement.
[765] Ibid., Article I (b).
[766] Ibid.

As mentioned above, there exist two transboundary procedures depending on whether the weather modification activities are carried out by a private or public person. If the actor is a private legal subject, the responsible agencies of the parties obtain their information through the reporting requirements provided in relevant internal laws and regulations.[767] The responsible agency is obligated to transmit information on the weather modification activities of mutual interest carried out by a private person only if the private legal subject has transmitted information to the responsible agency. This is indicated by the phrase 'acquired by a responsible agency', which implies that the responsible agency has no obligation to transmit information to another party if it has received none.[768] Any information acquired must be transmitted 'as soon as practicable to the responsible agency of the other Party'[769].

The Agreement states that 'whenever possible, this information shall be transmitted prior to the commencement of such activities'[770]. Does this amount to a legal obligation? This author would contend that although this norm is expressed in rather loose language, it nevertheless establishes an obligation of attempt. Article II (1) contains a norm on the exact time-frame for the notification: 'It is anticipated that such information will be transmitted within five working days of its receipt by a responsible agency'. The notification must contain copies of the reports provided to the responsible agency according to the internal laws of the respective party, and 'such other information and interpretation as the responsible agency might consider appropriate'[771]. Article V stipulates that consultations are to be held at the request of either party, with such consultations to be carried out 'promptly on the request of a

[767] Ibid. Article I (d) defines the concept of reporting requirements.
[768] Ibid., Article II (1).
[769] Ibid.
[770] Ibid.
[771] Ibid., Article II (2).

Party'[772]. In urgent cases, electronic means of communication are allowed in conducting consultations.[773]

The prior transboundary procedure between the two parties is more strictly regulated in relation to weather modification activities of mutual interest that are carried out by governmental authorities. In these cases, the notification is unconditional and always prior to the commencement of such activities.[774] It is even emphasized that 'every effort shall be made to provide such notice as far in advance of such activities as may be possible'[775]. As a legal obligation this amounts to a heightened duty to provide for timely notification. The other party must also be fully informed. The norms on consultation are the same as those in the first procedure.[776]

The 1987 Agreement Between the Government of Canada and the Government of the United States of America on the Conservation of the Porcupine Caribou Herd (hereinafter 'the 1987 Caribou Agreement')[777] defines the concept of 'Porcupine Caribou Herd' in Article 1 (a) in terms of knowledge of where these animals migrate normally. Their habitat is defined very broadly as covering 'the whole or any part of the ecosystem, including summer, winter and migration range, used by the Porcupine Caribou Herd during the course of its long-term movement patterns'[778]. If the Porcupine Caribou Herd or its habitat may be influenced by a proposed activity, the party initiating the activity is obligated to organize impact assessment and review 'consistent with domestic laws, regulations and processes'[779].

---

[772] Ibid., Article V.
[773] Ibid.
[774] Ibid. Article IV reads: 'each Party agrees to notify and to fully inform the other concerning any weather modification activities of mutual interest conducted by it prior to the commencement of such activities'.
[775] Ibid.
[776] Ibid., Article V.
[777] There are no limitations in the treaty on the exchange of information. The Agreement was signed on 17 July 1987, and entered into force by virtue of Article 8 on the same day. It is still in force.
[778] Article 1 (c) of the 1987 Caribou Agreement.
[779] Ibid., Article 3 (c).

The parties are obligated to assess the potential impacts, including cumulative impacts, on the Porcupine Caribou Herd, its habitat and affected users of porcupine caribou.[780] The other party must only be notified if 'an activity in one country is determined to be likely to cause significant long-term adverse impact on the Porcupine Caribou Herd or its habitat'[781]. The other party also has the right to be consulted prior to any licencing decision.[782] Another procedure is that carried out by the origin state and the International Porcupine Caribou Board (IPCB).[783] If an activity may significantly affect the conservation of the herd or its habitat, the origin state must notify the IPCB and 'provide an opportunity to the Board to make recommendations'[784]. The origin state must also take note of the advice and respond to the recommendations of the Board[785], providing written reasons if it has rejected in whole or in part the Board's recommendations.[786]

The 1991 Agreement Between the Government of the United States of America and the Government of Canada on Air Quality (hereinafter the 1991 Air Quality Agreement)[787] was preceded by *the 1980 Memorandum of Intent Between the Government of the United States of America and the Government of Canada*

---

[780] Ibid., Article 3 (g).
[781] Ibid., Article 3 (d).
[782] Ibid.
[783] Ibid. The Board is established in Article 4 (a and b): 'a. The Parties will establish an advisory Board to be known as the International Porcupine Caribou Board...b. The Parties will each appoint four members of the Board within a reasonable period following the entry into force of the present Agreement'.
[784] Ibid., Article 4 (e.2).
[785] Ibid., Article 4 (e.3).
[786] Ibid., Article 4 (e.4).
[787] The Agreement was signed on 13 March 1991, and, in accordance with Article XVI, it entered into force upon signature. It is still in force. The Agreement contains a provision limiting information exchange, Article VII (2), which reads: 'Notwithstanding any other provisions of this Agreement, the Air Quality Committee and the International Joint Commission shall not release, without the consent of the owner, any information identified to them as proprietary information under the laws of the place where such information has been acquired'. The geographical scope of the Air Quality Agreement is not confined to the parties' territorial sovereignty but covers the jurisdictional areas of states as well, as provided in Article 1 (2).

Concerning Transboundary Air Pollution[788], which already outlined a prior transboundary procedure. The Memorandum already contained some legally binding provisions, such as the one on prior transboundary procedure requiring the parties to provide advance notification and consultation on proposed actions 'involving a significant risk of or potential risk of causing or increasing transboundary air pollution'[789]. The Memorandum states that it 'will remain in effect until revised by mutual agreement'[790].

The prior transboundary procedure in the 1991 Air Quality Agreement is the following. The origin state is obligated to notify the potentially affected state of projects which are 'likely to cause significant transboundary air pollution'[791]. However, this obligation is qualified by a reference to the origin state's 'laws, regulations and policies'[792], meaning that only those activities must be notified which are both likely to cause significant transboundary air pollution and are covered by national prior evaluation procedures. This notification must take place 'as early as practicable in advance of a decision'[793]. If the affected state so requests, the origin state must organize consultations on the matter.[794] These consultations must commence 'thirty days from the date of the receipt of the request for consultations'[795] at the latest and 'include consideration of appropriate mitigation measures'[796].

The 1996 Administrative Arrangement for Cooperation and the Exchange of Information in Nuclear Regulatory Matters (hereinafter the 1996 Nuclear

---

[788] The Memorandum was signed on 5 August 1980, and, according to the Memorandum itself, it became effective upon signature.
[789] Article 3 of the Memorandum. The Memorandum provides prior transboundary procedure for not only concrete activities but also 'proposed major changes of policy, regulations or practices'.
[790] Ibid. Final clause of the Memorandum.
[791] Article V (1) of the 1991 Air Quality Agreement.
[792] Ibid.
[793] Ibid., Article V (2).
[794] Ibid.
[795] Ibid., Article XI.
[796] Ibid., Article V (4).

Agreement)[797] between the United States and Canada also contains provisions applicable to the planning phase of a stationary activity. The respective governmental branches are required to exchange information on the regulation of 'health, safety, security, safeguards, waste management, and environmental protection aspects'[798]. The exchange of information between the parties must take place also in respect of 'documents relating to significant licencing actions' in cases of 'siting, construction...nuclear facility in their respective jurisdictions' and 'actual and proposed civil uses of fissionable and radioactive substances, and the storage and disposal of spent fuel and byproduct materials resulting from such uses'[799].

The negotiations on a free-trade area comprising Mexico, Canada and the United States have also led to the adoption of a side-agreement on environmental cooperation between the parties.[800] The territorial scope of the 1994 Environmental

---

[797] The 1996 Administrative Arrangement for Cooperation and the Exchange of Information in Nuclear Regulatory Matters, with Appendices. The Agreement was signed on 15 August 1996 and entered into force on the same date. According to Article 10 (1) of the Agreement: 'This arrangement...will remain in effect for five years'. Appendix C contains a limitation on the exchange of information. Although the widest possible exchange of information is promoted, this is subject to any national protection provided for the proprietary information and intellectual property. In addition, Article 2.1. reads: 'To the extent that AECB [Atomic Energy Control Board now known as Canadian Nuclear Safety Commission] and USNRC [United States Nuclear Regulatory Commission] are permitted to do so under the laws, regulations, and policy directives of their respective countries...will exchange...technical information of the kinds described in Appendix A'. Articles 2.3 to 2.7 also provide limitations clauses. Agreements between administrative branches of states are normal international treaties. See Klabbers 1996 I, pp. 97-104.

[798] Article 2 (1) of the 1996 Nuclear Agreement.

[799] Ibid., Articles 2 (1.1) and 2 (1.2).

[800] Documents on the home page of the Department of Foreign Affairs and International Trade of Canada state (at http://www.dfait-maeci.gc.ca/nafta-alena/over-e.asp [21.4.2001]): 'The North American Free Trade Agreement (NAFTA) entered into force in Canada, the United States and Mexico on January 1994. Designed to foster increased trade and investment among the partners, the NAFTA contains an ambitious schedule for tariff elimination and reduction of non-tariff barriers, as well as comprehensive provisions on the conduct of business in the free trade area' It is also stated (at http://www.dfait-maeci.gc.ca/nafta-alena/side1-e.asp [21.4.2001]): 'in parallel to the NAFTA, the North American Agreements on Labour and Environmental Cooperation were designed to facilitate greater cooperation between the partner countries in those areas and to promote the effective enforcement of each country's laws and regulations'.

Cooperation Agreement[801] between the countries covers their continental shelf areas[802] and the sea areas outside their territorial sea up to the limit where their customs laws apply.[803] The Agreement contains two relevant provisions. Article 20 contains generally worded obligations. Each party must notify all others 'of any proposed or actual environmental measure that the Party considers might materially affect the operation of this Agreement or otherwise substantially affect that other Party's interests under this Agreement'[804]. Surely the plan for a project which may cause significant adverse transboundary impact comes under the operation of this Agreement since it clearly is in the interests of all parties under an agreement which deals with international environmental cooperation. However, the provision is qualified by the phrase 'to the extent possible'[805], requiring the contracting states only to attempt to fulfill this obligation. A more clearly worded obligation is found in paragraph 3:

> On request of any other Party, a Party shall promptly provide information and respond to questions pertaining to any such actual or proposed environmental measure, whether or not that other Party has been previously notified of that measure.[806]

The most important provision from the viewpoint of this study is Article 10 (7), which obligates the Council of the North American Commission for Environmental

[801] The Agreement was concluded in September 1993 and entered into force by virtue of Article 47 on 1 January 1994. The 1994 North American Agreement on Environmental Cooperation. Article 39 sets out limitations on provision of information. This exception to the information exchange rule is defined in very strict terms as 'nothing in this Agreement shall be construed to require a Party to make available or allow access to information' on the issues which affect its environmental law enforcement, or when its law prohibits the disclosure of such information. If confidential information is delivered, the potentially affected state must treat it as the origin state would have to treat it according to its national laws. The Agreement contains another provision on this issue under the heading 'national security', in which the provision of information is restricted on the basis of 'essential security interests'.

[802] Annex 45 (a, c (iii)) of the 1994 Environmental Cooperation Agreement.

[803] Ibid., Annex 45 (a, c (i)).

[804] Ibid., Article 20 (2).

[805] Ibid.

[806] Ibid., Article 20 (3).

Cooperation (NACEC) to consider and develop recommendations within three years of the signature of the Environmental Cooperation Agreement with a view to agreement on transboundary environmental impact assessment procedure. The Article also provides guidelines for a possible agreement, which suggest that such an agreement might contain an EIA procedure for proposed projects likely to cause significant adverse transboundary effects 'including a full evaluation of comments provided by other Parties and persons of other Parties'[807]. It then proposes that the procedure should contain a notification procedure as well as provision of relevant information and consultation between Parties.[808] From the material point of view, the provision also urges that mitigation of potential adverse effects should be organized.[809] It is clear that this provision only requires the Council of the CEC to consider and possibly develop recommendations within a certain time limit and with a view to agreement; there is no legal obligation for the parties to come up with an agreement of any kind.[810]

The Council of the NACEC has tried to implement the provision from the very beginning. Already in Council Resolution #95-7 the Council endorsed the so-called Transboundary Impact Assessment Overarching Principles, which are intended to inform and guide it in its task of developing more specific recommendations pursuant to Article 10 (7).[811] The first of these principles is that pollution respects no borders. The guidelines emphasize that the states parties are part of the same environment and encouraged them to communicate and share information on projects within their jurisdiction. The document also underscores that it is the affected state which is in the

[807] Ibid., Article 10 (7a).
[808] Ibid., Article 10 (7b).
[809] Ibid., Article 10 (7c).
[810] Ibid., Article 10 (7).
[811] Resolutions of the Council can be found on the NACEC home page at http://www.cec.org/who_we_are/council/resolutions/index.cfm?varlan=english [30.4.2001].

best position to determine the kinds of impacts that are of concern to itself and its citizens.[812]

The second principle invoked is good neighborliness as this is described in principle 2 of the Rio Declaration. Good neighborliness is seen as willingness to cooperate with neighboring states and willingness to address the concerns of those potentially impacted by the activities in another state. This principle is also seen as providing an opportunity for the potentially affected state to contribute comments and information to the environmental assessment process. The third principle is respect for national and subnational processes, which emphasizes that the transboundary impact assessment procedure should be organized subject to the constraints imposed by national and subnational processes. This principle is designed to accommodate the complex (con)federal structure of the states parties. The additional principles of cost-effectiveness, efficiency and complementarity discussed in the guidelines underline that no new procedures should be created; existing processes should be utilized instead. Finally, the Principles stress that the public should be entitled to take part in the transboundary EIA procedures.[813]

It is evident that the above-mentioned resolution is only a recommendation, the second of the two forms of decision which can be taken by the Council according to Article 9 (6 and 7). These paragraphs provide that the Council can also take decisions in addition to making recommendations, implying that the former are legally binding on the parties.[814] In fact, a couple of years later, the Council adopted a much stronger-worded obligation for the parties in Council Resolution #97-03:

---

[812] Resolution # 95-7 of the Council of the NACEC.
[813] Ibid.
[814] Article 9 (6 and 7) of the 1994 Environmental Cooperation Agreement reads: '6. All decisions and recommendations of the Council shall be taken by consensus, except as the Council may otherwise decide or as otherwise provided in this Agreement. 7. all decisions and recommendations of the Council shall be made public, except as the Council may otherwise decide or as otherwise provided in this Agreement'.

> The Council hereby decides that:
> The Parties, building upon the work of the North American Intergovernmental Group on Transboundary Environmental Impact Assessment agree to complete a legally-binding agreement consistent with their obligations under Article 10(7) of the North American Agreement on Environmental Cooperation by April 15, 1998. This agreement shall include, inter alia, provisions on assessment of transboundary environmental impacts, notification to the potentially affected Party, consideration of mitigation measures, and public participation.

This provision was binding on the parties, requiring them to conclude a treaty on Transboundary Environmental Impact Assessment by April 15, 1998. Considerable efforts were made to conclude such an agreement, progressing as far as a draft treaty that was even made available on the NACEC home page, i.e., the so-called North American Agreement on Transboundary Environmental Impact Assessment (hereinafter 'the draft TEIA')[815]. It seemed that a treaty would be concluded, but nothing has happened as yet,[816] meaning that the parties have not complied with

---

815 The home page of the NACEC ("latest news" at http://www.cec.org/news/announce/Data.cfm?varlan=english&vardate=9999&unique_no=85 [30.4.2001]), contains the following news: 'Full-Text of Proposed North American Transboundary Environmental Impact Assessment Agreement Now On-Line - Montreal, 2 December, 1997. The full text of the proposed Transboundary Environmental Impact Assessment Agreement being developed by Canada, Mexico and the United States is available in the CEC website. The objectives of the proposed Transboundary Environmental Impact Assessment Agreement would be to provide decisionmakers with timely information on the transboundary environmental consequences of proposed projects to ensure that decisions on such projects take into account such consequences and to provide a mechanism for potentially affected people and governments to participate in the process leading to a decision on the project. The CEC Council, the highest body of the CEC, has decided that the parties Canada, Mexico and the United States) would complete a legally-binding agreement by April 15, 1998. The Council also agreed that the work done by the group of experts be distributed publicly this fall'.

816 The NACEC home page also provides the following news ("latest news" at http://www.cec.org/news/announce/Data.cfm?varlan=english&vardate=9999&unique_no=114 [30.4.2001]): 'NAFTA partners Approaching Agreement on Transboundary Environmental Impact Assessment --Montreal, 22 February 1999. Recent announcements of progress in negotiations on a North American agreement on transboundary environmental impact assessment mark an important step in trinational cooperation to conserve and protect the North American environment, and avoid environment related border disputes. At a US-Mexican summit last Monday in Mérida, Mexico, President Bill Clinton and President Ernesto Zedillo expressed their satisfaction, "to be approaching

Decision #97-03. However, there are signs that negotiations on the draft TEIA will resume in the near future.[817] The draft treaty will be examined in chapter V (2).

4.1.3. Treaties Covering Both Geographical Regions

The treaties studied above are limited in their actual and potential application: the Nordic treaties and the 1992 EEA Agreement apply only and potentially to the Nordic Arctic, the Paris Conventions regulate only certain maritime areas, and the treaties covering the North American region are similarly limited in the number of parties involved. In contrast, the treaties to be studied in this chapter contain at least the potential for all the Arctic states to become parties to them.

Two of the largest inter-state conferences ever organized are relevant here: UNCLOS III[818] and the Rio Conference[819]. Both were organized under the auspices of the UN and participated in by almost all of the states in the world. The end-product of

---

agreement, along with Canada, on a trilateral arrangement to ensure the cross-border effects of many proposed projects and activities can be included in our respective environmental impact assessments, and that bordering states and their citizens will be able to participate in the assessment process." At the end of a Canada-Mexico bilateral meeting last Friday in Ottawa Canadian Environment Minister Christine Stewart and Mexican Secretary of Environment, Natural Resources and Fisheries Julia Carabias reiterated their commitment to finalize the negotiations on transboundary environmental impact assessment, within the framework of the environmental side agreement to the North American Free Trade Agreement (NAFTA). These efforts represent an important step in promoting cooperation on environmental assessment of many proposed projects'.

[817] According to an officer of the United States Fish and Wildlife Service, the new environmental minister of Mexico is aiming to have the TEIA negotiations commenced anew (e-mail communication 12.3.2001).

[818] The name 'The Third United Nations Conference on the Law of the Sea' derives from the fact that there had been two previous rounds of treaty negotiations on the law of the sea, the first in 1958, when four conventions were concluded, and the second in 1960, when no agreements were reached. In 1970, the UNGA decided to convene a third conference on the law of the sea in 1973. The negotiations lasted until 1982, with some 150 states participating. For a review of the negotiations by a scholar who participated in these negotiations as part of the Finnish Delegation, see Hakapää 1981, pp. 25-31.

[819] The United Nations Conference on Environment and Development (UNCED) was held in Rio de Janeiro, Brazil, from June 3 to 14 in 1992, with some 175 states, 1500 representatives of non-governmental organizations and 7000 journalists taking part. For a review, see Johnson 1993, pp. 3-10.

UNCLOS III was the 1982 United Nations Convention on the Law of the Sea (LOSC).[820] The Arctic parties to the LOSC are Finland, Iceland, Norway, the Russian Federation and Sweden; Canada and Denmark are signatories.[821] The scope of the prior transboundary procedure in the LOSC is very broad, including all planned activities under states' jurisdiction or control.[822] The LOSC does not require that the impact must be of a transboundary nature; it is enough that 'substantial pollution of or significant and harmful changes to the marine environment'[823] may take place. In fact, it is not even required that pollution be likely; reasonable grounds for believing that such might occur trigger the duty to assess potential effects on the marine environment.[824]

The prior transboundary procedure in the LOSC is an indirect one. Article 206 requires that states must assess the 'potential effects of such activities on the marine environment'. Since the provision speaks of effects on the marine environment in general, it means that states are obligated to conduct an assessment of the effects of activities taking place in their maritime jurisdiction on the marine environment located in other states' jurisdiction as well as on international areas. The assessment of transboundary impacts on the marine environment located in another state's jurisdiction cannot be very systematic, however, since there exist no provisions on how potentially affected states can contribute to an assessment. More importantly, the duty of assessment is qualified by the phrase 'as far as practicable'[825], which means

---

820 The 1982 United Nations Convention on the Law of the Sea. Article 30 of the LOSC contains a provision on limiting the exchange of information. According to the provision, which is drafted in rather general terms, a state party is not obligated to disclose information if the disclosure of that information 'is contrary to the essential interests of its security'. The LOSC was signed on 10 December 1982 and entered into force on 16 November 1994.

821 The following dates of entry into force are important: Finland (21.6.1996), Iceland (21.6.1985), Norway (24.6.1996), the Russian Federation (12.3.1997), Sweden (25.6.1996). The Convention has also entered into force for the European Community (1.1.1998). The United States has not signed the Convention. Information obtained from the UNTS.

822 Article 206 of the LOSC.

823 Ibid.

824 Ibid.

825 Ibid.

that the duty is only an obligation of attempt in international law. The results of assessments must be communicated to the competent international organizations, 'which should make them available to all States'[826]. A potentially affected state can thus obtain information through this channel.

At the Rio Conference, three instruments[827] were adopted and two conventions opened for signature, one of the latter being the 1992 Convention on Biological Diversity.[828] All the Arctic states signed the Convention and, with the exception of the United States, all are parties to it as well.[829] The origin state is obligated to provide prior notification in case of activities which may significantly affect the biological diversity of the environment of other states. The parties are also encouraged to exchange information and consult each other.[830] However, the provision setting out this obligation is written in rather vague language, using terms such as 'promote' and 'encourage', which clearly are legally indeterminate. In addition, the contracting parties are to observe the obligation 'as far as possible and as appropriate'[831]. On balance, the Convention thus seems to establish only obligations of attempt.

The most important inter-governmental organization from the viewpoint of the present study is the United Nations Economic Commission for Europe (UN ECE). All eight Arctic states are members of the organization, as one of whose aims is to

---

826 Ibid., Article 205.

827 These are the Rio Declaration on Environment and Development, Agenda 21 and the Non-Legally Binding Authoritative Statement of Principles for a Global Consensus on the Management, Conservation and Sustainable Development of All Types of Forests. For a review of the entire process of the Rio Conference and the documents adopted there, see Johnson 1993.

828 The 1992 Convention on Biological Diversity. The Convention does not contain any general provision limiting the exchange of information. The Convention was signed at the beginning of June 1992 and entered into force on 29 December 1993.

829 The following dates of entry into force are relevant: Canada (4.2.1992), Denmark (21.2.1993), Finland (27.7.1994), Iceland (12.9.1994), Norway (9.7.1993), the Russian Federation (5.4.1995), Sweden (16.12.1993) and the EC (21.12.1993). Information obtained from the UNTS.

830 Article 14 (1c) of the 1992 Convention on Biological Diversity.

831 Ibid., Article 14 (1).

promote measures to protect the environment.[832] The 1991 Espoo Convention[833] was signed by all eight Arctic states and thus provides potential for Arctic-wide regulation of prior transboundary procedures. Five of the Arctic states - Canada, Denmark, Finland, Norway and Sweden – are parties to the Convention.[834] Iceland is expected to become a party to the Convention in the near future.[835] Denmark made a note excluding Greenland from the scope of the Convention, but, according to the responsible ministry, Denmark sees that the Convention applies also to Greenland.[836] The Convention already covers almost all of the Nordic Arctic and will in all likelihood cover it totally in the near future when Iceland becomes a party. The situation in regard to the North American region is less encouraging, as only one of

[832] As provided on the home page of the UN ECE (at http://www.unece.org/oes/eceintro.htm [15.3.2001]): 'UN/ECE was set up in 1947 by ECOSOC. It is one of five regional commissions of the United Nations. Its primary goal is to encourage greater economic cooperation among its member States. It focuses on economic analysis, environment and human settlements, statistics, sustainable energy, trade, industry and enterprise development, timber and transport'.

[833] The 1991 Convention on Environmental Impact Assessment in a Transboundary Context. The Convention was adopted on 25 February 1991, signed mostly on 26 February 1991 and entered into force on 10 September 1997.

Article 2 (8) of the Convention provides for a limitation on the exchange of information: 'The provisions of this Convention shall not affect the right of Parties to implement national laws, regulations, administrative provisions or accepted legal practices protecting information the supply of which would be prejudicial to industrial and commercial secrecy or national security'. The Convention was adopted on 25 February 1991, signed mostly on 26 February 1991, and entered into force on 10 September 1997.

[834] The following dates of entry into force are relevant: Canada (13.5.1998), Denmark (14.3.1997), Finland (10.8.1995), Norway (23.6.1993), Sweden (24.1.1992). The Convention has also entered into force for the EC (24.6.1997). Information obtained from the UNTS.

[835] According to an Icelandic source, Iceland will become a party to the Convention in the very near future (e-mail communication on 26.3.2001). This is only natural since the EEA Agreement already requires largely the same measures of Iceland as the Espoo Convention would.

[836] Denmark delivered a note to the depositary upon signature of the Convention stating the following: 'Decision reserved as concerns the application of the Convention to the Faroese Islands and Greenland'. This piece of information was still on the record in the UNTS on 31 May 2001. Yet, according to an officer of the Danish Ministry of Environment and Energy, Spatial Planning Department, who is in charge of the tasks related to the Espoo Convention, Denmark filed a note when it signed the Convention but not when it ratified it. The understanding of the Ministry of Environment and Energy as well as the Foreign Ministry of Denmark is that since Denmark did not file the note a second time when it ratified the Convention, the Convention applies in its entirety to Greenland and the Faroe Islands. The Ministry of Environment and Energy plans to take action to remove the note from the UNTS. From the viewpoint of international law, the main thing is that Denmark itself is of the opinion that the note is not in force anymore. E-mail communication and telephone conversation on 31 May 2001.

the legal relations between the Arctic states in that region is governed by the Convention, that between Canada and Denmark (Greenland).

Since the Espoo Convention will be studied in detail in chapter V and the prior transboundary procedure in the 1992 Industrial Accidents Convention is almost identical, it will suffice to confine the examination here to the procedure in the latter instrument. Of the Arctic states, Finland, Norway, the Russian Federation and Sweden are parties to the Industrial Accidents Convention[837]; Canada, Denmark and the United States have signed it[838], although Denmark did so 'with reservation of application to the Faroe Islands and Greenland'[839].

According to the Convention, the origin state has a general obligation to 'identify hazardous activities within its jurisdiction and to ensure that affected Parties are notified of any such proposed...activity'[840]. If there exists disagreement as to whether the treaty is applicable, the concerned states are first obligated to discuss the matter

---

[837] The 1992 Convention on the Transboundary Effects of Industrial Accidents. The Convention was signed by most of the parties on 18 March 1992 and entered into force on 19 April 2000. The Convention contains a general clause on 'Limitations on the supply of information', Article 22: '1. The provisions of this Convention shall not affect the rights or the obligations of Parties in accordance with their national laws, regulations, administrative provisions or accepted legal practices and applicable international regulations to protect information related to personal data, industrial and commercial secrecy, including intellectual property, or national security. 2. If a Party nevertheless decides to supply such protected information to another Party, the Party receiving such protected information shall respect the confidentiality of the information received and the conditions under which it is supplied and shall only use that information for the purposes for which it was supplied'.

[838] The following dates of entry into force are relevant: Finland (13.9.1999), Norway (1.1.1993), the Russian Federation (1.2.1994), Sweden (22.9.1999) as well as the EC (24.1.1998). Information obtained from the UNTS.

[839] Information obtained from the UNTS.

[840] Article 4 (1) of the 1992 Industrial Accidents Convention. Interestingly, the first meeting of parties was to adopt a decision adding some further criteria on how to identify hazardous activities, e.g., distance from the international border. See Draft-decision (CP.TEIA/2000/7, 18 September 2000), which can be found on the home page of the Convention at http://www.unece.org/env/teia/english/homepage.html [1.3.2001]. No decision was reached on this issue, however. Information obtained from the Finnish Ministry of the Environment (telephone conversation on 22.3.2001).

and if no agreement is reached, submit the dispute to an inquiry commission.[841] Annex III provides that the inquiry commission procedure can be initiated and carried out without the consent of the other party or parties. Annex III also sets out the prior transboundary procedure.[842]

The origin state is obligated to initiate the prior transboundary procedure by furnishing the potentially affected state with the existing information on the hazardous activity.[843] It must also indicate at which time the potentially affected state must respond to the notification, confirming its participation, and may also, depending on the national procedure, add more comprehensive information on potential transboundary effects.[844] If the potentially affected state does not respond within the time specified by the origin state, or if it otherwise informs the origin state of its intention not to participate, the remainder of the prior transboundary procedure is not applied.[845] Upon an affirmative answer from the potentially affected state, the origin state must at least inform the affected state of the timetable for the national procedure and provide relevant information 'on the hazardous activity and its transboundary effects in the event of an industrial accident' as well as the 'opportunity to participate in evaluations of the information or any report demonstrating possible transboundary effects'[846]. The origin state is obligated to request information relating to the relevant area under the jurisdiction of the affected state and the latter is obligated to provide such information.[847]

[841] Ibid., Article 4 (2).
[842] Ibid., Article 4 (3).
[843] Ibid., Annex III (2).
[844] Ibid., Annex III (3).
[845] Ibid., Annex III (4 and 5).
[846] Ibid., Annex III (6).
[847] Ibid., Annex III (7).

The origin state is obligated to deliver the 'analysis and evaluation documentation' to the affected state.[848] The concerned states together are requested to have this information delivered to the public in areas reasonably likely to be affected. They must also arrange for the public to have an opportunity to make comments on and objections to the hazardous activity and to ensure that these observations are received by the competent licencing authority.[849] The origin state is obligated to initiate consultations with the potentially affected state on, *inter alia*, 'the transboundary effects of the hazardous activity in the event of an industrial accident, and measures to reduce or eliminate its effects'[850]. In the consultations, the origin state must take due account of the analysis and evaluation documentation, the comments of the affected state, and the comments of the public of that state.[851] Finally, the origin state is to deliver the licencing decision to the affected state 'along with the reasons and considerations on which it was based'[852].

Another important inter-governmental organization here is the International Atomic Energy Agency (IAEA). This organization is specialized in all aspects of the peaceful use of atomic energy and has all eight Arctic states as members.[853] Of the Arctic states, all except Iceland are parties to the 1994 Convention on Nuclear Safety,[854]

[848] Ibid., Annex III (8).
[849] Ibid., Annex III (9).
[850] Ibid., Annex III (10).
[851] Ibid., Annex III (11).
[852] Ibid., Annex III (12).
[853] The IAEA is a specialized agency within the United Nations system. According to the home page of the IAEA (at http://www.iaea.org/worldatom/About/ [15.3.2001]): 'By 29 July 1957, eighteen States ratified the Statute of the IAEA thus bringing it into force. As of 1 December 1999, the IAEA has 130 Member States', including all the eight Arctic states.
[854] The Convention was opened for signature on 20 September 1994 and entered into force on 24 October 1996. The Convention contains a very broad article on the limitation of information exchange, Article 27, which stipulates that the treaty does not 'affect the rights and obligations of the Contracting Parties under their laws to protect information from disclosure'. As examples of such information, the treaty mentions information relating to national security, physical protection of nuclear materials or personal data as well as information protected by intellectual property rights. The second paragraph relates directly to the information exchange in the context of transboundary procedures. If the above-mentioned information is protected, the 'information shall be used only for the purposes for which it has been provided and its confidentiality shall be respected'. Hence, the

although Denmark declared that '[u]ntil further notice the Convention shall not apply to Greenland and the Faroe Islands'[855] and thus excluded Greenland from the scope of the Convention for the time being. The material scope of the treaty has been defined primarily through the concept of 'nuclear installation', i.e., 'any land-based civil nuclear power plant under its jurisdiction including such storage, handling and treatment facilities for radioactive materials as are on the same site and are directly related to the operation of the nuclear power plant'[856].

The prior transboundary procedure is regulated in Article 17 of the Convention.[857] All the contracting parties are obligated to 'take the appropriate steps to ensure that appropriate procedures are established and implemented'[858]. Hence, the prior transboundary procedure provisions aim at providing guidelines for more detailed procedures to be negotiated between the contracting states; in other words, the parties have an obligation to negotiate a prior transboundary procedure of a certain kind, inasmuch as the Convention does not provide directly applicable rules for these situations.[859] The procedure starts if the origin state finds that another contracting

---

origin state is not entitled to withhold such information, but the affected state is obligated to make sure that the confidentiality of such information is maintained. The following dates of entry into force are relevant: Canada (24.10.1996), Denmark (11.2.1999), Finland (24 10.1996), Norway (24.10.1996), the Russian Federation (24.10.1996), Sweden (24.10.1996), the United States (10.7.1999). Information obtained on the home page of the IAEA at http://www.iaea.org/worldatom/Documents/Legal/nukesafety.shtml [5.2.2001]. Last change of status 31.1.2000.

855 Information obtained on the home page of the IAEA at http://www.iaea.org/worldatom/Documents/Legal/nukesafety.shtml [11.5.2001].

856 Article 2 (i) of the 1994 Convention on Nuclear Safety, which also provides: 'Such a plant ceases to be a nuclear installation when all nuclear fuel elements have been removed permanently from the reactor core and have been stored safely in accordance with approved procedures, and a decommissioning programme has been agreed by the regulatory body'.

857 Ibid. In paragraph (v) of the Preamble, the parties state that they are '[a]ware that accidents at nuclear installations have the potential for transboundary impacts'.

858 Ibid., Article 17.

859 Ibid. Article 17 (iv) reads: 'for consulting Contracting Parties in the vicinity of a proposed nuclear installation, insofar as they are likely to be affected by that installation and, upon request providing the necessary information to such Contracting Parties, in order to enable them to evaluate and make their own assessment of the likely safety impact on their own territory of the nuclear installation'.

state is likely to be affected by the nuclear installation to be operated in its jurisdiction. The origin state is then obligated to consult the potentially affected state and, if the latter so requests, to provide necessary information.[860]

The 1997 Joint Convention[861] will enter into force in the near future. Of the Arctic states, five have ratified, accepted or approved the instrument, i.e., Canada, Denmark, Finland, Norway and Sweden.[862] Denmark made a declaration that '[u]ntil further notice the Convention shall not apply to Greenland and the Faroe Islands'[863]. The Russian Federation and the United States have signed the Convention. Iceland has taken no action in relation to the instrument.

The Joint Convention applies to the management of both spent fuel and radioactive waste, with these activities treated in separate sections of the Convention. Spent fuel management is defined as 'all activities that relate to the handling or storage of spent

---

[860] Ibid.

[861] The 1997 Joint Convention on the Safety of Spent Fuel Management and on the Safety of Radioactive Waste Management. The Convention was adopted on 5 September 1997 and opened for signature on 29 September 1997. The Convention will soon to enter into force, because as of 5 January 2001, 23 of the 25 contracting parties required had already become parties to it. To be exact, certain additional conditions are also required, as stipulated in Article 40 (1): '[the Convention] shall enter into force on the ninetieth day after the day of deposit with the Depositary of the twenty-fifth instrument of ratification, acceptance or approval, including the instruments of fifteen States each having an operational nuclear power plant'. The Convention contains also a provision on the limitation on information exchange. In Article 36 it is stipulated that the treaty does not 'affect the rights and obligations of the Contracting Parties under their laws to protect information from disclosure'. As examples of such information, the treaty mentions information relating to national security, physical protection of nuclear materials or personal data as well as information protected by intellectual property rights. The second paragraph relates directly to the information exchange in the context of transboundary procedures. If the above-mentioned information is protected, the 'information shall be used only for the purposes for which it has been provided and its confidentiality shall be respected'. Hence, the origin state is not entitled to withhold such information, but the affected state is obligated to make sure that the confidentiality of such information is maintained.

[862] The following dates of deposit are relevant: Canada (7.5.1998), Denmark (3.9.1999), Finland (10.2.2000), Norway (12.1.1998), Sweden (29.7.1999). Information obtained on the home page of the IAEA at http://www.iaea.org/worldatom/Documents/Legal/jointconv.shtml [5.2.2001]. Last change of status on 5 January 2001.

[863] Information obtained from the home page of the IAEA, see previous footnote.

fuel, excluding off-site transportation'[864] and radioactive waste management as 'all activities, including decommissioning activities, that relate to the handling, pretreatment, treatment, conditioning, storage or disposal of radioactive waste, excluding off-site transportation'[865]. An exception is made for spent fuel or radioactive waste 'within military or defence programmes', in that the Convention does not apply unless the contracting party makes a declaration to this effect in a certain specific case.[866] If materials are transferred permanently to civilian facilities, however, the Convention applies.[867] The prior transboundary procedure is identical for both spent fuel and radioactive waste facilities.[868] The relevant articles only provide for obligations to negotiate a certain kind of prior transboundary procedure,[869] with the origin state obligated to consult the potentially affected state if that state is 'likely to be affected'[870]. If the potentially affected state so requests, the origin state is obligated to provide it with 'general data relating to the facility to enable [it] to evaluate the likely safety impact of the facility upon [its] territory'[871].

There exist many conventions that include all or almost all of the Arctic states and provide for a prior transboundary procedure, i.e., the 1982 LOSC, the 1992 Convention on Biodiversity, the 1994 Convention on Nuclear Safety as well as a convention soon to enter into force, the 1997 Joint Convention. However, the Biodiversity Convention only contains obligation of attempt, and the obligations in the two IAEA Conventions are merely obligations to negotiate a prior transboundary procedure for the situations defined. The LOSC contains an indirect procedure for these situations. Given the limitations of these instruments, more promising prospects

---

864 Article 2 (o) of the 1997 Joint Convention.
865 Ibid., Article 2 (i).
866 Ibid., Article 3 (3).
867 Ibid.
868 Ibid., Articles 6 (iv) and 13 (iv).
869 Ibid. Articles 6 and 13 both start with the following sentence: 'Each Contracting Party shall take the appropriate steps to ensure that procedures are established and implemented'.
870 Ibid., Articles 6 (iv) and 13 (iv).
871 Ibid.

are offered by the 1992 Industrial Accidents Convention and especially the 1991 Espoo Convention; indeed, these have already demonstrably influenced prior transboundary procedures in the Arctic and will do so even more in the future.

4.2. Conclusion

Two main sets of norms regulate prior transboundary procedures between the Arctic states, treaties and customary law, and it is necessary to examine whether there exist any conflicts between the two or between treaties relating to the same issue. The due diligence principle of general international law is of a high order of generality and consequently it is not easy to determine in the abstract what material and procedural requirements it establishes for the Arctic states. Precisely because of its high order of generality, its content remains very much context-dependent, although it can be suggested that it requires notification and transboundary impact assessment of the states and imposes an obligation for the concerned states to consult each other. Few treaties provide for prior transboundary procedures that fulfill these obligations. Those that do include the 1991 Espoo Convention, the 1992 EEA Agreement (EIA Directive) and the 1992 Industrial Accidents Convention.

Yet, as was also argued above, these preventive duties are not independent duties but guidelines urging states to observe their duty of due diligence in customary law. The prior transboundary procedures provided for by treaties and the prior transboundary procedure required by the due diligence principle operate at different levels because treaties create independent obligations in international law that can be invoked in the planning stage of an activity, whereas the preventive duties required by the due diligence principle are only guidelines for the *ex post facto* determination of whether an origin state has observed due diligence or not. In this perspective, there exists no conflict between the applicable treaties and the principle of due diligence.

In the Eurasian region, it has been established that there exist four treaties which provide for prior transboundary procedures between the Nordic states: the 1974 NEPC, the 1976 Nordic Guidelines Treaty, the 1992 Paris II Convention and the 1992 EEA Agreement. The EEA Agreement (hereinafter 'the EIA Directive') presently requires the states parties to implement the prior transboundary procedure stipulated in the Amendment Directive. This obligation to implement the EIA Directive concerns all five Nordic Arctic states - Finland, Denmark, Sweden, Iceland and Norway - with the exception of Greenland and the Svalbard Islands.[872] Can it be concluded that the Nordic Arctic states have implemented the requirements of the EIA Directive through the three above-mentioned conventions?

In principle, all three conventions can be seen as implementing the requirements of the EIA Directive because they include some form of prior transboundary procedure and govern legal relations between the Nordic Arctic states. There is overlap between these three treaties. The legal relation between the NEPC and the Nordic Guidelines Treaty is governed by an explicit provision of the NEPC (Finland, Sweden, Norway and Denmark).[873] When the Nordic Guidelines Treaty applies, it overrides the NEPC but the latter treaty, as a treaty having a broader scope, is applicable to other environmentally harmful activities potentially causing transboundary impact. The relation between Iceland and other Nordic Arctic states, which falls outside of the NEPC and the Nordic Guidelines Treaty, is covered by the more extensive Paris II Convention, although the prior transboundary procedure of the Convention applies only to marine pollution.

---

[872] see *supra* chapter IV (4.1.1.).

[873] The relationship between the NEPC and the Nordic Guidelines Treaty is governed by Article 1 (2) of the NEPC: 'The Convention shall not apply insofar as environmentally harmful activities are regulated by a special agreement between two or more of the Contracting States'. According to Article 30 of the Vienna Convention as well as the customary law of treaties, if states have agreed on the legal relationship between treaties, it is this agreement which governs their relationship.

It seems, however, that the three conventions do not fully implement the requirements of the EIA Directive, which is based on the requirements of the 1991 Espoo Convention. The prior transboundary procedure of the Espoo Convention requires much more extensive procedures than those provided for in the treaties studied, and thus the EIA Directive requires the Nordic Arctic states to become parties to the Espoo Convention. It should still be remembered that this obligation does not pertain to Greenland and the Svalbard Islands. This obligation has been observed well by the Nordic Arctic states since the Espoo Convention currently applies in all areas required by the EIA Directive except Iceland, which will become a party to the Convention in very near future.[874] On balance, the Nordic Arctic is well regulated where prior transboundary procedures are concerned, in sharp contrast to the legal relations between the Russian Federation and the Nordic states, for which no prior transboundary procedures exist. Yet, pending Iceland's ratification of the Espoo Convention, parties must rely on other Nordic Arctic treaties in some instances of potential transboundary pollution in the Nordic Arctic.[875]

Hence, the Nordic Guidelines Treaty applies whenever a nuclear installation is proposed, and the NEPC applies otherwise as between parties.

[874] According to an officer involved in the implementation of the Espoo Convention in the Arctic, Iceland may become a party in 2001 (e-mail communication, 26.3.2001). Indeed, since Iceland has already been deemed by the EFTA surveillance authority to have implemented the Amendment Directive, it is only a matter of time before it becomes a party to the Espoo Convention, inasmuch as the Amendment Directive provides a prior transboundary procedure that is based on the Convention.

[875] What is the legal relationship between the Espoo Convention and the other treaties governing prior transboundary procedure between the Nordic Arctic states when the Espoo Convention becomes applicable in the whole Nordic Arctic? The basic rule in the customary law of treaties is that if a specific agreement exists between parties with regard to earlier and later treaties, as is the case between the NEPC and the Nordic Guidelines Treaty, it is this agreement which governs the relation between the treaties. The Espoo Convention also contains a specific provision to this effect: 'The Parties may continue existing or enter into new bilateral or multilateral agreements or other arrangements in order to implement their obligations under this Convention. Such agreements or other arrangements may be based on the elements listed in Appendix VI' (Article 8 of the Espoo Convention). The most important phrase in this Article is 'in order to implement their obligations under this Convention', as this means that the criterion by which other, overlapping treaties are assessed is whether they provide for the same level of protection as the Espoo Convention; if they do, these treaties remain operative as implementation conventions of the Espoo Convention. As was examined above, the three treaties applicable in the Nordic Arctic cannot be seen as fulfilling the requirements of the Espoo Convention but they do apply in situations not covered by the

The basic rules governing the prior transboundary procedures that apply mainly between the Nordic Arctic states have been changed by the two UN ECE Conventions, the 1991 Espoo Convention and the 1992 Industrial Accidents Convention. Since some of the Nordic Arctic states are parties to both of these conventions, it is necessary to clarify the legal relation between the two. The scope of these conventions covers largely the same activities. According to a study made by the UN ECE:

> [the] application [of the conventions] to new activities largely overlaps, as most of the proposed activities within the scope of the Industrial Accidents Convention are hazardous activities that are capable of causing transboundary effects, and so are likely to require an EIA under the EIA Convention.[876]

The relationship between the Espoo Convention and the 1992 Industrial Accidents Convention[877] is governed by an explicit provision in Article 4 (4) of the latter. The Espoo Convention is clearly the more important convention for the situations studied here, as the 1992 Industrial Accidents Convention provides that if the prior transboundary procedure carried out in accordance with the Espoo Convention

---

Convention. The other treaties remain in force since it cannot be established on the basis of Article 59 of the Vienna Convention that the parties had any intention to terminate them or suspend their operation, particularly as they regulate also issues other than the Espoo Convention. For an in-depth analysis of the problems related to multilateral treaties with overlapping scope, see Vierdag 1988, pp. 75-111.

876 Paragraph 9, p. 4 of Annex to the Decision II/5 'Recent EIA development and links with other ECE Conventions' adopted by the second meeting of parties of the Espoo Convention (ECE/MP.EIA/5). The document can be found on the home page of the UN ECE at http://www.unece.org/env/eia/2p.htm [24.4.2001].

877 Article 24 of the 1992 Industrial Accidents Convention also provides a provision on conflict of treaties: 'The Parties may, in order to implement their obligations under this Convention, continue existing or enter into new bilateral or multilateral agreements or other arrangements'. The provisions of this Convention shall not affect the right of Parties to take, by bilateral or multilateral agreement where appropriate, more stringent measures than those required by this Convention'.

'fulfill[s] the relevant requirements of this Convention',[878] it is the Espoo Convention that takes precedence.

However, this provision requires that the transboundary EIA procedure of the Espoo Convention '[include] an evaluation of the transboundary effects of industrial accidents from the hazardous activity which is performed in conformity with the terms of this Convention'[879]. In other words, there are certain requirements set out for the procedure of the Espoo Convention for it to be able to replace the procedure of the 1992 Industrial Accidents Convention. The evaluation of whether the Espoo Convention meet these requirements will depend on the given circumstances, but it can be presumed that in most cases the Espoo Convention will fulfill the requirements of the 1992 Industrial Accidents Convention since it is specially designed for situations necessitating prior transboundary procedures.

As stated above, there are no specific treaties providing for a prior transboundary procedure between the Russian Federation and its Nordic neighbors. This lack of a treaty has been significantly mitigated by the entry into force of the 1992 Industrial Accidents Convention, however, for until the Russian Federation becomes a party to the Espoo Convention, this sets out a basic prior transboundary procedure to be used between the Russian Federation and its three Nordic neighbors, Finland, Norway and Sweden. The applicability of the Convention derives from its definition of 'hazardous activity' as 'any activity in which one or more hazardous substances are present or may be present in quantities at or in excess of the threshold quantities listed in Annex I to the Convention and which is capable of causing transboundary effects'[880], which

[878] Ibid., Article 4 (4).
[879] Ibid., Article 4 (4).
[880] Ibid., Article 1 (b).

encompasses most large-scale industrial activities. However, there is large list of exclusions from the scope of the Convention.[881]

In the North American Region, the starting point is that the 1975 Weather Agreement, the 1987 Caribou Agreement, the 1991 Air Quality Agreement and the 1996 Nuclear Agreement apply to certain specific situations only and that their scopes do not overlap. The NACEC partners have not been able to conclude a treaty on Transboundary Environmental Impact Assessment (TEIA).[882] The only general treaty governing almost all possible situations of potential transboundary pollution between Canada and the United States is the 1994 Environmental Cooperation Agreement, but the notification and information exchange duties it contains are very general.

There are no applicable treaties between the United States and the Russian Federation or between Canada and the Russian Federation. It would be of the utmost importance for Canada and the United States to become parties to the 1992 Industrial Accidents Convention or for the United States and the Russian Federation to ratify the 1991 Espoo Convention. The 1984 Marine Agreement between Canada and Denmark (Greenland) provides for a prior transboundary procedure, albeit for limited sets of activities, i.e., offshore hydrocarbon activities. Significantly, the Espoo Convention covers the legal relation between Denmark (Greenland) and Canada. In case of conflict between the 1984 Marine Agreement and the Espoo Convention, the latter prevails as *lex posteriori*.

---

[881] Ibid. Article 2 (2) reads: 'This Convention shall not apply to: (a) Nuclear accidents or radiological emergencies; (b) Accidents at military installation; (c) Dam failures, with the exception of the effects of industrial accidents caused by such failures; (d) Land-based transport accidents with the exception of: (i) Emergency response to such accidents; (ii) Transportation on the site of the hazardous activity; (e) Accidental release of genetically modified organisms; (f) Accidents caused by activities in the marine environment, including seabed exploration or exploitation; (g) Spills of oil or other harmful substances at sea'.

4.3. The Prior Transboundary Procedures Applicable to Offshore Hydrocarbon Exploitation

From the viewpoint of the Arctic environment offshore hydrocarbon exploitation is the most problematic of all the stationary activities. First of all, as can be seen on the map below, there exists a significant number of hydrocarbon deposits dispersed around the region's waters (see figure 6).[883] Second, although these deposits have been exploited already for some time, exploitation has not been very extensive,[884] and it is likely that commercial exploitation of hydrocarbon resources will quicken in the near future. Untapped deposits have become increasingly tempting since they are located in areas that are politically stable, unlike the Middle East.[885]

---

[882] See *supra* chapter IV (4.1.2.).

[883] As provided in the AMAP Assessment Report (1998 (10.2.), p. 664): 'Exploration for oil and gas resources in recent years has indicated that the Arctic potentially contains some of the world's largest oil and gas reserves'. A good example here is the Shtockmanovskoye gas field in the Barents Sea, which is estimated to fulfill the demand for natural gas in Europe for over 25 years. At the moment, it seems that the exploitation of the gas field will not start until 2010 at the earliest, and several options still exist as to which route would be commercially and politically the most feasible to transport the gas to Europe. See the report of Corke and Holmes (2000) on the home page of the Finnish Ministry of Trade and Industry at http://www.vn.fi/ktm/eng/3/maakaasu.htm [6.3.2001].

[884] For a brief history of Arctic on- and offshore hydrocarbon exploitation, see Jumppanen 1990, pp. 73-74. For a review of the present situation, see AMAP Assessment Report 1998, pp. 664-667.

[885] Jumppanen 1990, pp. 76-77.

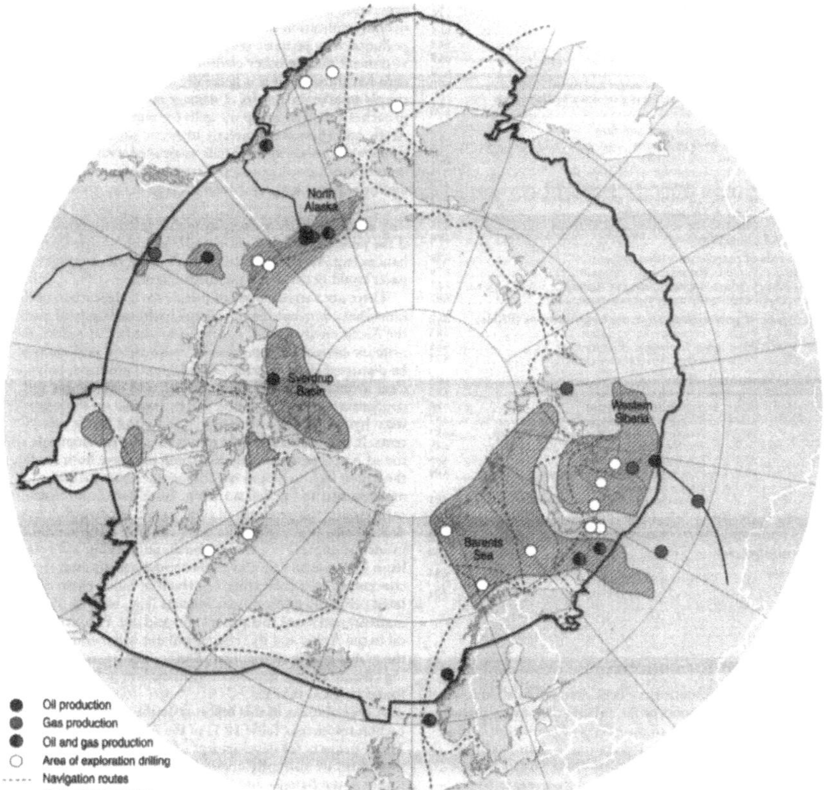

Figure 10·1. Major areas of oil and gas development and potential development in the Arctic, and major shipping routes and possible new routes through Arctic waters.

Figure 6 (source: AMAP Assessment Report 1998, p. 662)

The phenomenon of creeping coastal state jurisdiction has been very pronounced in the Arctic, especially with respect to the continental shelf areas.[886] As the Arctic waters are naturally very shallow, states are entitled to claim extensive areas of the shelf as part of their continental shelf,[887] which has lead to overlapping claims to the shelf between the states.[888] In general, the resolution of these disputes has brought the international jurisdictional borders of the states closer to each other, increasing in turn the likelihood of prior transboundary procedures being triggered in connection with offshore hydrocarbon exploitation.

---

[886] From the viewpoint of environmental protection, the phenomenon of creeping jurisdiction of maritime zones has been beneficial. The expanding coastal state EEZ and continental shelf jurisdiction has subjected larger areas of sea-bed and water column to the environmental protection standards of states. This is clearly a situation preferable to their still being under the high seas regime since at least now there is a body – a national government - in charge of environmental protection. Benvenisti (1996, pp. 384-415) argues (p. 395): 'The private property approach is not alien to the management of resources previously regarded as shared under international law. An excellent example is the assignment of private property rights to continental shelves and exclusive economic zones. This is an exemplary translation of the insights of economic analysis to the law governing international marine resources'. If there were no EEZ and continental shelf zones, states would be more inclined to pollute the marine environment than if they acted as guardians of the environment located in their continental shelf and EEZ. In addition, the establishment of the ISA has placed an international organ in charge of exploitation activities taking place in the Area.

[887] The largest areas of continental shelf fall under the jurisdiction of the Russian Federation. This is due to number of factors. First, the Russian Arctic waters are naturally very shallow. Second, the continental part of the Russian Arctic coastal areas is continued by its northern islands, which extend the shelf to its natural limits and even further since the present law of the sea allows coastal states to include parts of the ocean floor as their continental shelf area if it falls within the limits of their EEZ. The northern islands of the Russian Federation are so close to the ocean floor of the Arctic Ocean that parts of the ocean floor are part of Russia's continental shelf.

[888] As was discussed above, in some cases the Arctic states have reached an agreement on the location of the shelf, e.g., the Russian Federation with the United States over the shelf in Chukchi and Bering Seas, Canada and Denmark (Greenland) over the shelf of Labrador, Baffin and Lincoln Seas, Iceland with Denmark (Greenland) and Norway (the Jan Mayen Island) over the shelf of Greenland and Iceland Seas. Interestingly, some of these treaties contain common exploitation of hydrocarbon deposits in the border areas. See e.g., Article V of the 1973 Agreement between Canada and Denmark (Greenland) and Article 2 of the 1998 Agreement between Iceland and Denmark (Greenland). There are also disputes between the Arctic states over the location of their shelf boundaries, e.g., between the United States and Canada over the shelf of the Beaufort Sea and the Russian Federation and Norway over the shelf of the Barents Sea.

Although pollution from hydrocarbon exploitation is estimated to be the greatest threat to the Arctic environment in the near future, it is not a serious problem as yet. As concluded by the 1998 AMAP Assessment Report:

> Petroleum hydrocarbons have been detected in all biotic and abiotic compartments of the circumpolar Arctic environment. However, with the exception of the concentrations in a few areas exposed to catastrophic releases from oil pipeline ruptures or prolonged chronic releases, hydrocarbon levels in the Arctic associated with anthropogenic inputs have been found to be relatively low and not of ecological consequence.[889]

The Assessment Report nevertheless emphasizes the problems that these petroleum hydrocarbons may have in the Arctic. The special vulnerability of the Arctic ecosystems is stressed, and the problems in taking remedial measures in the severe conditions of the region are underscored.[890] The main reason why these activities in particular are studied is that these resources are likely to be exploited extensively in the near future, a development that will introduce problems of potential transboundary pollution.

4.3.1. Applicable Prior Transboundary Procedures

The principal instruments providing for prior transboundary procedures are two UN ECE Conventions: the 1991 Espoo Convention and the 1992 Industrial Accidents Convention; however, the latter excludes from its scope 'spills of oil or other harmful substances at sea'[891] and thus excludes all proposed offshore hydrocarbon activities. The Espoo Convention includes '[o]ffshore hydrocarbon production'[892] as one category of activities in its Appendix I, but, curiously, not onshore exploitation. Since

---

[889] AMAP Assessment Report 1998, p. 693 (Chapter 10.8 (1)).
[890] Ibid., Chapter 10.8. (8 and 10).
[891] Article 2 (2g) of the 1992 Industrial Accidents Convention.
[892] Appendix I (15) of the Espoo Convention.

the transboundary EIA procedure of the Espoo Convention uses the general term 'production', it includes the exploration phase of offshore hydrocarbon exploitation[893], which can also cause environmental impacts.

As stated above, the Espoo Convention covers all of continental shelves around Jan Mayen, Svalbard, the Norwegian mainland, Greenland and the Canadian Arctic archipelago. However, the continental shelves of Iceland, the United States (Alaska) as well as the extensive shelf of the Russian Federation are at present excluded from the scope of the Convention. The Espoo Convention covers as yet only two legal relations between neighboring Arctic states, that between Canada and Denmark (Greenland) and between Denmark (Greenland) and Norway (Jan Mayen Land). Also the NEPC applies to offshore hydrocarbon exploitation. Article 13 of the instrument stipulates that: 'This Convention shall also apply to the continental shelf areas of the Contracting States'. However, all the legal relations that were previously covered by the NEPC are now regulated by the Espoo Convention as a treaty *lex posteriori*. The legal relations between Iceland and Greenland and between Iceland and Jan Mayen Land are still unregulated, a situation that will be rectified in the near future when Iceland becomes a party to the Espoo Convention.[894]

The 1992 Paris II Convention also provides important guidance here since its prior transboundary procedure is triggered if the interests of other contracting states are affected by, for instance, the potential transboundary marine pollution from offshore hydrocarbon exploitation. Since Iceland is party to this Convention, it applies to the legal relation between Iceland and Norway (Jan Mayen Land) and Iceland and Denmark (Greenland). The problem with the 1992 Paris II Convention is that it contains only a right to consultations for the potentially affected state. There will be

---

[893] For a presentation of the different phases of oil exploitation (exploration, construction, production and decommissioning) and the possible environmental impacts related to each stage, see the AMAP Assessment Report 1998, p. 663.
[894] See also *infra* chapter V (3.).

no agreement covering the legal relation between the Russian Federation and Norway in these situations until the Russian Federation becomes a party to the Espoo Convention.[895]

The situation is worse in the North American region, where only one of the legal relations between the neighboring states will be covered by the Espoo Convention, that between Canada and Denmark (Greenland). The only agreement focusing on the potential transboundary impact of offshore hydrocarbon exploitation is the 1983 Marine Agreement between Denmark (Greenland) and Canada, which, as discussed above, provides for a detailed prior transboundary procedure.[896] Yet, in the situations of potential transboundary impact from offshore hydrocarbon production between Canada and Denmark (Greenland), it is now the Espoo Convention that is the applicable treaty as *lex posteriori* to the Marine Agreement. The legal relation between Canada and the United States in the Beaufort Sea is governed by the general provisions of the 1994 Environmental Cooperation Agreement, Article 20 (3) of which provides:

> On request of any other Party, a Party shall promptly provide information and respond to questions pertaining to any such actual or proposed environmental measure, whether or not that other Party has been previously notified of that measure.

Since the Agreement applies also to continental shelf areas, this rule provides that Canada and the United States must at least exchange information and respond to one another's questions in a case of potential transboundary impact of offshore hydrocarbon activities in the Beaufort Sea. The most problematic legal relations are those of the Russian Federation vis-à-vis Canada and the United States since no

[895] It is, of course, possible that Norway and the Russian Federation commence negotiations outside the system of the Espoo Convention. However, before the unresolved boundary dispute, this development is not likely to take place.
[896] See *supra* chapter IV (4.1.2.).

treaties will govern these situations until and unless the Russian Federation and the United States become parties to the Espoo Convention or the parties conclude special agreements covering transboundary impacts of proposed offshore hydrocarbon activities.

4.4. Potential Treaties

Apart from the numerous legally relevant documents providing for prior transboundary procedures,[897] there are a number of instruments that may be negotiated in the future and become international treaties with the potential of providing binding prior transboundary procedures in the Arctic. As the draft TEIA Treaty will be studied in the next chapter,[898] the main process studied here is the ILC project 'International Liability for Injurious Consequences Arising Out of Acts not Prohibited by International Law,' which has proceeded to the phase of draft

[897] There are many documents that are legally relevant and applicable to the situations studied here and can be found in the Basic Documents of International Law 1992, Vol 1, the important declarations. The main international organizations in respect of prior transboundary procedures are the United Nations General Assembly (UNGA), the United Nations Environment Programme (UNEP), the Organization for Economic Cooperation and Development (OECD), the United Nations Economic Commission for Europe (UN ECE) and the Council of Europe. See, e.g., the following UN instruments: United Nations General Assembly Resolution 2995 (XXVII), Co-operation Between States in the Field of the Environment; 1974 Charter of Economic Rights and Duties of States (Art.3); 1987 UNEP Goals and Principles of Environmental Impact Assessment; Provisions for Co-operation Between States in Weather Modification. For the OECD, see, e.g.,: 1979 OECD recommendation on the Assessment of Projects with significant Impact on the Environment; 1974 OECD Recommendation on the Analysis of Environmental Consequences of Significant Projects; 1974 OECD Principles Concerning Transfrontier Pollution (paras. 6,8); Equal Right of Access in Relation to Transfrontier Pollution in relation to implementation; Strengthening International Co-operation on Environmental Protection in Transfrontier Regions; Provision of Information to the Public and Public Participation in Decision-making Processes Related to the Prevention of, and Response to Accidents Involving Hazardous substances. See, e.g., the following UN ECE instruments: Declaration of Policy on Prevention and Control of Water Pollution, Including Transboundary Pollution; Decision on International Co-operation on Shared Water Resources; Decision on Principles Regarding Co-operation in the Field of Transboundary Waters, 1987. A relevant instrument concluded by the Council of Europe is the Air Pollution in Frontier Areas. See also Chapter 8 of Agenda 21, especially Section 8 (5b). Agenda 21 can be found from Johnson 1993, pp. 125-508.

[898] See *infra* chapter V (1.).

articles.[899] ILC has provisionally adopted draft articles on 'prevention of transboundary damage from hazardous activities'. It is not yet sure whether these articles will form a basis for an international treaty or be adopted in another form. In any case, the draft articles will have major consequences for the development of customary law on the subject because they have already been adopted by the ILC and are general in scope.

The draft articles apply to a broad set of situations. They include all activities which 'involve a risk of causing significant transboundary harm through their physical consequences'[900] and are not prohibited by international law. The required physical consequences are not confined to likely impacts; only the existence of risk is required. According to Article 2 (a), this includes both 'a low probability of causing disastrous harm and a high probability of causing other significant harm'. 'Harm' is defined to include harm to 'persons, property or the environment'[901]. The geographical scope of the draft articles is broad since the link between the activity and the origin state is established through the criterion of 'jurisdiction or control', which includes the areas of jurisdictional competence of states.[902] A potentially

---

[899] The text of the draft articles provisionally adopted by the Commission on first reading can be found on the home page of the Commission at http://www.un.org/law/ilc/reports/1998/chp4.htm#_C [6.2.2001]. See the history of this project in the present Special Rapporteur's first report (A/CN.4/487 and Add.1). The following was stated on the home page of the ILC (at http://www.un.org/law/ilc/sessions/52/52sess.htm#Unilateral_acts [6.2.2001] from its 52nd session (2000)): 'With regard to the topic of "International liability for injurious consequences arising out of acts not prohibited by international law (prevention of transboundary damage from hazardous activities)", the Commission established a Working Group to examine the comments and observations made by States on the draft articles on the sub-topic of prevention which had been adopted on first reading by the Commission in 1998 (See: Report of the Secretary-General (A/CN.4/509)). On the basis of the discussion in the Working Group, the Special Rapporteur presented his third report (A/CN.4/510) containing a draft preamble and a revised set of draft articles on prevention, along with the recommendation that they be adopted as a framework convention'.

[900] Article 1 of the draft articles.

[901] Ibid., Article 2 (b).

[902] Ibid., Article 2 (d).

affected state is also defined inclusively as any state 'which has jurisdiction or control over any other place where such harm is likely to occur'[903].

The states are required to establish prior procedures - both impact assessment[904] and licencing procedures - for the environmentally harmful activities.[905] The origin state must notify the potentially affected state if the assessment indicates a risk of significant transboundary harm.[906] This notification must be done in a timely manner, and the origin state must transmit to the potentially affected states 'the available technical and other relevant information on which the assessment is based'[907]. The potentially affected state must respond to this notification within a reasonable time.[908]

If the origin state does not notify the potentially affected state, and the affected state has reasonable grounds to believe that the draft articles apply, the latter can make a request for the origin state to apply Article 10.[909] If the origin state in its turn finds that the draft articles are not applicable, it must 'so inform the other State within a reasonable time, providing a documented explanation'[910]. In such an eventuality, the potentially affected state has a right both to enter into consultations[911] and to request that the origin state take measures to minimize the risk or 'suspend the activity in question for a period of six months unless otherwise agreed'[912]. If the origin state refuses to apply the draft articles, Article 17 (2) provides that if the concerned states have not reached an agreement within six months, the potentially affected state can

903 Ibid., Article 2 (e).
904 Ibid. Article 8 provides: 'Any decision in respect of the authorization of an activity within the scope of the present draft articles shall be based on an evaluation of the possible transboundary harm caused by that activity'.
905 Ibid., Article 7.
906 Ibid., Article 10 (1).
907 Ibid.
908 Ibid., Article 10 (2).
909 Ibid., Article 13 (1).
910 Ibid., Article 13 (2).
911 Ibid., Article 11 (1).
912 Ibid., Article 13 (3).

initiate a fact-finding procedure. The report of the fact-finding commission has only an advisory character.[913] The origin state must ensure that the public likely to be affected is provided with information relating to the activity, the risk involved and the harm that may ensue.[914] In addition, the public is guaranteed an undefined right to make its views known.[915]

5. EIA Norms in Transboundary Watercourses

Before proceeding to consider the EIA norms relating to transboundary watercourses, it will be necessary to define what is meant by the term. Transboundary watercourses are well defined in the 1992 Transboundary Watercourses Convention:

> "Transboundary waters" means any surface or ground waters which mark, cross or are located on boundaries between two or more States; where ever transboundary waters flow directly into the sea, these transboundary waters end at a straight line across their respective mouths between points on the low-water line of their banks.[916]

As discussed above, the principle of due diligence applies to all transboundary pollution and thus includes transboundary pollution within a transboundary watercourse. However, the law relating to transboundary watercourses has for a long time been considered lex specialis. The 1966 'Helsinki Rules on the Uses of the Waters of International Rivers' (hereinafter 'the Helsinki Rules')[917] are widely taken as representing the basic principles governing the use of international rivers.[918]

---

[913] Ibid., Article 17 (2).
[914] Ibid., Article 9.
[915] Ibid.
[916] Article 1 (1) of the 1992 Transboundary Watercourses Convention.
[917] The Helsinki Rules can be obtained in Vol. 1, Basic Documents of International Environmental Law, p. 227. These rules were adopted by a non-governmental expert body of international lawyers, the International Law Association (ILA). The ILA has continued to work within this field; see Manner and Metsälampi 1988.
[918] Article I of the Helsinki Rules declares: 'The general rules of international law as set forth in these chapters'. Thus, the intent was to codify the law relating to transboundary watercourses.

Chapter 2 in particular of these rules, which relates to the principle of equitable utilization, has received the status of *lex lata* and been codified in the recent *1997 Convention on Non-Navigational Uses of International Watercourses (hereinafter 'the 1997 Watercourses Convention')* prepared by the ILC.[919]

Article 7 of the 1997 Watercourses Convention refers to prevention of 'causing of significant harm to other watercourse States'[920] and has provoked a debate on whether the Convention has changed the principle of equitable utilization.[921] Those who argue that such a change has occurred point out that the legality of the use of transboundary waters is determined by the principle of equitable use except when the use may cause significant harm to the other riparian states, a situation governed by the principle of due diligence and its prohibition against causing significant harm to other states.[922] It seems that it is too early to speak of such a change especially in light

---

Although the term 'international river' is used in the name of these rules, they apply to 'international drainage basin', a term defined in Article II of the Helsinki Rules as follows: 'An international drainage basin is a geographical area extending over two or more States determined by the watershed limits of the system of waters, including surface and underground waters, flowing into a common terminus'.

[919] Articles 5 and 6 of the 1997 Watercourses Convention. In the 1966 Helsinki Rules, equitable utilization is defined in the following manner: Article IV provides: 'Each basin State is entitled, within its territory, to a reasonable and equitable share in the beneficial uses of the waters of an international drainage basin'; the beginning of Article V adds: 'What is a reasonable and equitable share within the meaning of Article IV is to be determined in the light of all the relevant factors in each particular case'; and, finally, Article VI states: 'A use or category of uses is not entitled to any inherent preference over any other use of category of uses'. Article 6 of the 1997 Watercourses Convention is the most relevant here. Paragraph 1 of the Article provides: 'Utilization of an international watercourse in an equitable and reasonable manner within the meaning of Article 5 requires taking into account all relevant factors and circumstances'. Paragraph 3 adds: 'The weight to be given to each factor is to be determined by its importance in comparison with that of other relevant factors. In determining what is a reasonable and equitable use, all relevant factors are to be considered together and a conclusion reached on the basis of the whole'.

[920] Article 7 (1) of the 1992 Transboundary Watercourses Convention.

[921] For instance, Brunnee and Toope (1994, p. 64) argue that a significant harm resulting from a failure to exercise due diligence violates the principle of equitable utilization. Hence, the principle of due diligence, if it is applicable, overrides the principle of equitable utilization.

[922] See, e.g., the Third Report of the Special Rapporteur Schwebel, Article 8 (1), ILC Yearbook 1982, Vol. II, Part one, p. 103. For a contrary view, see Fuentes 1998, pp. 119-200. Fuentes states (p. 171): 'The conclusion, therefore, should be that the environmental impact of the use of an international watercourse is one factor that should be taken into account, among others, in the establishment of an equitable regime for the utilization of international watercourses...In the

of the ICJ's decision in the *Gabcíkovo-Nagymaros* case, which relied on the principle of equitable utilization.[923]

The 1997 Watercourses Convention in general, and Part IV of the Convention in particular, entitled 'Protection, Preservation and Management', introduced a new environmental perspective on transboundary watercourses law, i.e., that vital human needs[924], the ecosystems[925] and the sustainability of water use[926] can be protected from the open-ended balance of interests embodied in the principle of equitable utilization if seen in terms of collective interests rather than bilateral interests of riparian states.[927] Yet, even proponents of this approach view it mainly as something for the future[928] and the present author prefers to see the principle of equitable utilization as a principle of *lex lata*[929] and the new environmental perspectives as considerations entailed by the goal of sustainable development.[930]

What are the requirements of the principle of equitable utilization in the situations studied in this thesis? The Helsinki Rules did not set out the preventive duties of states although these could be derived from the general obligation in Article X (1a) to

---

process of weighing up the various relevant factors for the establishment of an equitable regime for the utilization of international watercourses, no factor has inherent priority over the others'.

923 For an analysis of the effect of the *Gabcíkovo-Nagymaros* case in relation to the law relating to transboundary watercourses, see Bourne 1997, pp. 6-12. Bourne argues (p. 11): 'The judgment of the ICJ in the *Gabcikovo-Nagymaros* case, then, marks a milestone in the evolution of international water law and international environmental law. It establishes that the protection of the environment has no absolute priority over other considerations'. For a more 'environmental' reading of this judgment, see Castro 1997, pp. 21-31.

924 Article 10 (2) of the 1992 Transboundary Watercourses Convention.

925 Ibid., Article 20.

926 Ibid., Article 5 (1).

927 For an analysis, see Nollkaemper 1996, pp. 39-73, especially pp. 60-69.

928 Ibid., p. 69.

929 For an analysis of the criteria for equitable utilization, see Fuentes 1996, pp. 337-412.

930 For a similar account, see Fuentes 1998. Fuentes argues (p. 199): 'equitable utilization and sustainable development differ in that the first is an existing principle of general international law while the second is a goal to be achieved at the national and international levels, and which aims at the harmonization of existing and emerging principles of international law'.

prevent any new form of water pollution.[931] This situation changed with the 1997 Watercourses Convention, Part III of which sets out preventive duties and can be seen as reflecting the state of customary law. The difference here to preventive duties required by the due diligence principle is clear: the preventive duties relating to transboundary watercourses are obligations of customary law that are independent of the general principle of equitable utilization.[932]

5.1. Applicable Treaty Regimes

The transboundary watercourses of the North American continent in general have been regulated extensively,[933] the principal exception being the watercourses crossing the international boundary between the United States (Alaska) and Canada (the Yukon Territory), which are governed by a single treaty. The transboundary watercourses of the Eurasian region have been well regulated through many bilateral agreements. Two treaties have been concluded that potentially apply to both regions.

---

[931] Article X (1a) of the Helsinki Rules provides: 'Consistent with the principle of equitable utilization of the waters of an international drainage basic, a State must prevent any new form of water pollution or any increase in the degree of existing water pollution in an international drainage basin which would cause substantial injury in the territory of a co-basin State'. This general obligation must be read in light of the principle of equitable utilization, and thus becomes part of the balancing of interests. This principle can also be interpreted as requiring the prior duties of notification, transboundary impact assessment and consultation in the case of a new plan likely to cause water pollution.

[932] In a recent general study on procedural environmental duties, Okowa (1996, pp. 275-336) comes to the conclusion that no preventive duties exist independently of the general principle of due diligence except in the law of international watercourses. As argued by Okowa (pp. 321-322): 'given the extensive State practice in the context of international watercourses, it may be argued with some degree of confidence that procedural duties in that context have passed into the corpus of general international law. But it would be unsound to rely on that practice to support the existence of customary norms in other areas such as the protection of species or atmosphere, or the regulation of the transboundary movement of hazardous waste'.

[933] See, e.g., the 1961 Agreement between the United States and Canada regarding the Columbia River Basin: Cooperative Development of Water Resources; the 1984 Treaty relating to the Skagit River and Ross Lake, and the Seven Mile Reservoir on the Pend d'Oreille River, with Annex; the 1988 Agreement between the United States and Canada Concerning Cooperation in Flood Control.

A relevant distinction can be drawn here between agreements on transboundary watercourses, which regulate only a specific watercourse, and framework treaties,[934] which aim at introducing new principles to existing specific regimes and potentially apply to varying kinds of transboundary watercourses. Common to the former category of agreements is that they establish a joint organ to administer and/or to resolve disputes between the parties. In considering the planning phase of environmentally harmful stationary activities, it is useful to make a distinction between those joint organs that are competent to make a decision on whether a certain activity may proceed and those that are not. None of the joint organs to be studied below has absolute decision-making power in all cases of potential negative effect on the transboundary watercourse; it is more useful to see them as falling on a scale from purely cooperative bodies to organs that have decision-making power in almost all cases.

The 1929 Agreement Between Sweden and Norway Concerning Certain Issues Over Water Rights[935] does not establish any permanent joint organ, but both Sweden and Norway are entitled to invoke the commission procedure whereby a dispute is

[934] For a brief overview of the phenomenon of framework treaties, see Kiss 1993, pp. 792-797.

[935] The 1929 Water Rights Agreement was signed on 11 May 1929. According to Article 35 of the Agreement, in order for the Agreement to enter into force both parties needed to ratify it and exchange notes to confirm this state of affairs; this took place on 2 July 1931. The Agreement entered into force 2 August 1931 and is still in force, although some modifications between the parties are presently in progress. The Water Rights Agreement contains another preventive procedure. If negative effects will occur on one side of the transboundary watercourse, consent must be obtained (Article 12 (1)), but, if not, the affected state is not entitled to object to the development of the activity (Article 12 (2)). The procedure starts with the submission of an application by a public or private person who is to exercise the functions covered by the Water Rights Agreement (Article 14 (1)). The application must contain the descriptions and assessments necessary for the evaluation of the effects of the project on the waters of both states (Article 14 (2)). When the application arrives at the competent authority in the state of origin, copies of the application and its annexes must be sent to the other state (Article 14 (3)). The public of the origin state is informed as provided for by the law of the origin state and the appropriate authority of the affected state may, if it sees fit, also inform its public of the proceedings taking place in the origin state (Article 15). Each state is entitled to obtain the necessary information from the competent authorities of the other in order to make the investigations required to determine the potential effects of the proposed activity (Article 16). (Author's translation from the Swedish).

submitted for comment by a commission composed of an equal number of members from both states.[936] The commission procedure and the Agreement apply to all watercourses, rivers, lakes and other watercourses in which the international border of the states runs, or which are intersected by the international border.[937] All construction works or other measures by one state party which may cause a change of any kind in the composition and functioning of or the movement of fish in the waters of the other fall within the scope of the procedure.[938] The Commission's task is to investigate the matters brought before it and it has a right to obtain necessary assistance from both states.[939] Article 19 stipulates that the Commission must provide an opinion on the whole content of the licence for the activity, this non-binding opinion then being submitted to both states for consideration.[940] The final licencing decision must also be delivered to the potentially affected state.[941]

Two of the international joint organs are quite similar in that they have been established as permanent organs of cooperation serving merely advisory and fact-finding tasks. The International Joint Commission (IJC), a joint organ of Canada and the United States established by *the 1909 Treaty Between the U.S. and Great Britain Relating to Boundary Waters, and Questions Arising Between the United States and Canada (hereinafter 'the 1909 Boundary Waters Treaty')*[942], can be considered as having competence also in transboundary watercourses crossing the northern boundary between the two states.[943] The Commission has powers over alteration of

[936] Ibid., Article 17.
[937] Ibid., Article 1 (2 and 3).
[938] Ibid., Article 1 (1).
[939] Ibid., Article 18 (1).
[940] Ibid., Article 19 (2).
[941] Ibid., Article 21 (2).
[942] The 1909 Treaty Between the U.S. and Great Britain Relating to Boundary Waters, and Questions Arising Between the United States and Canada.
[943] The Treaty was signed in 1909 and entered into force the following year. It is still in force. The Treaty did not initially apply in the Arctic since Alaska was not admitted to the Union until 1959. Accordingly, the Preliminary Article provides: 'boundary waters are defined as the waters from main shore to main shore of the lakes and rivers and connecting waterways' (Preliminary Article). However, since the international boundary of the two states was extended when Alaska became part

the physical condition of the boundary waters, which are defined as 'lakes and rivers and connecting waterways...including all bays, arms, and inlets thereof, but not including tributary waters'[944].

The Treaty permitted the existing uses of the boundary waters at the time but provided that 'no further or other uses or obstructions or diversions, whether temporary or permanent...affecting the natural level or flow of boundary waters on the other side of the line shall be made except...with the approval...[of] the International Joint Commission'[945]. The Commission was also given the power to look into the facts and circumstances of the particular questions and matters referred to it.[946] In other words, in those issues which are under its mandate, and also in those cases submitted to it directly by the parties, the Commission is competent to study the environmental impacts of a proposed activity.

The Borderwater Commission, established in Articles 1 and 3 of *the 1981 Agreement Between Finland and Norway on Finnish Norwegian Borderwater Commission*[947],

---

of the United States, the Treaty came to apply to the boundary waters between Alaska and Yukon. The basic principle governing these situations can be found in the Vienna Convention on State Succession, which provides that in the case of territorial extension, treaty obligations extend correspondingly unless explicitly agreed otherwise. As stated by Klein (1984, p. 473): 'It is commonly accepted as a rule of customary international law that...territorial changes alter the treaty borders; but the régime itself of already existing treaties is not affected'. When Alaska became part of the United States, this territorial change also affected the coverage of the Treaty since the parties had an additional boundary in the Arctic. The parties could have agreed not to include the Arctic border waters but opted not do so. The same view is held by an officer of the IJC (e-mail communication 14.12.2000). The mandate of the IJC has been broadened from what it was in the 1909 Boundary Waters Treaty through the 1978 Agreement between the United States and Canada on Great Lakes Water Quality (Article VII) and the 1991 Air Quality Agreement (Article IX). For an analysis of the IJC and the Great Lakes Agreements in Finnish, see Kuokkanen 1988, pp. 77-136.

[944] The Preliminary Article of the 1909 Boundary Waters Treaty.

[945] Ibid., Article III. See also Article IV.

[946] Ibid., Article IX.

[947] The Agreement was adopted through an exchange of notes on 1 April 1981 and entered into force on 1 May 1981. It is still in force. See also, e.g., the 1989 Agreement on Joint Fishing Regulations for the Tana River Fishing District (with annexes). The Agreement was signed on 1 March 1989 and entered into force on 1 January 1990.

oversees all rivers which the international border of Finland and Norway traverses, as well as lakes, rivers and brooks which are intersected by the international border.[948] The Commission, which is composed of three members from each state, has a mandate to make motions, present initiatives and give statements on preventing pollution of the borderwaters, on construction on the borderwaters, and on regulation of the river flow. It does not have any binding decision-making power; i.e., it is purely a cooperative body. The basic obligation for the parties is that the relevant officials of the states must inform the Commission if they become aware of plans which may crucially affect the borderwaters.[949] The Commission is entitled to make preventive investigations and thus also impact assessment studies as well as receive assistance from the officials of both parties.[950]

The Joint Finnish Russian Commission on the Utilization of Frontier Watercourses, established by *the 1964 Agreement Concerning Frontier Watercourses,*[951] has advisory tasks for the most part but in some limited cases the power to make a final licencing decision as well. Its mandate covers all those lakes, rivers and streams which are crossed by the international border between the states or through which the border runs.[952] The scope of the Agreement is defined mainly by the effects on the transboundary waters. First, there are the negative effects defined in Article 2, which are related mainly to changes in the composition or flow of the river water itself.[953] Second, there are negative effects produced by polluting activities, which mainly involve the introduction of harmful substances into transboundary waters.[954] The basic procedure to be followed by the states in these situations obligates the origin

948 Ibid., Article 2.
949 Ibid., Article 1 (2).
950 Ibid., Article 5.
951 The 1964 Agreement Between the Finnish Republic and the Soviet Union Concerning Frontier Watercourses (translation in the UNTS). The Agreement was signed on 24 April 1964 and came into force on 6 May 1965. Since the Russian Federation took over the treaty commitments of the Soviet Union, the Agreement remains in force between the parties.
952 Ibid., Article 1.
953 Ibid., Article 2.

state to notify the potentially affected state when it is about to carry out any measures on its side of the border which may have effects described in Articles 2 and 4.[955] The potentially affected state has two months to comment on the planned activity, after which the states may decide to submit the matter to the Finnish Soviet Borderwater Commission[956] for an advisory opinion or a final decision.[957]

The Commission has the power to make any necessary investigations, and thus also impact assessments, and to obtain assistance from the officials of both of the parties in this task, regardless of whether it has only advisory powers or final decision-making power.[958] When the Commission has final decision-making power, it must observe the national laws of both parties as well as request comments from those officials who, according to national law, are entitled to submit such comments.[959] If it can not reach a decision, or if the case is of a serious nature, the Commission must submit the matter for decision by the respective governments.[960] If the Commission reaches a unanimous decision, and no complaints against the decision are made by either party, the decision becomes binding on the parties.[961]

The mandate of the Finnish Russian Borderwater Commission was expanded by *the 1992 Action Program Between Finland and Russian Federation with a view to*

---

[954] Ibid., Article 4 (1).
[955] Ibid., Article 9.
[956] Ibid. The Commission is established in Articles 6 and 7.
[957] Ibid.
[958] Ibid., Articles 10 (1) and 12 (1). A good example here is the mining project (phosphorite ore) in Sokli in northern Finland above the Arctic Circle and very near the northern Russian border. Although the EIA procedure was not in force at the time in Finland, the EIA was applied to the project and since the project was to operate so near the Russian border, the Borderwater Commission was also notified. The Commission discussed the matter and even prepared a report before the developer, Kemira Ltd., decided to discontinue the project for economic reasons. For a general review of the environmental impact assessment process applied to Sokli, see Mähönen 1994. Information on the Borderwater Commission procedure obtained from the Finnish Ministry of the Environment (author's personal copy).
[959] Ibid., Article 10.
[960] Ibid., Article 10 (4).
[961] Ibid., Article 11 (1).

Reduce Pollution and Implement Water Protection in the Baltic Sea Area as well as Other Areas Near the Border of Finland and Russian Federation (hereinafter 'the 1992 Action Treaty')[962]. The Treaty covers all of Finland but only the districts in Russia along the Finnish border and includes both water and sea areas near the border;[963] only the former fall under the mandate of the Commission however. The material scope of the Treaty is very wide. The origin state is obligated to notify 'in good time before any planned action which may have effect on the condition of those water and sea areas near the border'[964]. This notification must contain an evaluation of the project's environmental impacts in general and its transboundary effects in particular on the water and sea areas near the border.[965] In making the notification, the origin state must take into account the 1992 Transboundary Watercourses Convention as well as the 1991 Espoo Convention.[966] Interestingly, the Action Treaty contains a special procedure for particularly serious cases - a joint impact assessment.[967] In cases where the potential impact affects transboundary watercourses, the body in charge is the Borderwater Commission. The parties have also agreed to develop the regime outlined in paragraph 3 of the Treaty.

---

[962] The Action Treaty was signed on 26 May 1992 and entered into force on the same day as stipulated in Article 8 of the Treaty. The Treaty is still in force and does not contain any provisions limiting the exchange of information. The Joint Finnish Russian Commission on the Utilization of Frontier Watercourses is referred to in Chapter 4 (1). Interestingly, Finland and the Russian Federation included a provision on EIA in their 1992 Agreement Between the Government of the Republic of Finland and the Russian Federation on Cooperation in the Field of Environmental Protection. According to Article 3, the cooperation also relates to environmental impact assessment. Finland and the Russian Federation have been negotiating a bilateral agreement on transboundary EIA on the basis of the Espoo Convention since 1996. The negotiations have been at standstill for a while but will start again at the end of June 2001 (e-mail communication with an officer of the Finnish Ministry of the Environment 3.4.2001). (Author's own translation).

[963] Ibid., Section 1.

[964] Ibid., Section 4 (6.1).

[965] Ibid.

[966] Ibid., Section 4 (6.2).

[967] Ibid., Section 4 (6.1).

The Transfrontier River Commission, established by *the 1971 Agreement Concerning Frontier Rivers*[968], has the largest mandate of the joint bodies operating in the Arctic where final licencing decisions are concerned. The mandate of the Commission covers the frontier rivers of Finland and Sweden.[969] From the viewpoint of the situations studied here, the scope of the Agreement is mainly governed by two chapters: Chapter 3, 'Hydraulic construction works', and Chapter 6, 'Protection against water pollution'. Chapter 3 (1) enumerates all the activities which are considered as coming under the chapter, although not exhaustively. These are mainly hydraulic works such as the erection of installations in any waters, the regulation of water level or flow, diversion of water from a body of water.[970] According to Article 13 of the Agreement, a licence must be obtained for these activities from the Frontier River Commission.[971] If the activity is of minor scope, however, no licence is required.

---

[968] The 1971 Agreement Concerning Frontier Rivers, with annexes (English translation in the UNTS). The Agreement was signed on 16 September 1971 and came into force on 1 January 1972. It is still in force as adopted, although an amendment is underway. For an overview in Finnish of the changes that have to be made in the Agreement because of several EC directives and international conventions, see Hepola 1999, pp. 145-174. The latest development, on 8 March 2001, is that initial agreement has been reached within the Finnish government to have part of the Torne and Muonio river area included as part of the ecological network of special areas of conservation, known as Natura 2000 (Article 3 of the Council Directive 92/43/EEC of 21 May 1992 on the Conservation of Natural Habitats and of Wild Fauna and Flora). The final decision will be made by the Finnish government. Sweden has already included the Torne and Kalix rivers in its proposal. Information obtained from the home page of the Finnish Ministry of the Environment at http://www.vyh.fi/ajankoht/tiedote/ym/tied2001/ym0132.htm [9.3.2001]. On the Agreement, see Fitzmaurice 1992, pp. 33-67. See also, e.g., Exchange of Letters Constituting an Agreement Concerning the Construction and Maintenance of a Bridge Over the River Torne at Pello (Overtornea), which entered into force on 31 October 1968; Exchange of Notes Constituting an Agreement on Building a Bridge Across River Muonio, which entered into force on 1 July 1985.

[969] Ibid. The following water areas are under its competence: - The Rivers Könkämä and Muonio and the portion of the River Torne and the lakes through which the State frontier between Finland and Sweden runs "frontier rivers": - The special effluents formed by the various branches at the mouth of the River Torne; - The part of the Gulf of Bothnia lying between the Finnish and Swedish parishes of Lower Torne (Section 1 (1)).

[970] Ibid. The regulation of water flow is specifically regulated in Section 4, but almost all of the norms in Section 3 also apply to these hydraulic works. In addition, planned construction which may affect the movement of fish requires a permit from the Frontier River Commission.

[971] Ibid. The Frontier River Commission is established in Section 2.

Chapter 6 (3) states: 'The provisions of articles 4-14 shall apply to use of land, buildings or installations in a way that may cause pollution of waters where such use does not constitute hydraulic construction'. The implementation of a stationary activity which may pollute the water area is governed by Chapter 6. The difference between the activities in this chapter and those listed in Chapter 3 is that the former introduce alien particles into the water while the latter alter the water area itself in some way. The material scope of the Agreement is specified in Annex C, Article 2, which lists 54 categories of activities. If these activities may cause pollution of the water area, they require a licence from the Transfrontier River Commission.

A natural or legal person who is planning to undertake any of the activities mentioned in the Agreement must obtain a licence from the Transfrontier Rivers Commission.[972] The Commission is composed of three members from each state and thus represents the interests of both states.[973] In certain issues, the Commission is obligated to submit the permissibility of the activity to the governments of the states before it makes its licencing decision. In relation to construction on the water area, such issues are:

1. The construction of a dam, or major change to a dam.
2. Regulation of the river flow if such an activity may have essential impact on the water conditions in the water area.
3. The construction may 'result in a substantial deterioration in the living conditions of the population or cause a permanent change in natural conditions' as defined in Article 3 (2) of Chapter 3.[974]

The submission to the governments may also take place if the construction is viewed by the Commission as requiring such consideration or one of the governments so requests.[975]

---

[972] Ibid., Sections 3 (13) and 6 (4).
[973] Ibid., Section 2 (2).

There exist certain exceptions to the licencing power of the Commission also with respect to pollution of the water area as regulated in Chapter 6. If the Commission so determines, it may submit a matter for consideration by the national governments. The reasons for this may be, for instance, that the nature or the impacts of the activity require such submission. The Commission is obligated to make such a submission if either of the national governments so requests.[976]

No explicit rules exist on evaluating the impacts of the proposed activity on the water area. However, according to Chapter 2 (3):

> [The Commission shall] pursuant to its own decisions, institute such inquiries and investigation as are required in order to enable it to accomplish its tasks in accordance with this Agreement. The Commission may enter into direct contact with the authorities of either State and may call upon them for assistance in obtaining any necessary information and arranging for any necessary consultations. The Commission may, where necessary, employ the services of experts for special investigations.

Since the procedure in all cases begins with the licence application to the Commission, the latter is in a position to evaluate the potential impacts on the water area. Inasmuch as the licencing procedure is governed by the respective national laws of the contracting states, and both Finland and Sweden have introduced EIA legislation, the Commission is normally required to undertake some sort of impact assessment procedure.

---

974 Ibid., Section 3 (13).
975 Ibid.
976 Ibid., Section 6 (4).

There exists one framework treaty, the 1992 Transboundary Watercourses Convention[977], that applies potentially to all transboundary watercourses located in the Arctic. From the viewpoint of the joint organs studied above, this treaty requires the introduction of new principles that govern the management of a specific transboundary watercourse. The Convention presently applies to Denmark, Finland, Norway, the Russian Federation and Sweden.[978] The geographical scope of the Convention covers all of the 'transboundary waters' defined in Article 1 (1).[979] The principal importance of this Convention is that it harmonizes the basic principles governing the protection and use of transboundary watercourses.[980] It obligates the riparian states to conclude more specific cooperation agreements in regard to transboundary watercourses not governed by any treaty and requires the partners of an existing treaty to reform their joint measures to better accommodate the principles of the Convention.[981]

It is important from the viewpoint of the present study that the Convention favors a preventive approach. This is emphasized in Article 2 (1): '[t]he parties shall take all

977 The 1992 Convention on the Protection and Use of Transboundary Watercourses and International Lakes. The Convention was adopted in March 1992, opened for signature from 17 to 18 March 1992 and entered into force on 6 October 1996. The Convention was adopted by the Senior Advisers to the Economic Commission for Europe Governments on Environmental and Water Problems at their Resumed Fifth Session held at Helsinki from 17 to 18 March 1992. The Convention was opened for signature at Helsinki from 17 to 18 March 1998 and had been open for signature at United Nations Headquarters since 18 September 1992. Information obtained from the UNTS. According to Article 8, the parties are entitled to withhold information: 'related to industrial and commercial secrecy, including intellectual property, or national security'.

978 The dates of ratification, accession, acceptance or approval by the Arctic states are the following: Denmark (28.5.1997), Finland (21.2.1996), Norway (1.4.1993), the Russian Federation (2.11.1993), Sweden (5.8.1993) and the EC (14.9.1995). Denmark made a note 'with reservation of application to the Faroe Islands and Greenland'. Information obtained from the UNTS. The parties to the Convention have already met twice. The first meeting took place in Helsinki, Finland, from 2 to 4 July 1997, and adopted the rules of procedure (ECE/MP.WAT/2). The second meeting, held in the Hague, Netherlands, from 23 to 25 March 2000 adopted terms of reference for the International Water Assessment Centre (IWAC), which are set out in Annex V(ECE/MP.WAT/5). Both documents can be found on the home page of the UN ECE at http://www.unece.org/env/water/topfra3.htm [9.3.2001].

979 Article 1(1) of the 1992 Transboundary Watercourses Convention.

980 Ibid., Articles 2 and 3.

appropriate measures to prevent, control and reduce any transboundary impact'. Measures to this effect are to be taken at the source of pollution.[982] One of the principles laid down to guide the measures is the precautionary principle.[983] This and the remaining principles are specified in Article 3 and require the parties, 'as far as possible', to harmonize their national measures, for instance, through prior licencing of waste-water discharges[984] and environmental impact assessment.[985]

Part II of the Convention concerns the cooperation between the riparian parties. The parties are required to 'enter into bilateral or multilateral agreements or other arrangements, where these do not exist, or adapt existing ones, where necessary to eliminate the contradictions with the basic principles of this Convention'[986] Significantly, the Convention requires the parties to establish joint bodies[987] and also outlines the basic tasks of these bodies. Of importance here is that these joint bodies are to 'serve as a forum for the exchange of information on existing and planned uses of water and related installations that are likely to cause transboundary impact'[988]. They are also to participate in the implementation of EIA procedures relating to transboundary waters.[989] The 1999 Protocol on Water and Health will also provide relevant regulation when it enters into force.[990]

---

981 Ibid., Article 9.
982 Ibid., Article 2 (3).
983 Ibid. Article 2 (2a) provides: 'The precautionary principle, by virtue of which action to avoid the potential transboundary impact of the release of hazardous substances shall not be postponed on the ground that scientific research has not fully proved a causal link between those substances, on the one hand, and the potential transboundary impact, on the other hand'.
984 Ibid., Article 3 (b).
985 Ibid., Article 3 (h).
986 Ibid., Article 9 (1).
987 Ibid., Article 9 (2).
988 Ibid., Article 9 (2h).
989 Ibid., Article 9 (2j).
990 The Protocol on Water and Health to the 1992 Convention on the Protection and Use of Transboundary Water courses and International Lakes. The Protocol was adopted on 17 June 1999 on the occasion of the Third Ministerial Conference on Environment and Health at London but has not yet entered into force. The Protocol was opened for signature in London on 17 June 1999 and thereafter at United Nations Headquarters until 18 June 2000, the potential signatories being states members of the UN ECE, states members of the Regional Committee for Europe of the World

Potential future regulation applicable in the Arctic is also to be found in the 1997 Watercourses Convention,[991] the end-product of the ILC project Non-Navigational Uses of International Watercourses. Of the Arctic states, Finland, Sweden and Norway have all accepted the Convention as binding.[992] The Convention is applicable to all international watercourses, with concept of watercourse defined as 'a system of

---

Health Organization (WHO), states having consultative status with the UN ECE and regional economic integration organizations which are either members of the UN ECE or members of the Regional Committee for Europe of the WHO. Information obtained from the UNTS. The document from the first meeting of signatories to the Protocol (MP.WAT/AC.2/2000/2-EUR/ICP/5021651/2) can be found on the home page of the UN ECE at http://www.unece.org/env/water/topfra3.htm [9.3.2001]. The 1999 Protocol on Water and Health to the Convention has been signed by the Nordic Arctic states, but the only one that has accepted the Convention as binding is the Russian Federation. The Russian Federation accepted the Protocol on 31 December 1999. The Convention was designed especially to prevent, control and reduce water-related diseases (Article 1) The scope of this Protocol covers not only surface freshwater, groundwater and estuaries but also some other forms of water not usually covered by agreements relating to transboundary watercourses (Article 3). As explained above, the parties are obligated to establish extensive systems for access to information by the public and the participation of the public in prior procedures; in addition, they are required to assess any proposed measure in respect of human health, water resources and sustainable development. This assessment must take into account 'the differing new impacts of any proposed measure on the differing environmental mediums' (Article 4). These assessment responsibilities also relate to the potential transboundary impact of the proposed measure (Article 6). The Protocol also ensures that in the case of a conflict between the Protocol and the 1991 Espoo Convention, the 'compliance by public authorities of that Party with the requirements of that Convention in relation to a proposed action shall satisfy the requirement under paragraph 6 of this article in respect of that action' (Article 7). It also specifies these transboundary obligations. In Article 12 (vii) it is provided: 'authorities of states parties are required to notify other parties of: significant threats of such outbreaks and incidents'. Importantly, in Article 13 (1d), the parties are also required to 'consult each other, at the request of any one of them, on the significance of any adverse effect on human health which may constitute a water-related disease'. In the case of transboundary impacts, the transboundary procedure of the 1992 Transboundary Watercourses Convention takes priority over the Protocol, as is stipulated in Article 13 (2).

[991] The 1997 Convention on Non-Navigational Uses of International Watercourses. The Convention was adopted by the UNGA on 21 May 1997 and opened for signature at the same day. The Convention was adopted by resolution A-RES-51-229 of 21 May 1997 of the UNGA at its 51st session. The Convention became open for signature at the Headquarters of the United Nations on 21 May 1997 and remained open to all states and regional economic integration organizations until 21 May 2000. Information obtained from the UNTS. The Convention has not yet entered into force, since only 16 states have signed it. It does not seem likely that the Convention will enter into force in the foreseeable future since thirty-five instruments of ratification, acceptance approval or accession are needed for it to enter into force (Article 36).

[992] The following dates of ratification, acceptance, accession or approval are relevant: Finland (23.1.1998), Norway (30.9.1998) and Sweden (15.6.2000). Information obtained from the UNTS.

surface and ground waters'[993]. The watercourse becomes an 'international watercourse' if its parts are 'situated in different States'[994].

The Convention also provides general-level guidance for the states parties, encouraging them to conclude watercourse agreements[995], exchange data[996] as well as protect and manage the ecosystems[997]. It also contains basic rules on the rights and obligations of the riparian parties[998] as well as participation rights for private legal subjects of the potentially affected state.[999] Of particular importance from the viewpoint of this study is that it contains a specific Part III on planned measures. The origin state is obligated to notify the potential affected state if planned measures 'may have a significant adverse effect upon other Watercourse States'[1000]. The Convention contains a procedure for situations in which there exists disagreement between the states as to whether a significant adverse effect will follow from the planned measure. If the origin state does not initiate the transboundary assessment procedure, the potentially affected state has a right to request it to notify.[1001] This request must provide a 'documented explanation setting forth its grounds'[1002]. If the origin state still determines that the procedure is not applicable, it must notify the potentially affected state and furnish a documented explanation.[1003] In such cases, the potentially

993 Article 2 (a) of the 1997 Watercourses Convention.
994 Ibid., Article 2 (b).
995 Ibid., Articles 3 and 4.
996 Ibid., Article 9.
997 Ibid., Part IV.
998 Ibid., Articles 5, 6, 7 and 8.
999 Ibid. Article 32 provides a norm on non-discrimination. According to this Article, private legal subjects 'who...are under serious threat of suffering significant transboundary harm as a result of activities related to an international watercourse' must be allowed access also to judicial or other procedures in the state of origin. Yet, if the states concerned have otherwise agreed 'for the protection of the interests' of private legal subjects, it is these measures of protection that prevail over the norm established in Article 32. It should be noted that this other agreement must have been made for the protection of the interests of the private legal subjects, a condition constituting a certain standard of protection from this form of agreement.
1000 Ibid., Article 12.
1001 Ibid., Article 18 (1).
1002 Ibid.
1003 Ibid., Article 18 (2).

affected state has a right to enter into consultations with the origin state.[1004] If the potentially affected state so requests, the origin state must refrain from implementing the planned measures for a period of six months.[1005]

The notification must contain at a minimum 'available technical data and information, including the results of any environmental impact assessment'[1006]. The contracting states are thus not obligated to establish national EIA procedures; but if such a procedure exists and has been applied to a given case, the information produced by it must be included in the notification.[1007] The origin state must 'allow the notified States a period of six months within which to study and evaluate the possible effects of the planned measures and to communicate the findings'[1008] back to the origin state. This period can be extended.[1009] During this period, the origin state must provide additional data and information, if available, on request by the potentially affected state and refrain from implementing the proposed activity.[1010]

As mentioned above, the potentially affected state must provide its reply within six months of the date on which it receives the notification. It also has a right to have this deadline extended for a period of six months if the 'evaluation of the planned measures poses special difficulty'[1011]. In its reply to the origin state, the potentially affected state must indicate the way in which the planned measures breach the material principles of the Convention outlined in Articles 5 to 7.[1012] If the potentially affected state does not reply within the specified time, the origin state can go ahead

---

1004 Ibid.
1005 Ibid., Article 18 (3).
1006 Ibid., Article 12.
1007 Ibid.
1008 Ibid., Article 13 (a).
1009 Ibid., Article 13 (b).
1010 Ibid., Article 14.
1011 Ibid., Article 13 (b).
1012 Ibid., Article 15.

with its implementation of the proposed activity. Failure to reply also has effects on the possible question of compensation from transboundary damage.[1013]

If the potentially affected state has replied to the notification and stated that the planned measures are in breach of the material principles of the Convention, the concerned states must enter into consultations 'with a view to arriving at an equitable resolution of the situation'[1014]. If the potentially affected state so requests, the origin state must refrain from implementing the planned measures for a period of six times.[1015] The Convention also contains a special procedure for planned measures which must be urgently implemented in order to protect public health, public safety or other equally important interests.[1016] If this criterion is fulfilled, the origin state can implement the planned measures immediately 'subject to Articles 5 and 7'.[1017] The origin state must notify the potentially affected states of such a case 'with the relevant data and information'[1018]. The potentially affected states also have a right of consultation in these cases.[1019]

5.2. Conclusion

Since the border between Alaska and the Yukon Territory does not include transboundary waters forming an international boundary, it is not surprising that no specific agreements have been concluded between the United States and Canada on the waters in the region. There is only one joint organ, the IJC, that states may turn to when pollution problems appear in the northern transboundary waters of these states.

[1013] Ibid., Article 16.
[1014] Ibid., Article 17 (1).
[1015] Ibid., Article 17 (3).
[1016] Ibid., Article 19 (1).
[1017] Ibid.
[1018] Ibid., Article 19 (2).
[1019] Ibid., Article 19 (3).

The principle of equitable utilization provides guidance in disputes over potential transboundary impacts of a proposed activity.

As in the case of prior transboundary procedures, there are new possibilities for Arctic-wide considerations in relation to Arctic transboundary waters. The 1992 Transboundary Watercourses Convention and its Protocol, as well as the 1997 Watercourses Convention, are available to all the Arctic states. Yet, it seems that only the Eurasian Arctic states are interested in making these Conventions operable. In fact, since all the Eurasian states are parties to the 1992 Transboundary Watercourses Convention, they are obligated to adapt the existing joint organs and management agreements to conform to the modern principles of environmental protection introduced by the Convention, and, some measures have already been taken.[1020] There exists only one agreement between Norway and the Russian Federation, the 1957 Agreement on the Utilization of Water-Power on the Pasvik (Paatso) River[1021], but this neither provides any environmental protection standards nor establishes any joint organ. As parties to the 1992 Transboundary Watercourses Convention, however, the Russian Federation and Norway have been obligated to establish a joint organ to manage this frontier river[1022] that is capable of implementing the environmental protection standards of the Convention.[1023]

[1020] Article 9 (1) of the 1992 Transboundary Watercourses Convention. For an overview of the effects of the 1992 Transboundary Watercourses Convention on the amendment of the 1971 Agreement Concerning Frontier Rivers between Finland and Sweden in Finnish, see Hepola 1999, pp. 161-165. According to an officer of the Swedish Environmental Protection Agency, Sweden and Norway are planning to study how the 1928 Agreement on Water Rights should be amended (e-mail communication 27.4.2001). According to an official at the Finnish Ministry of the Environment, the principles of the Convention have already been implemented in the practice of the Finnish Norwegian Borderwater Commission (e-mail communication 14.5.2001). According to an official at the Finnish Ministry of Agriculture and Forestry, the principles of the Convention have already been implemented in the practice of the Joint Finnish Russian Commission on the Utilization of Frontier Watercourses (e-mail communication 14.9.2001).

[1021] The Agreement (with annexed charts) on the Utilization of Water-Power on the Pasvik (Paatso) River was signed at Oslo on 18 December 1957 and came into force on 27 June 1958.

[1022] There is also another river forming an international boundary between Norway and the Russian Federation, the Jakobsriver (Jakobselva). According to an officer of the County Environmental

6. EIA Norms in International Areas

Pollution affecting international areas may have its origin in the area of jurisdiction of Arctic states. As provided by a valid principle of international law, the due diligence principle: 'States have...the responsibility to ensure that activities within their jurisdiction or control do not cause damage to the environment of other States or of areas beyond national jurisdiction'[1024]. As discussed above,[1025] the due diligence principle expresses states' obligations towards 'the environment of other States and of areas beyond national jurisdiction'. Since there are great difficulties in determining the legal status of the areas of jurisdictional competence as either part of state territory or as part of international areas, this formulation describes what the due diligence principle requires: since the term 'jurisdiction' is used, areas beyond states' national jurisdiction are clearly those areas that are outside the areas of the jurisdictional competence of states, not merely outside their areas of territorial sovereignty. Hence, in the context of the principle of due diligence international areas mean only those areas that are outside the areas of states' national jurisdiction.

The same preventive duties required by the due diligence principle in regard to inter-state pollution are applicable in situations of potential transboundary pollution originating in the area of jurisdiction of states and affecting international areas. From the material point of view, the Arctic states are required to prevent potential transboundary impacts on the environment of international areas, mainly the water column of the Arctic Ocean and the international air space lying outside the states'

Deparment of Finnmark, fishing rules have been adopted by Norway and the Russian Federation over the Jakobsriver (e-mail communication 28.5.2001).

[1023] Article 9 (1 and 2) of the Transboundary Watercourses Convention.

[1024] Principle 21 of the Stockholm Declaration and principle 2 of the Rio Declaration. For a view that the due diligence principle also requires measures protecting areas beyond national jurisdiction, see Charney 1991, pp. 162-163, and Leigh 1992, pp. 134-135.

[1025] See *supra* chapter IV (4.).

areas of sovereignty. From the procedural perspective, the principle of due diligence requires the origin state to notify the potentially affected state, to conduct a transboundary impact assessment of environmental effects and to consult the potentially affected state. The problem here is that there exists only one organ with which states can communicate about potential transboundary impacts,[1026] the International Sea Bed Authority (ISA), and in practice it is not the deep sea-bed that normally suffers from the pollution effects.[1027] The due diligence principle thus mainly requires that states conduct a transboundary impact assessment of potential effects on the environment of international areas and take measures to mitigate these effects such that they remain below the threshold of significance.

There exists only one treaty procedure which creates – albeit in a limited fashion - a prior transboundary procedure for international areas.[1028] Article 206 of the LOSC obligates the states to assess the environmental effects of planned actions affecting the entire marine environment, thus including international areas. The results of these studies must be communicated to competent inter-governmental organizations, one of which in this case could be the International Maritime Organization (IMO). The organization(s) informed are then obligated to communicate the information received to all states.

Stationary activities as defined in this study may also take place in international areas. Since the study here concentrates on environmentally harmful stationary activities, the main area of concern here is the exploitation of hydrocarbon resources deposited in the deep sea-bed (the Area) of the Arctic Ocean. Only a general

[1026] For an analysis, see Rosas 1991, p. 42.
[1027] Charney (1991, pp. 149-177) asserts that it is the specially affected states, third-states, that can most appropriately take action on behalf of the violation of rights 'owned' by international areas. It could thus be argued that, e.g., in the case of the potential pollution of the Arctic Ocean, the origin state would be obligated to inform the other littoral states.

overview is necessary here because it is not likely that offshore hydrocarbon exploitation will begin in the Area in general in the near future, not to mention the Arctic Ocean, which poses particularly difficult operating conditions.

It is only the ISA[1029] and states that can authorize activities to operate in the Area. The basic requirements of the goal of sustainable development are that the ISA and the Arctic states establish EIA procedures for offshore hydrocarbon exploitation activities taking place in the Area of the Arctic Ocean, e.g., the procedures described in the 1997 Offshore Guidelines.[1030] The principle of due diligence applies only to potential transboundary pollution from hydrocarbon exploitation activities in the Area. The international legal subjects, i.e., states and the ISA, are required to notify the Arctic states and the littoral states in particular if it is likely that a significant adverse transboundary environmental impact will result from offshore hydrocarbon exploitation activities in the Area. The subjects are also obligated to conduct a transboundary impact assessment and to consult the potentially affected states. Clearly, the LOSC is the most important treaty in this context. Article 209 provides basic rules relating to pollution from activities operating in the Area.

Article 209 is divided into two parts. The first states: 'international rules, regulations and procedures shall be established in accordance with Part XI to prevent, reduce and

---

1028 Interestingly, as will be shown below, one legally relevant document contains such a duty. Principle 12 of the UNEP Guidelines for EIA contain an obligation to evaluate the potential effects on international areas. See infra chapter V (1.).

1029 The principle of due diligence applies to legal subjects other than states, in this case the ISA and its organs. As provided by Jennings and Watts (1992, Oppenheim's International Law, p. 16): 'To the extent that bodies other than states directly possess some rights, powers and duties in international law they can be regarded as subjects of international law, possessing international personality. It is a matter for inquiry in each case whether - and if so, what - rights, powers and duties in international law are conferred upon any particular body'. The ISA is clearly an international legal subject, and since it governs a certain physical area, is obligated by the principle of due diligence. To be sure, the formulation of the due diligence principle refers only to states, but this is because it is primarily states which are obligated by the principle. The goal of sustainable development also provides normative guidance here although not in strict legal terms.

1030 See *supra* chapter IV (3.3.1.).

control pollution of the marine environment from activities in the Area'[1031]. These rules and regulations must also be re-examined from time to time. Part XI provides rules for adopting norms in the case of pollution from activities in the Area. Article 145 of the LOSC requires measures to be taken in order to ensure effective environmental protection from activities taking place in the Area. The ISA is required to adopt rules and procedures for the prevention of pollution to the marine environment and for conserving the natural resources of the Area.[1032] This provision is contained in Annex III of the LOSC as well as the 1994 Implementation Agreement.[1033]

The 1994 Implementation Agreement is of importance here, especially Paragraph 7 of Section 1. The plans of work submitted by the qualified applicants must specify

[1031] Article 209 (1) of the LOSC.
[1032] Ibid. Article 145 reads: '(a) the prevention, reduction and control of pollution and other hazards to the marine environment, including the coastline, and of interference with the ecological balance of the marine environment, particular attention being paid to the need for protection from harmful effects of such activities as drilling, dredging, excavation, disposal of waste, construction and operation or maintenance of installations, pipelines and other devices related to such activities; (b) the protection and conservation of the natural resources of the Area and the prevention of damage to the flora and fauna of the marine environment'.
[1033] The prevailing governance regime of the deep sea-bed consists of the LOSC and the 1994 Implementation Agreement (hereinafter 'the regime of the deep sea-bed'). The LOSC rules and principles were modified quite a bit through the Implementation Agreement in order to make the regime of the deep sea-bed acceptable and ensure that the many industrialized states would ratify the LOSC. Where the exploitation of the sea-bed resources is concerned, the basic idea of the regime is that companies registered in the contracting states, the so-called qualified applicants, are entitled to prospect, explore and exploit the sea-bed for commercial purposes, but under the rules and principles laid down in the regime and under the control of the ISA. The prime organs of the ISA are the plenary body, the Assembly representing the contracting states, the Council which is responsible for the implementation of the regime, the Secretariat which takes care of the day-to-day administration of the ISA, and finally the Enterprise, a commercial entity which is supposed to engage in sea-bed mining on behalf of the ISA. Another important organ is the Legal and Technical Commission, which is responsible for many different kinds of tasks, for instance, examining the applications for exploration. Prospecting, i.e., a general search for sea-bed resources, requires only notification to the ISA. In addition, a written notification must be provided confirming that the environmental protection rules of the regime have been observed. No exclusive rights over a certain area are created through the notification. The situation is different with regard to the next phases of the process: exploration and exploitation. A specific authorization must be requested from the ISA, and, if given, exclusive rights are obtained by the company in question. For an overview, see Churchill and Lowe 1999, pp. 248-252.

two sites of equal estimated commercial value, one of which must be reserved for the exploitation of the Enterprise of the Authority for a certain period of time. In all cases, the Legal and Technical Commission of the ISA is the first body to examine the proposed plan. If the Commission recommends approval to the Council, which decides these issues, the plan is, as a rule, approved unless specific grounds are adduced for rejecting it.[1034] The applications must be accompanied by an assessment of the potential environmental impact of the proposed activities:

> An application for approval of a plan of work shall be accompanied by an assessment of the potential environmental impacts of the proposed activities and by a description of a programme for oceanographic and baseline environmental studies in accordance with the rules, regulations and procedures adopted by the Authority.[1035]

Clearly, these assessments must be of a wide scope since Article 145 requires preventive measures with regard to all areas of the marine environment, both international and national. Since the exploitation of the deep sea-bed has thus far been mainly confined to polymetallic nodules[1036], the Assembly of the ISA has approved 'Regulations on Prospecting and Exploration for Polymetallic Nodules in the Area'[1037], which contain rules on EIA as well as environmental protection.[1038]

[1034] The beginning of paragraph 5 (a), Section 1 of the 1994 Implementation Agreement, reads: 'An application for approval of a plan of work for exploration shall be considered by the Council following the receipt of a recommendation on the application from the Legal and Technical Commission'.

[1035] Ibid., Paragraph 7, Section 1.

[1036] These have been defined in Article 3 (d) of the Regulations referred to in the next footnote: 'any deposit or accretion of nodules, on or just below the surface of the deep seabed, which contain manganese, nickel, cobalt and copper'.

[1037] Decision of the Assembly relating to the regulations on prospecting and exploration for polymetallic nodules in the Area (ISBA/6/A/18). The document can be found on the home page of the ISA at http://www.isa.org.jm/en/default.htm [15.2.2001].

[1038] Regulation 18 and Part V of the Regulations. See also the Report of the Deep-Seabed Polymetallic Nodule Exploration (20 November 2000). Development of Environmental Guidelines (ISA 99/02). Part 3 contains draft guidelines for the EIA procedure: 'Chapter 9, Guidelines for the Assessment of the Environmental Impacts from the Exploration for Polymetallic Nodules in the Area'. Information obtained from the home page of the ISA at http://www.isa.org.jm/en/default.htm [14.2.2001].

Second, the deep sea-bed regime of the LOSC ensures that not only the ISA but also the states parties are obligated to protect the environment from activities taking place in the Area. According to Article 209 (2), states parties are required to adopt regulations to prevent pollution of the marine environment from activities in the Area undertaken by a state. These regulations must be as strict as the ones adopted by the ISA.[1039] States which are not yet parties to the LOSC regime are not bound by these regulations.[1040] Of the Arctic states, Canada and Denmark have not yet ratified the LOSC, and the United States has not even signed it. These states are still bound by the principle of due diligence and directed by the goal of sustainable development.

[1039] Article 209 (2) of the LOSC reads: 'Subject to the relevant provisions of this section, States shall adopt laws and regulations to prevent, reduce and control pollution of the marine environment from activities in the Area undertaken by vessels, installations, structures and other devices flying their flag or of their registry or operating under their authority, as the case may be. The requirements of such laws and regulations shall be no less effective than the international rules, regulations and procedures referred to in paragraph 1'. For an account of the negotiating history of this Article, see Hakapää 1981, pp. 124-125.

[1040] It can be argued that since by customary law the Area is nowadays part of the common heritage of mankind, states exploiting its resources unilaterally are violating these basic principles. On the other hand, it can also be argued that certain states, e.g., the United States, have consistently objected to the formation of the deep sea-bed regime, and can thus be considered as persistent objectors and not obligated by this regime. Brownlie argues (1998, pp. 253-254): 'the dissident states, such as the United States, might be said to have the status of persistent objectors, having opposed the new customary rule while it was in the process of formation; but to claim this status would involve conceding the existence of the new rule'. The applicability of these general principles is not, however, dependent on whether a state is exercising legal or illegal rights: the principle of due diligence obligates them irrespective of this question, even where the state is perceived as breaching the law. Only in war time does the principle of due diligence lose its applicability, with different sets of environmental protection standards becoming applicable. In the law of war, or as it is nowadays called 'international humanitarian law', there are certain principles and rules that protect the environment, for instance, rules prohibiting the use of environmental modification techniques. See the 1977 United Nations Convention on the Prohibition of Military or

7. Conclusion on the EIA Norms in the Arctic

It will be useful here to make a distinction between the goal of sustainable development and the principles of due diligence and equitable utilization on the one hand and applicable international treaties on the other. Where the former group is concerned, international law lays down a strict set of EIA norms, which also apply in the Arctic. The goal of sustainable development directs the Arctic states and the ISA to conduct EIA procedures and provide for prior licencing procedures with regard to all planned environmentally harmful stationary activities in the Arctic, as well as to enhance public participation in general and the participation of indigenous peoples in particular. The principle of due diligence requires the Arctic states and the ISA to provide for prior notification to the potentially affected state, conduct transboundary EIA and consult the potentially affected state in situations of potential transboundary pollution in the Arctic. The principle of equitable utilization imposes the same duties of notification, transboundary EIA and prior consultations on riparian states of transboundary Arctic watercourses.

International legal subjects are not obligated to implement these principles through international treaties, but it is obviously very important that they attempt to do so. Without any treaties on the subject, the likelihood that a state will observe general standards decreases. International treaties serve to operationalize these standards, and, as has been shown above, some aspects of the standards have been quite well operationalized in this way. An increasing number of treaties provides for prior procedures, which formerly were seen as belonging to the exclusive jurisdiction of states. The ISA procedures are also well advanced in observing the goal of sustainable development and the requirements of the principle of due diligence.

Any Other Hostile Use of Environmental Modification Techniques, Documents on the Laws of War, pp. 377-386.

Similarly, quite a number of agreements provide for EIA norms related to transboundary watercourses in the Arctic. For example, the 1992 Transboundary Watercourses Convention sets out an obligation for states in the Eurasian region to amend existing agreements on watercourses.

Increasingly, situations of potential transboundary pollution between states are governed by prior transboundary procedures of high quality. The most important rules here are those created by the two UN ECE Conventions, the 1991 Espoo Convention and the 1992 Industrial Accidents Convention. These replace the older treaties in their own field of application, but leave the earlier treaties to operate in some situations of potential transboundary pollution from stationary activities in the Arctic. The same Conventions are also very important where there is a risk of potential transboundary pollution from offshore hydrocarbon exploitation in the Arctic waters. Yet, of all the forms of potential transboundary pollution, only inter-state pollution is governed by international treaties. Pollution spreading from the areas under the jurisdiction of states to international areas is problematic because there is only one treaty provision regulating such situations.

The goal of sustainable development and the principles of due diligence and equitable utilization do not make any distinctions between categories of activities as long as they are environmentally harmful. The most problematic of these activities are military activities, since they have almost always been explicitly or implicitly excluded from the scope of the applicable treaties. As is evident from the treaties studied, military activities are still largely excluded from the area of environmental protection. Here, general standards are important despite the problems related to their implementation and application,[1041] for they provide at least some requirements for the planning phase of the implementation of a stationary military activity. It seems, however, that a more useful way to combat the environmental hazards presented by military activities, in particular those in Northwest Russia, is financial and technological assistance[1042] as provided, e.g., through the Arctic Military Cooperation (AMEC)[1043] and the EC's Northern Dimension programs[1044].

There are also some prospects for universal treaty regulation that would extend its influence into the Arctic. The 1997 Watercourses Convention, the end-product of an ILC project, provides the possibility of a universal treaty to govern environmental protection of transboundary watercourses, including the issue of prevention of environmental impact. Another ILC project has adopted draft articles entitled 'Prevention of Transboundary Damage From Hazardous Activities', and the Special Rapporteur has recommended that a framework treaty be negotiated on this basis. It seems, however, that the success of both initiatives as international treaties is likely to

---

1041 See generally in Finnish Koivurova 2000, pp. 470-472.

1042 For an analysis, see Stokke 1994, pp. 145-158; Ojala 1997, pp. 153-158; and, in Finnish, Heininen 1999, pp. 270-274.

1043 An important cooperation program is the Arctic Military Cooperation Program (AMEC) between the United States, Norway and the Russian Federation. As stated on the home page of the United States Environmental Protection Agency (at http://www.epa.gov/oia/amec.htm [21.4.2001]): 'AMEC Projects 1.1 and 1.1-1 address specific environmental issues associated with the removal and storage of spent nuclear fuel from nuclear submarines being decommissioned and dismantled in Northwest Russia'.

1044 The EC's Northern Dimension Program has also paid attention to the problem of cleaning the nuclear reactors which have been in military use in northwestern Russia. In a conference organized in Helsinki, Finland, on 9 March 2001, various international financing institutions and the Commission of the EC made long-term plans to combat this problem. Information obtained from the Finnish newspaper, Lapin Kansa (10.3.2001). The EC's Northern Dimension Program is presented well on the home page of the Northern Dimension (at http://europa.eu.int/comm/external_relations/north_dim/index.htm [12.3.2001]): 'the northern dimension in the external and cross-border policies of the European Union covers the Baltic Sea region, Arctic Sea region and North West Russia. It addresses the specific challenges of those regions and aims to increase cooperation between the EU member states, the EU applicant countries and Russia. The Northern Dimension is implemented within the framework of the Europe Agreements with the Baltic States, the Partnership and Cooperation Agreement with Russia and the European Economic Area regulations. The areas for cooperation under the Northern Dimension are the environment, nuclear safety, energy cooperation, Kaliningrad, infrastructure, business cooperation, Justice and Home Affairs, social development and others. The Northern Dimension operates through the EU's financial instruments available for the region: Phare, Tacis and Interreg'. One Tacis project completed in 2000, aimed at assessing potential sites for the disposal of low- and medium-level radioactive waste in the Murmansk and Archangelsk Oblasts and the Novaya Zemlja

remain limited, although they are still important from the viewpoint of the development of customary international law.[1045]

From a chronological viewpoint, it can be observed that the 1990s provided new opportunities for Arctic-wide perspectives in comparison to earlier region-specific prior transboundary procedures that operated mainly between the Nordic states on the one hand and between the United States and Canada on the other. During the 1990s, important treaties were negotiated which were open to all eight Arctic states. It is the regional cooperation developed in the International Atomic Energy Agency (IAEA) but especially within the United Nations Economic Commission for Europe (UN ECE), that has contributed most to the development of Arctic-wide perspectives in the field of EIA norms. The prior transboundary procedures stipulated by the 1991 Espoo Convention and the 1992 Industrial Accidents Convention, as well as the preventive approach provided by the 1992 Transboundary Watercourses Convention, have made it possible to have Arctic-wide regulation operable, if not at present at least in the near future. In addition, the 1998 Aarhus Convention, when it comes into force, will provide the civic society with a new kind of opportunity to participate in prior procedures applicable in the Arctic. Unfortunately from the viewpoint of indigenous peoples of the Arctic, not even this Convention accords them a special status in prior procedures.[1046]

---

area in Northwest Russia. Information obtained from the home page of the Northern Dimension at http://europa.eu.int/comm/external_relations/north_dim/examples.htm [12.3.2001].

1045 Interestingly enough, the second meeting of the parties to the Espoo Convention decided to open the possibility for all member states of the United Nations to become a party to the instrument (new Article 17 (3) of the Convention). For further information, see infra chapter V (2.1.). For a general account of the influence of treaties on the development of custom, see Jennings and Watts 1992, Oppenheim's International Law 1992, pp. 33-36.

1046 See *supra* chapter IV (3.3.2.).

8. EIA Norms in Action in the Arctic: the Case of the Vuotos Reservoir

The debate on the construction of the Vuotos Reservoir has been a protracted one. The historical basis for the harnessing of the Kemi River for hydropower generation was an Act passed in 1952 permitting the systematic exploitation of the hydropower resources of the river and establishing a state-owned company for the purpose, Kemijoki Ltd.[1047] The initial plans for the Vuotos reservoir, or as it was known then 'the Kemihaara project', by Kemijoki Ltd. were completed in the 1960s, with the reservoir to be located in the area of the municipalities of Savukoski, Salla and most extensively Pelkosenniemi. The project was suspended for environmental reasons in 1974, and in 1982 the Finnish government made an informal decision not to proceed with the construction of the reservoir. The project was resumed as the 'Vuotos project' by the middle of the 1980s and the preparatory work was completed when Kemijoki Ltd. applied to the Water Court of Northern Finland on 25 September 1992 for a permit to construct the Vuotos reservoir and dam.[1048] (See figure 7).

[1047] The shareholders of Kemijoki Ltd. are presented on the home page of the company at http://www.kemijoki.fi/kemijoki_eng/omistuse.htm [8.3.2001].

[1048] For a review of the history of the Vuotos project in English by an employee of Kemijoki Ltd., see Hellsten 1994, pp. 179-184. For a critical historical account in Finnish, see Mustavuori 1995, pp. 295-296. For an in-depth analysis, see Suopajärvi 2001, pp. 30-55. Another project proposed to be implemented near the Finnish Arctic was the enlargement of the plants of Outokumpu Polarit Ltd. and Outokumpu Chrome Ltd. in Tornio, Finland. As the site is very near the Swedish border, the Swedish governmental authorities and the Swedish public were involved in the EIA procedure. Information can be found on the home page of the Finnish Ministry of the Environment at http://www.vyh.fi/poltavo/yva/arkisto/lap/okumpu/index.htm [20.3.2001]. For an analysis of the transboundary EIA procedure in this case, see Korhonen 1998.

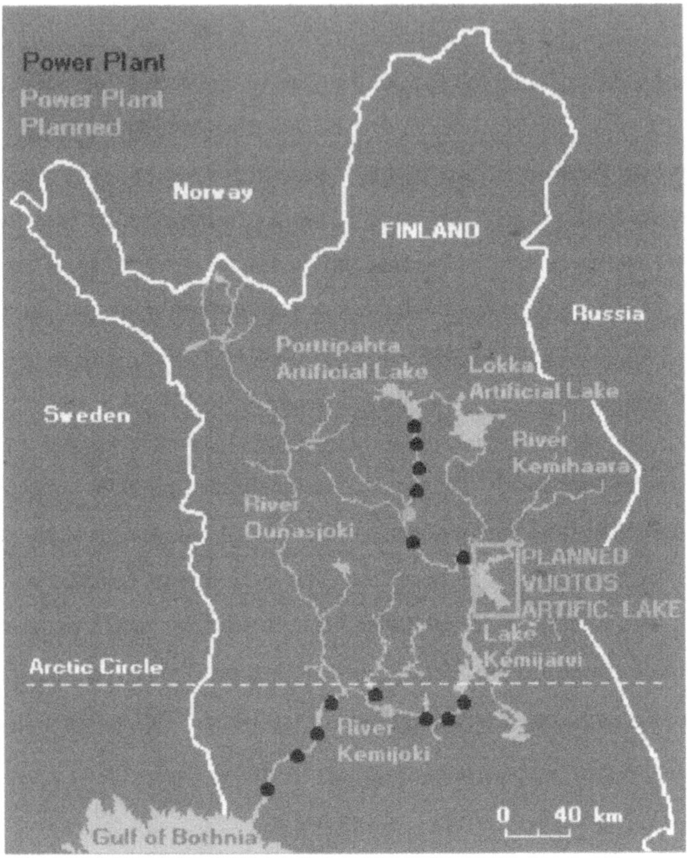

Figure 7 (source: home page of the Kemijoki Ltd. at http://www.kemijoki.fi/vuotos_eng/vsiie.htm)

As presented on the home page of Kemijoki Ltd., the purpose of the reservoir is to function as a water reserve for all the existing and future power plants on the main course of the Kemi River (see figure 7). The Vuotos reservoir would 'increase power generation of all power plants located below it. Furthermore, it would make economical sense to build additional machinery at the existing plants and two new power plants. The addition to the energy production would be about 560 GWh/a'[1049]. The size of the reservoir 'would be about 0.25 % of the area of the province of Lapland'[1050].

The main focus here is on how the potential transboundary impacts of the construction of the Vuotos Reservoir have been dealt with in the Finnish national procedure.[1051] To illustrate how this complicated procedure has advanced, it is important to distinguish two procedures: the procedure stipulated in the Finnish Water Act and that coordinated by the Finnish Ministry of the Environment (hereinafter 'the FME'). The traditional procedure, which relied on the Finnish Water Act and the Nordic Environment Protection Convention (NEPC), was considered deficient by the FME, especially in light of the requirements of the Espoo Convention

As Finland's national EIA legislation was not yet in force when Kemijoki Ltd. applied for the permit from the Water Court of Northern Finland (hereinafter 'the Water Court'), the Finnish Water Act provided the applicable internal permit procedure for Vuotos.[1052] The procedure was the following. Kemijoki Ltd., the

---

1049 Information obtained from the home page of Kemijoki Ltd. at http://www.kemijoki.fi/vuotos_eng/vuotose.htm [8.3.2001].

1050 Information obtained from the home page of Kemijoki Ltd. at http://www.kemijoki.fi/vuotos_eng/vuotose.htm [8.3.2001].

1051 Some aspects of this chapter rely on an article by the author published in the Nordic Journal of International Law, see Koivurova 1997 I, pp. 516-522.

1052 Kemijoki Ltd. proposed that a voluntary environmental impact assessment of the Vuotos project should be carried out. It even had an environmental impact statement drawn up by the consulting firm EKONO. Subsequently, the Water and Environment District of Lapland proposed to the Finnish Water and Environment Council that Vuotos could be handled through a voluntary EIA (27.8.1993). It also proposed that an ad hoc coordination authority be formed, inasmuch as the

proponent of the activity, filed its application with the Water Court on 25 September 1992 for constructing the Vuotos reservoir and dam (hereinafter 'Vuotos').[1053] According to the Water Act, all large-scale projects have to be studied by the implementation of inspection proceedings which determine the impact of the project and lead to the drawing up of a proposal for consideration by the Water Court. To this end, the Water Court appointed a panel consisting of a proceedings engineer and two trustees for the Vuotos on 11 May 1993.[1054]

Since the Finnish Water Act and the NEPC require that a procedure to be carried out by the origin and the potentially affected states be commenced only after the impact studies have been completed, the FME proposed that Finland apply the Espoo Convention although it was not binding on Finland at the time. Indeed, when signing the Convention, the contracting states had adopted a resolution affirming their 'resolve to strive for the entry into force of the Convention as soon as possible and to seek to implement the Convention to the maximum extent possible pending its entry into force'[1055].

The FME initiated its own procedure by first discussing informally with the Water Court how the potential notification to Sweden should be organized. The Water Court, the competent authority to notify by virtue of the NEPC, was of the opinion that in the early phase of the procedure, the FME could notify Sweden with reference

internal EIA legislation was still in preparation. This suggestion was rejected by the Finnish Water and Environment Council after consultation with the Finnish Ministry of the Environment (1 July 1993). (Author's personal copy of letters exchanged by the Finnish Water and Environment Council and Water and Environment District of Lapland).

1053 Information obtained from the Water Court of Northern Finland (via telephone conversation 2.12.1996). In the application, there was no reference to any significant transboundary impact on Swedish maritime areas. (Interview with the head of the legal department and the director of environmental affairs for Kemijoki Ltd. 2.12.1996).

1054 An additional proceedings engineer was appointed for the Vuotos procedure because of its large scale. Information obtained from the Water Court (via telephone conversation 16.12.1996).

1055 Paragraph 1 of the resolution of the signatories of the Espoo Convention (ECE/ENVWA/19).

to the Espoo Convention and the resolution of the signatories.[1056] According to the EIA Act, which entered into force on 1 September 1994,[1057] the FME must consult with the Finnish Ministry for Foreign Affairs,[1058] and this consultation took place before the FME notified Sweden about the potential transboundary impact of the Vuotos. On the basis of this informal consultation, it was agreed that the Espoo Convention should be invoked provided that the obligations it imposed did not exceed the level of those set forth in a valid international treaty, the NEPC.[1059]

Notification to Sweden took place on 22 June 1995, long before the inspection proceedings were able to produce the results of impact studies for the Water Court.[1060] In the notification, the FME informed the Swedish point of contact, the Swedish Environmental Protection Agency (hereinafter 'the SEPA'), that the construction of the Vuotos might result in an overload of phosphorus in the maritime area of Sweden in the Gulf of Bothnia.[1061] While the FME referred primarily to article 3 of the Espoo Convention and the resolution of the signatories as the legal bases of the notification, it also invoked the NEPC.[1062] The SEPA was asked to respond before 1 September 1995 and did so on 30 August 1995, requiring that environmental studies be done covering the possible impact on Swedish maritime areas.[1063] The

[1056] Information obtained from the FME (via telephone conversation 16.12.1996).
[1057] Since Finland was a party to the EEA Agreement, which entered into force on 1 January 1994, it was obligated to implement the original EIA Directive 85/337/EEC as part of the EEA requirements. For further information, see *supra* chapter IV (4.1.1.).
[1058] Section 14 (3) of the EIA Act reads: 'The Ministry of the Environment shall request a Ministry for Foreign Affairs opinion on the matter'.
[1059] Information obtained from the FME (via telephone conversation 16.12.1996).
[1060] Author's personal copy received from the FME.
[1061] Author's personal copy from a letter from the FME to the SEPA.
[1062] The bases of the notification are described as follows: 'The notification is carried out in accordance with Article 3 of the Convention on Environmental Impact Assessment in a Transboundary Context adopted in the United Nations Economic Commission for Europe (Convention E/EEC/1250) and in accordance with declaration on the application of the Convention before it enters into force, signed on 25 February 1991, and in accordance with the Nordic Environment Protection Convention (19.2.1974)'. Author's translation from the Swedish.
[1063] Author's personal copy received from the FME.

FME responded to Swedish concerns by collecting expert opinions on the matters identified.[1064]

The inspection proceedings stipulated by the Water Act included several studies of the impact of the activity, with the panel drawing conclusions as to whether a permit should be granted for the activity and under what conditions.[1065] This occurred on 24 May 1996, when the persons responsible submitted their proposal to the Water Court. They concluded that no significant transboundary impact of the construction of the Vuotos on the maritime area of Sweden was likely[1066] and recommended to the Water Court that the Vuotos be constructed. After the inspection proceedings, various private persons and governmental authorities had an opportunity to present their comments on and objections to the conclusions and proposals of the inspection proceedings. The deadline for submitting these comments expired on 11 September 1996.[1067] Because of the action taken by the FME also Swedish authorities and private legal subjects submitted their comments and objections to the Water Court.

The Water Court faced a difficult task. According to Article 5 of the NEPC, as the examining authority, the Water Court was obligated to inform the supervisory authority of the other Nordic states if the project in question 'may entail nuisance of

[1064] The FME delivered the application of Kemijoki Ltd. and the conclusions arrived at by the inspection panel to the SEPA (10.6.1996) (Information obtained from the FME via telephone conversation 16.12.1996). According to the decision of the Water Court, the information provided by the Swedish authorities to the FME was delivered to the Water Court on 13 September 1996. In their deposition, Swedish authorities requested further information on the project, which was delivered on 25 June 1997 to the SEPA by the FME. The SEPA specified its comments in a deposition delivered to the Ministry of the Environment on 14 September 1997 and referred to the application of the Nordic Environment Protection Convention in the procedure at the Water Court. Finally, the FME delivered the information collected from the Swedish authorities to the Water Court on 2 October 1997.

[1065] For an explanation of the procedure in the case of the Vuotos project, see Hellsten 1994, pp. 179-187.

[1066] Information obtained from the Environment Centre of Lapland (via telephone conversation 16.12.1996).

[1067] Information obtained from the Water Court (via telephone conservation 16.12.1996).

significance in another Contracting State'[1068]. However, the inspection proceedings indicated that the construction would cause no significant transboundary impact. If the procedure stipulated by the valid standards at the time - the Water Act and the NEPC - had been followed, the Water Court would not even have had to consider whether the Swedes had to be involved in the procedure; since no nuisance of significance was found by the inspection proceedings, no need to notify the Swedish authorities existed. However, the measures taken by the FME placed the Court in a new position, for once various parties from Sweden had presented their objections, the Court was bound to take them into consideration in making its decision.

This did not change the decision of the Water Court, however, since it came to the conclusion, on 29 February 2000, that no significant transboundary impact would follow from the construction of the Vuotos and thus the applicable treaty, the NEPC, did not in fact apply to the case. The Court made a decision not to study the demands by the Swedish authorities and private legal subjects:

> According to the NEPC...if the activity may entail a nuisance of significance in another Contracting State, the examining authority shall...inform the supervisory authority of the other state of the proposed activity. According to the inspection proceedings, the effects of the project on the area of Sweden would be minor, an evaluation which was confirmed by the Ministry of the Environment in an oral hearing. Because of the insignificance of the effects of the activity on the Swedish territory, the Water Court, as the examining authority, has not informed the supervisory authority of Sweden.[1069]

[1068] Article 5 of the NEPC.

[1069] Author's own translation. This part of the decision can be found on the home page of the FME at http://www.vyh.fi/hoito/vesirak/vuotos/psvrat01.htm [21.4.2001]. The Court also stated: 'The Ministry of the Environment has on 22 June 1995 notified the Swedish government as stipulated by the Convention on Environmental Impact Assessment in a Transboundary Context, the so-called Espoo Convention. The purpose of the notification was to inform the Swedish authorities about the phase of the planning process and reserve a possibility to participate in the pending licencing procedure before the permit was given. The Ministry of the Environment has on 13 September 1996 delivered to the Water Court the information provided by the Swedish authorities. In their deposition, Swedish authorities requested further information on the project, information which was delivered on 25 June 1997 to Naturvårdsverket by the Ministry of the Environment.

The potential effects on the Gulf of Bothnia were not taken into consideration in the licencing decision, because the Water Court based its decision on the conclusion of the inspection proceedings and, significantly, also on an alleged statement by the FME in an oral hearing. It was thus no wonder that the complainants to the next stage of administrative judicial review, the Administrative Court of Vaasa, included the SEPA.

The Water Court gave the Kemijoki Ltd. a licence to construct the Vuotos reservoir and dam in accordance with the proponent's application. Kemijoki Ltd. had also applied for extensive rights to go ahead with necessary work related to the construction of the reservoir. The Water Court permitted the company to clear the forests under its ownership but provided that 'the permit to initiate work does not apply to the clearing of the shores of lakes'[1070]. This right was revoked by the Administrative Court of Vaasa (hereinafter 'the Administrative Court'), to which several parties had appealed against the Water Court's decision, in an initial ruling pending its final decision on the matter. The Court held an interesting public hearing in Pelkosenniemi from 6 to 8 February 2001, which was attended by the representatives of the SEPA, Norbottens provincial government and private legal subjects.[1071] All of the private and public actors from Sweden were of the opinion that the Vuotos project would have a variety of negative impacts.

Importantly, the FME argued that the Water Court had misunderstood its presentation when the Court concluded that the FME had determined that no significant impact on

---

Naturvårdsverket has specified its comments in a deposition delivered to the Ministry of the Environment on 14 September 1997, and referred to the application of the Nordic Environment Protection Convention in the procedure at the Water Court. The Ministry of the Environment has on 2 October 1997 delivered the above-mentioned deposition to the Water Court' (author's own translation).

[1070] Ibid. Author's own translation.

[1071] The decision by the Administrative Court, pp. 35-43. The decision was received from the Administrative Court by the author. Other private legal subjects were the Nature Protection Associations of Norrbottens province and Haparanda.

the area of Sweden would result.[1072] Another interesting aspect was that Kemijoki Ltd. wanted to argue explicitly that Sweden and the Swedish authorities could not be considered injured parties but that they could only be heard during the process. Both the FME and SEPA demanded that Swedish authorities be given the status of injured parties in the process. The chairman of the Administrative Court stated that, according to the law, the Swedish authorities could be given their say but that their status as injured parties would be decided when the matter was dealt with by the Court. He also stated that the Swedes and Kemijoki Ltd. could challenge the decision by making a complaint to the Supreme Administrative Court of Finland.[1073]

Quite unexpectedly, the Administrative Court, on 14 June 2001, revoked the licencing decision made by the Water Court and rejected the permit application by Kemijoki Ltd. on the basis of Article 2 (5) of the Finnish Water Law.[1074] Since Kemijoki Ltd. has appealed against this decision to the Supreme Administrative Court, it will be useful to examine the parts of the decision by the Administrative Court which pertain to the transboundary process with Sweden. It is anticipated that the Supreme Administrative Court will hand down its decision by the end of the year 2002.[1075]

1072 Information obtained from the FME via e-mail (8.3.2001).

1073 See the decision by the Administrative Court, pp. 35-43. See also Finnish newspaper, Lapin Kansa, p. 8 (7.2.2001).

1074 The decision of the Administrative Court, pp. 55-58. Copy of the decision obtained from the Administrative Court.

1075 Information obtained from the Supreme Administrative Court via telephone conversation (21.8.2001). The construction of the Vuotos has caused some problems also from the viewpoint of EC law. The Commission of the EC started proceedings (on 26 August 2000) against Finland in the European Court of Justice for Finland's failure to classify special protection areas under the Wild Birds Directive. The Commission's position is that Finland should have designated the mires of Kemihaara as one such special protection area. See the Europa home page at http://curia.eu.int/en/act/act00/0028en.htm [22.4.2001] (Case C-240/00). Moreover, the Commission decided to send a letter of formal notice (on 13 September 2000) to Finland for failure to protect the mires of Kemihaara. The Commission points out that the permit given to the Vuotos project threatens the bird species listed in Annex I to the Wild Birds Directive (1979 Directive on the Conservation of Wild Birds). Information obtained from the home page of the Europa site, press releases at http://europa.eu.int/rapid/start/cgi/guesten.ksh?p_action.gettxt=gt&doc=IP/00/1008|0|AGED&lg=E

Kemijoki Ltd. claimed that SEPA cannot be considered an injured party in the procedure and that the latter's appeal to the Administrative Court should therefore not be considered.[1076] However, the Administrative Court opined that the construction of the Vuotos Reservoir may entail damaging effects on the Swedish side of the Gulf of Bothnia, and, accordingly, SEPA, as a supervisory authority in the meaning of Article 4 of the NEPC, could be considered an injured party in the procedure and thus have a right to make an appeal against the licencing decision of the Water Court.[1077] In its appeal, SEPA states that the permit can only be given if heavy metal and phosphorus emissions can be controlled through the permit conditions in such a way that no danger is caused to Swedish territory and the Swedish environment. SEPA also notes that if the permit is given, a separate condition for the permit must be that Kemijoki Ltd. is to regularly deliver monitoring information to the government of the Norrbottens District.[1078]

N [22.4.2001]. Recently, the Commission decided to give a second written warning, a reasoned opinion (on 4 April 2001), to Finland on the same issue; see at http://europa.eu.int/rapid/start/cgi/guesten.ksh?p_action.gettxt=gt&doc=IP/01/505|0|RAPID&lg=EN [22.4.2001]. According to Article 226 of the Treaty Establishing the European Community, the Commission must give the state concerned 'the opportunity to submit its observations' after a reasoned opinion. If the state does not comply, the Commission can 'bring the matter before the Court of Justice'.

1076 Kemijoki Ltd. also claimed that the appeal of Statens Naturvårdsverket should not be examined because its complaint was received by the Water Court after the final date for filing such complaints according to Finnish procedure. The Administrative Court rejected this argument because at least some parts of the complaint were received by the Water Court. Decision of the Administrative Court, pp. 49-51.

1077 Kemijoki Ltd. also disputed the right of appeal from the local district and county representatives on the Swedish side on the basis that these are not supervisory authorities in the meaning of Article 4 of the NEPC. The Administrative Court, however, decided that the arguments of these agencies can be taken into account as addenda to the appeal of the Statens Naturvårdsverket. Per Lundbäck, a fisherman from Sweden, was also considered an injured party by the Administrative Court although his appeal arrived too late to the Water Court. This was due to the fact that he was not given appropriate information to safeguard his rights in the procedure. On the other hand, the Administrative Court did not consider the association 'Länsförbundet Naturskyddsföreningen i Norrbottens Län' as an injured party whose appeal also had arrived too late. Decision of the Administrative Court, pp. 51-52.

1078 Ibid., p. 18.

8.1. Analysis of the Vuotos Process

Before the decision of the Water Court, the problems in the Vuotos process were not so much international as domestic. From the viewpoint of international law and Sweden, Finland performed its duties in an exemplary manner. Surely international law does not prevent the premature application of the Espoo Convention if this discharges Finland's international duties. Besides, it can be argued that the NEPC, a valid international treaty, provided a legal basis for the FME's position. The NEPC requires states parties, more precisely the examining authorities, to 'send as soon as possible a copy of the documents of the case to the supervisory authority of the other State, and afford it the opportunity of giving its opinion'[1079]. Clearly the actions of the FME, also justified in that they were in accordance with the NEPC, were also covered by the NEPC. It seems only reasonable to assume that the Water Court and the FME decided to fulfill their obligations according to the NEPC and transfer the matter to the Ministry in order for it to act 'as soon as possible'[1080].

However, the internal competence to execute the obligations set forth in the NEPC was transferred to an organ other than that provided for by the NEPC. The NEPC accords this competence to the examining authority, in this case the Water Court, whereas the Espoo Convention accords it to the points of contact, i.e., in Finland to the FME. Kemijoki Ltd. argued that this transfer of internal competence was illegal as the Espoo Convention was not yet in force.[1081] But is the coming into force of the Espoo Convention really that decisive in this context? As international law seems to commend Finland for the way it performed its international duties, attention turns to

---

[1079] Article 5 of the NEPC.
[1080] Ibid.
[1081] Interview with the head of the legal department and the director of environmental affairs for Kemijoki Ltd. (2.12.1996).

Finnish domestic legislation. The pivotal question is: who has the internal competence to notify Sweden according to the NEPC? According to the text of the NEPC itself, this authority rests with the Water Court, but the Water Court saw fit to have the FME handle the international process, for which the Ministry clearly had better resources and know-how. This functional justification, however, does not constitute a legal basis for the FME's exercising authority which, according to the NEPC, belongs to the Water Court.

Section 14 of the Act on Environmental Impact Assessment Procedure would grant the competence to notify the other state to the FME. However, section 27 of the same Act restricts the application of the Act to those activities initiated after 14 January 1994, and, as noted above, the Vuotos was not among these. But does this exclusion apply to international procedure as well? It seems that although the Finnish EIA Act is not applicable to the Vuotos procedure internally, it is internationally. This is clearly shown by the wording of section 14:

> The provisions of this Act shall also apply...if enforcement of an international agreement binding on Finland requires an environmental impact assessment procedure to be arranged in cooperation with another state in the case of a project being carried out in Finland. The Ministry of the Environment shall see to notification and consultation functions connected with environmental impact assessment.

It is essential to realize that the Act accords the FME competence with regard to not only the Espoo Convention but other international EIA conventions as well.[1082] Thus,

---

[1082] In the Government Bill for an Act on Environmental Impact Assessment Procedure, two international conventions are mentioned in addition to the Espoo Convention: 'These kinds of conventions would be those referring to the environmental impact assessment procedure. The central convention is the Espoo Convention, but the Convention on the Transboundary Effects of Industrial Accidents and the Convention on the Protection of the Baltic Sea may also be regarded as examples of such conventions' (author's own translation). The argument advanced here is that as a Convention dealing with the international environmental impact assessment procedure, the NEPC is also to be included. To be sure, these two conventions are only examples, as is evident from the text of the Government bill.

after the coming into force of the Act, all those international treaties which require 'an environmental impact assessment procedure to be arranged in cooperation with another state'[1083] come within the competence of the FME. The NEPC is such a treaty. Consequently, the internal competence of notifying Sweden by virtue of the NEPC was in the hands of both the FME and the Water Court. As the two agreed in the beginning of the procedure, it was the FME which was to act in the first phases of the procedure with Sweden.

The international problems in the Vuotos process appeared when the decision was made by the Water Court. An interesting aspect of the licencing decision of the Water Court was that it did not study the comments made by Swedish authorities and private legal subjects since: 'According to the inspection proceedings, the effects of the project on the area of Sweden would be minor, an evaluation which has been confirmed by the Ministry of the Environment in an oral hearing'[1084]. Thus, in principle, the Water Court acted in accordance with the NEPC since it determined that no significant impacts on the area of Sweden would follow from the construction of the Vuotos and thus no notification on the basis of Article 5 of the NEPC was necessary. Also the private legal subjects from Sweden, who are guaranteed the same rights as their Finnish counterparts by Article 2 of the NEPC, could not be considered potential injured parties. However, the FME argues that it was misunderstood by the Water Court in an oral hearing and now maintains that the potential impact on Swedish territory may be significant. The same argument has been presented by the Swedish authorities and private parties.

The Administrative Court approached the matter from a different perspective than the Water Court. Since the Swedish parties were already involved in the procedure and certain evaluations existed about the potential impacts on the Swedish area, there was

---

[1083] Section 14 (1) of the Act on Environmental Impact Assessment Procedure.
[1084] The decision of the Water Court. See supra footnote 1069.

no need to initiate the procedure on the basis of Article 5 of the NEPC, i.e., to determine whether the effects on the Swedish area may be significant. The Court studied whether SEPA as a supervisory authority in the meaning of the NEPC can be considered as having the right to participate in the licencing procedure as an injured party. Article 4 of the NEPC stipulates that the supervisory authority has a right to 'appeal against the decision of the Court or the Administrative Authority'[1085] in accordance with the rules of the state of origin. The Administrative Court applied these rules to the Swedish parties and decided that SEPA can be considered an injured party and thus having a right to appeal against the decision. The northern local authorities of Sweden were not seen as independent injured parties but their statements could be studied as part of SEPA's claims. This is an important decision since even though the Administrative Court denied Kemijoki Ltd. a permit to build the Vuotos Reservoir and dam, the case will still be heard by the Supreme Administrative Court of Finland.

If the Supreme Administrative Court overturns the decision made by the Administrative Court and permits the construction of the Vuotos Reservoir and dam, it will in all likelihood take note of the statements made by SEPA. As discussed above, SEPA demanded and will in all likelihood demand in the Supreme Administrative Court that if the permit is granted, two conditions are to be stipulated with respect to the Swedish areas: the first is that heavy metal and phosphorus emissions are kept under control through the permit conditions and the second that monitoring arrangements be organized by Kemijoki Ltd. to supervise the state of the waters lying on the Swedish side. If the Supreme Administrative Court overturns the decision of the Administrative Court, it is likely that the statements made by SEPA will influence the final permit decision.

[1085] Article 4 (2) of the NEPC.

V TRANSBOUNDARY EIA IN THE ARCTIC

From the perspective of international law, the most important EIA norms are those that deal with the potential inter-state transboundary impacts of proposed large-scale stationary activities. Although international law already regulates certain aspects of how states' national environmental protection systems must operate, its main task, at least in the Arctic context, is still the resolution of transboundary problems. As was shown in the previous chapter, the most important environmental management procedures for such situations in the Arctic are the two UN ECE Conventions, the more significant being the Espoo Convention.[1086] Before studying more specifically the content of the Espoo Convention, it will be necessary to examine the development of the idea of a transboundary EIA procedure. Thereafter, the chapter studies scope and content of the Espoo Convention in detail, and, finally, discusses the future prospects for the Espoo Convention in the Arctic.

1. The Development of the Idea of a Transboundary EIA Procedure

It was the spread of national EIA procedures that made it possible to conclude legally relevant documents and international treaties on transboundary EIA. The concept of transboundary EIA is used here to denote those legally relevant documents and international treaties that try to integrate foreign impact and foreign actors into the operation of the national EIA procedure of the origin state in cases where transboundary impact is likely. States cannot do this unilaterally since their jurisdiction is confined to certain pieces of territory, and thus international treaties are the only means to do this.

In the previous chapter, the concept of prior transboundary procedures was used to refer to all treaty procedures carried out between states by which they obligate themselves to exchange information on the proposed activity. As examined, there exist many prior transboundary procedures that do not make any requirements of the national law of the states parties but only require that information is to be exchanged between the parties.[1087] The basic difference between prior transboundary procedures and 'transboundary EIA procedures' lies in the fact that the operation of transboundary EIA procedures is linked to the operation of national EIA procedures. The transboundary EIA treaties try to ensure that foreign impact and foreign actors are included in the operation of the national EIA procedure of the respective origin state. The quality of the information exchanged between the states parties thus increases because of the high quality of the information produced by the national EIA procedures in general.

Treaty procedures or legally relevant documents that try to organize a transboundary EIA procedure must take into account the basic fact that states have differing national EIA procedures. From the viewpoint of the functioning of a transboundary EIA procedure, the ideal situation would obviously be one in which all the contracting states had an identical EIA procedure, for the potentially affected state and its public would know when and how to participate in the EIA procedure taking place in accordance with the EIA procedure of the origin state. As will be discussed below, one of the main issues is how much the contracting states' EIA procedures can and must be harmonized in order to achieve smooth transboundary EIA procedure.[1088] It should be noted that there are many treaties that provide for transboundary EIA procedures but as part of a regime consisting of a wide variety of measures of

---

[1086] See *supra* chapter V (7.).
[1087] See *supra* chapter IV (4.1.).
[1088] See *infra* chapter V (2.5.1.).

international environmental protection.[1089] In order to study the special problems related to the transboundary EIA procedure, and especially the Espoo Convention, the examination is confined here only to those normative instruments that concentrate solely on transboundary EIA procedure.[1090]

The historical development of the idea of a transboundary EIA procedure is clearly connected to the idea of the EIA in general because a transboundary EIA procedure is only a means of extending the national EIA procedure to foreign impacts and foreign actors. As was mentioned, the first national EIA procedures appeared already at the end of 1960s and especially in the beginning of the 1970s.[1091] However, the first transboundary EIA procedure having an international importance, the EIA Directive on EC, was not introduced until 1985. This Directive was, however, preceded by an interesting resolution of the United States Senate in 1978 containing the outline of a general convention for a transboundary EIA procedure and urging other nations to participate in the negotiations for such a convention.[1092] Although the resolution did not lead to any treaty negotiations, it is useful to examine it here as a first outline of a convention on the matter.

[1089] See e.g., Convention on the Protection of the Marine Environment of the Baltic Sea Area of 1992 (Article 7) and 1991 Protocol to the Antarctic Treaty (Article 8 and Annex I). For an analysis, see Sands and Werksman 1995, pp. 187-198, Kuokkanen 1996, pp. 165-172 and Dzidzornu 2001, pp. 15-26. EIA procedures have also been used increasingly in international development projects. See Klein-Chesivoir 1990, pp. 517-551.

[1090] Although the 1974 Nordic Environment Protection Convention (NEPC) also contains some kind of transboundary EIA procedure, the transboundary EIA procedure is not based on the national EIA procedure but on the licencing procedure, and thus the procedure of the NEPC is excluded from the treatment here.

[1091] See *supra* chapter IV (1.).

[1092] Senate resolution 49 passed the Senate on July 21, 1978. It is reproduced in 4 I.L.M. 1082 (1978). In the following year, the Council of the OECD made a recommendation to the member governments on the 'Assessment of Projects with Significant Impacts on the Environment'. However, this recommendation only referred to 'environmental assessment procedures for actions that might have significant transboundary effects' (paragraph 8). Recommendation adopted on 8 May 1979 (C (79) 116). The recommendation can be found in the Basic Documents of International Environmental Law, Vol. 1, 1992, pp. 400-401.

The scope of the proposed convention was broad since all major projects that might reasonably be expected to have adverse effects on the environment of other states or of international areas were to be assessed, and the results of such 'international environmental assessments' were to be communicated to the potentially affected state as well as the United Nations Environment Program (UNEP), especially in cases of potential damage to the environment of international areas.[1093] In case the potentially affected state or, in the case of global commons, the UNEP so requested, the origin state was to consult these affected subjects 'with a view toward preventing or minimizing any potential adverse environmental consequences beyond its territory'[1094]. If the potentially affected state or UNEP so requested, the origin state was to 'refrain from initiating the project or activity for a period of not more than 90 days after submission of an international environmental assessment'[1095]. However, if the postponement of the project '[involved] serious risks to the safety of life or property or would otherwise be clearly infeasible', the origin state could go ahead with the project.[1096]

As indicated above, the 1985 EIA Directive placed more emphasis on harmonizing the national EIA procedures of the member states of the EC than on extending this procedure to foreign impacts and actors.[1097] Article 7 of the Directive provides only that if the origin state or the potentially affected state so wishes, the documents of the EIA study must be delivered to the affected state. The potentially affected state is only given a right to consultations. Upon closer scrutiny, then, it can be seen that the EIA Directive of 1985 does not fulfill the minimum requirements of the Espoo Convention since, for instance, there are no rules entitling the public of the potentially affected state to participate in the EIA procedure of the origin state or

---

1093 Article I of the resolution.
1094 Ibid., Article II (1).
1095 Ibid., Article II (2).
1096 Ibid.
1097 See *supra* chapter IV (3.3.1. and 4.1.1.).

rules requiring that the information obtained from the potentially affected state be taken into account in licencing procedures. Nevertheless, the Directive provided a first example of a procedure which at least connected the potentially affected state in some way to the EIA procedure of the origin state.

A more extensive legally relevant document was the 1987 Goals and Principles of Environmental Impact Assessment adopted by the UNEP.[1098] Like the Directive on EIA of 1985, this document placed much more emphasis on the harmonization of national EIA procedures than on the involvement of foreign actors and impacts in the procedure[1099], but it provided more refined rules for the extension of the EIA procedure to foreign impact and actors. First of all the states are generally required to conclude arrangements for cooperation not only with regard to potential transboundary impacts of an inter-state character but also, interestingly, with regard to potential pollution affecting 'areas beyond national jurisdiction'[1100]. Principle 12 of the Goals and Principles requires that if the national EIA procedure indicates that the environment of another state is 'likely to be significantly affected by a proposed activity'[1101], the origin state should notify the potentially affected state and 'transmit to the potentially affected State any relevant information from the EIA'[1102]. If an agreement is reached between the states, timely consultations should take place between them. The principal importance of the document is that it was used as a model in the negotiations for the Espoo Convention.[1103]

[1098] Reproduced in Basic Documents of International Environmental Law, Vol. 1, 1992, pp. 187-190.

[1099] This is evident from the fact that out of the 13 principles of the document, only three are concerned with transboundary extension of the EIA procedure (Principles 1, 11 and 12), whereas 11 principles regulate the harmonization of national EIA procedures.

[1100] Principle 11 of the Goals and Principles.

[1101] Ibid., Principle 12.

[1102] Ibid.

[1103] For an analysis of this connection, see Connelly 1999, p. 38.

The first full-fledged transboundary EIA procedure was not concluded until 1991, with the conclusion of the Espoo Convention. Since the Espoo Convention will be studied in detail below, it is enough to state here that the Convention provides basic rules on how to connect the foreign impact and foreign actors to the national EIA procedure of the origin state. The Convention only requires that the national EIA procedure of the origin state contain a procedure for public participation and that the EIS meet the requirements set out in Appendix II of the Convention. When the EC became a party to the Espoo Convention, it became necessary to amend the EIA Directive of 1985, which, as was shown above, did not fulfill the requirements set out in the Convention. The Council of Ministers adopted the amendment to the EIA Directive in 1997, and the member states were obligated to implement the Amendment Directive by the year 1999.[1104]

This Amendment Directive does fulfill the requirements of the Espoo Convention. Since the Directive harmonizes the national EIA procedures of the contracting states extensively, it can also provide for a strict transboundary EIA procedure. A potentially affected state is entitled to be informed by and to consult with the origin state.[1105] Not only the private legal subjects of the potentially affected state but also the governmental bodies having environmental expertise are entitled to participate in the EIA procedure of the origin state.[1106] The procedure implements the duty of the Espoo Convention to ensure that the potentially affected state furnishes information on the environmental conditions of the area likely to be affected in its territory. The final decision must take into account the comments of both the potentially affected state and its public.[1107]

[1104] See more closely *supra* chapter IV (4.1.1.).
[1105] Article 7 of the Amendment Directive.
[1106] Ibid., Article 7 (3a).
[1107] Ibid., Article 8.

A transboundary EIA procedure, the draft Transboundary Environmental Impact Assessment Agreement (draft TEIA), has been negotiated under the umbrella of Northern American Commission for Environmental Cooperation (NACEC), even though two of the parties are also signatories to the Espoo Convention. Although the negotiations between Canada, the United States and Mexico produced a draft agreement, it has not yet been accepted by the parties. The draft TEIA is different in structure from the transboundary EIA procedure of the Espoo Convention,[1108] as the national EIA procedures of both Canada and the United States differ generally from those in Europe. The main difference is that the European EIA procedures, built as they are upon the EIA Directive of the EC, are triggered if objective criteria are fulfilled, viz. a list of large scale environmentally harmful activities. If the procedure is initiated, it normally continues until the licencing decision is made. The Northern American EIA procedures contain more flexibility, as was noted above.[1109] Especially in the federal EIA procedure of the United States, the agency in charge can decide to make a preliminary impact study after which it may decide whether a full-blown impact study is required or whether the project can proceed.[1110]

The draft TEIA contains two separate procedures for transboundary EIA procedure: a notification procedure and a transboundary environmental impact assessment procedure. The origin state is obligated to notify the potentially affected state if certain objective criteria are fulfilled. If the proposed project is closer than 100 kilometers to the international border of the states, or if it belongs to the classes of

---

[1108] The Espoo Convention has been the basis for the amendment of the EIA Directive of 1997 as well as the transboundary EIA procedure contained in the 1992 Industrial Accidents Convention (Article 4 and Annex II).

[1109] See *supra* chapter IV (1.).

[1110] For a general presentation of the EIA procedures prevailing in the United States and Canada, see Gilpin 1995, pp. 113-123. A good overview of the EIA procedures prevailing in these states can also be obtained from the home page of the NACEC, more specifically, in relation to Canada at http://www.cec.org/pubs_info_resources/law_treat_agree/summary_enviro_law/publication/ca07.cfm?varlan=english [19.2.2001] and with respect to the United States at http://www.cec.org/pubs_info_resources/law_treat_agree/summary_enviro_law/publication/us07.cfm?varlan=english [19.2.2001].

activities defined in Appendix I of the TEIA, notification must be carried out by the origin state.[1111] This is a very objective criterion for initiation of the notification procedure, even more objective than that of the Espoo Convention, in which the notification duty is based on two criteria, i.e., that the proposed activity must be among the categories of activities listed in Appendix I of the Convention and that it must be likely to cause significant adverse transboundary impact on the environment of other contracting states. However, the notification duty of the Espoo Convention triggers the full transboundary EIA procedure, whereas in the draft TEIA it is only a duty to notify other contracting states about the proposed activity, and nothing else.

In the draft TEIA, the duty to conduct a transboundary impact assessment is separated from the duty of notification. At this stage, the origin state is left with much larger discretion as to whether to initiate a transboundary impact assessment: the only criterion is whether significant adverse transboundary environmental impact is likely from the proposed activity. The determination of whether such effects will ensue from the proposed project is made by the origin state alone, guided by the criteria in Appendix III of the instrument.[1112] The potentially affected state has no right to dispute the findings of the origin state other than by appeal to the general dispute settlement provisions of the instrument.[1113]

Although the United States and Canada signed the Espoo Convention, and Canada has even ratified it, the transboundary EIA procedure of the Espoo Convention seems to have been modelled on the national EIA procedures prevalent in Europe: once the public has been informed, the origin state is obligated to initiate the transboundary EIA procedure, and the procedure must continue until the final decision is made concerning the future of the proposed activity. On the other hand, the federal EIA procedures of Canada and the United States are quite flexible as to their national

[1111] Article 2 of the draft TEIA.
[1112] Ibid., Article 10.

application, especially with regard to their international application, and thus are clearly capable of fulfilling the obligations of the Espoo Convention as well.

2. The Espoo Convention

The Espoo Convention is the most promising Convention focusing on potential transboundary pollution from planned activities in the Arctic. This is due to the fact that it has been signed by all eight Arctic states. Moreover, it covers the most important activities carried out in the Arctic and is of a high quality in comparison with the older prior transboundary procedures.

2.1. The Espoo Process

The starting point in the United Nations Economic Commission for Europe (UN ECE) for the procedure leading to the signing of the Espoo Convention can be traced back to a decision by the Group of Experts to hold a seminar on the topic in Warsaw in 1987. This seminar prompted the Group of Experts to recommend to the Senior Advisers of the ECE governments the development of a framework agreement on EIA in the transboundary context. The actual negotiations were initiated by the Senior Advisers through a decision to convene an *ad hoc* meeting to negotiate a draft framework agreement. Thereafter, six meetings of a working group[1114] established for this purpose were held between October 1988 and September 1990.[1115]

[1113] Ibid. Article 19 was only named but not elaborated by the Group of Experts.

[1114] The following reports are important: Report of the *ad hoc* meeting to elaborate a draft Framework Agreement on Environmental Impact Assessment in a Transboundary Context (Geneva 18-21 October 1988). ENVWA/AC.3/2 27 October 1988; Report of the second meeting of the *ad hoc* Working Group to elaborate a draft Framework Agreement on Environmental Impact Assessment in a Transboundary Context (Geneva 16-19 May 1989). ENVWA/AC.3/4 26 May 1989; Report of the third meeting of the *ad hoc* Working Group to elaborate a draft Agreement on Environmental Impact Assessment in a Transboundary Context (Hague, 28 November to 1 December 1989). ENVWA/AC.3/6 19 December 1989; Report of the fourth meeting of the *ad hoc* Working Group to elaborate a draft Agreement on Environmental Impact Assessment in a Transboundary Context (Geneva 12-16 February 1990). ENVWA/AC.3/8 23 February 1990; Report

The Convention was finally adopted by the Senior Advisers to ECE Governments on Environmental and Water Problems at their fourth session, which was convened in Espoo, Finland.[1116] The Convention was open for signature at Espoo from 25 February to 1 March 1991 and thereafter at the United Nations Headquarters in New York until 2 September 1991. After the adoption of the Convention, it was already known that its entry into force would take a few years because of the requirements laid down in the Convention.[1117] Interestingly, when signing the Convention, the contracting states adopted a resolution affirming their 'resolve to strive for the entry into force of the Convention as soon as possible and to seek to implement the Convention to the maximum extent possible pending its entry into force'[1118].

The resolution requires measures from the contracting states in two respects: to ratify the Convention 'as soon as possible' and to implement the Convention before it enters into force. Could this resolution be considered an agreement between the negotiating states for provisional application of a treaty as provided for in Article 25 of the Vienna Convention? According to Article 25:

---

of the fifth meeting of the *ad hoc* Working Group to elaborate a draft Convention on Environmental Impact Assessment in a Transboundary Context (Geneva 11-15 June 1990). ENVWA/AC.3/10 28 June 1990; Report of the sixth meeting of the *ad hoc* Working Group to elaborate a draft Convention on Environmental Impact Assessment in a Transboundary Context (Geneva 24-28 September 1990). ENVWA/AC.3/12 15 October 1990.

1115 The history of the Espoo Convention is carefully presented by Robert Connelly, who acted first as a rapporteur for the Warsaw seminar and later as chairman of the Working Group to elaborate the Draft Convention on Environmental Impact Assessment in a Transboundary Context. See Connelly 1999, pp. 37-46.

1116 In the Report of the fourth session of Senior Advisers to ECE Government on Environmental and Water Problems, as adopted by the Senior Advisers on 1 March 1991, Part III (10) reads: 'The Senior Advisers adopted the Convention on Environmental Impact Assessment in a Transboundary Context (E/ECE/1250) and the accompanying Resolution (ECE/ENVWA/19)'. ECE/ENVWA/18 18 March 1991.

1117 Article 18 (1) of the Espoo Convention reads: 'This Convention shall enter into force on the ninetieth day after the date of deposit of the sixteenth instrument of ratification, acceptance, approval or accession'.

1118 Paragraph 1 of the resolution of the signatories of the Espoo Convention (ECE/ENVWA/19).

> A treaty or a part of a treaty is applied provisionally pending its entry into force if: (a) the treaty itself so provides; or (b) the negotiating States have in some other manner so agreed.

It is point (b) which is relevant here, since the Espoo Convention does not so provide. Since the resolution was adopted at the time of the signing of the Convention, it can be taken as a presumption that provisional application was envisioned. The wording of the resolution does not, however, express that direct orders have been created. Both requirements are expressed as obligations of attempt, e.g., the parties strive or seek to achieve a goal 'as soon as possible' and 'to the maximum extent possible'. This obligation of attempt has been taken seriously since Finland, for example, applied the Convention three times before its entry into force.[1119] Also, as will be shown below, the signatories to the Convention acted on the obligation to strive for the speedy entry into force of the Convention by developing a procedure to that end.

The parties to a treaty are obligated not to frustrate the object and purpose of the treaty before its entry into force. According to Article 18 of the Vienna Convention, and the identical norm in the customary law of treaties:

> A state is obligated to refrain from acts which would defeat the object and purpose of a treaty when: (a) it has signed the treaty...until it shall have made its intention clear not to become a party to the treaty.

Thus, signature already requires some measures of the contracting states. In the case of the Espoo Convention, this means that at the very least some kind of notification must be made to the potentially affected state even if no comprehensive transboundary EIA procedure is organized; otherwise the object and purpose of the

[1119] See Koivurova 1997 I, pp. 515-522 and 1997 II, pp. 194-197. For further information concerning the application of the Espoo Convention, see Furman et al. 1999 and the final report of the workshop 'Practical application of the UN/ECE Convention on Environmental Impact Assessment in a Transboundary Context', 2000.

treaty, 'to prevent, reduce and control significant adverse transboundary environmental impact from proposed activities'[1120], would clearly be frustrated.

Pending the entry into force of the Espoo Convention, the states which had signed the Convention started to meet each other as urged by the resolution.[1121] The first meeting of the signatories was organized from 2 to 4 December 1991 in Geneva and was documented in the form of a report.[1122] This procedure, the meeting of the signatories, was of much importance in urging the states parties to ratify the Convention as soon as possible and to implement and apply it pending its entry into force. In the second meeting[1123], the work on listing the points of contact, which had

---

[1120] Article 2 (1) of the Espoo Convention.

[1121] The resolution of the signatories provides that the signatories: 'resolve to strive for the entry into force of the Convention as soon as possible and to seek to implement the Convention to the maximum extent possible pending its entry into force; 2. Consider that, pending the entry into force of the Convention, the necessary authority should be given to the Economic Commission for Europe and to its Executive Secretary to provide for a sufficient secretariat and, in the framework of the existing budgetary structure, for the appropriate financial means; 3. Continue to co-operate in bringing closer together their policies and strategies in relation to environmental impact assessment'.

[1122] ENVWA/WG.3/2 16 December 1991. The first meeting studied the implementation situation of the Convention and reported on some ratifications of the instrument (paragraph 8 of the report). Significantly, some of the signatory states were prepared to initiate preparatory work on many aspects of the Convention to prepare for its entry into force. Annex I to the report provides for work plans for the interim implementation of the Convention in accordance with the Resolution. Four subjects were chosen: legal and administrative aspects; methodological aspects; preparation of draft rules of procedure for the Meeting of parties; and policies and strategies promoting environmental impact assessment. This included a report on 'the practical problems encountered and the solutions found regarding the implementation of the provisions of the Convention' (paragraph 9) as well as the establishment of a task force on legal and administrative aspects (paragraph 19). In addition '[t]he delegation of Sweden offered to host a workshop…on methodological aspects' (paragraph 20). The meeting also 'requested the secretariat to prepare a list of points of contact regarding notification' (paragraph 21) and 'requested government rapporteurs with the assistance of the Secretariat to prepare draft rules of procedures' (paragraph 18) for the future meeting of parties.

[1123] ENVWA/WG.3/4 17 December 1992. The second meeting was convened in Rome, from 1 to 4 December 1992. In the meeting, the implementation situation of the Convention was reviewed and the parties encouraged speedy ratification of the Convention. The work on listing the points of contact, which had started at the first meeting, was continued: 'It was the understanding of the Meeting that, in the absence of a different bilaterally agreed arrangement, a notification transmitted to the relevant authority mentioned in the list would be sufficient to comply with the relevant obligations of the Convention' (paragraph 11 of the report). The parties were also in agreement that if no point of contact had been designated, the Ministry of Foreign Affairs of the party should be used as one (paragraph 11) and requested the Secretariat to prepare for the third meeting 'an

already started in the first meeting, was continued. It was also important that the draft rules of procedure for the future meeting of parties were elaborated at this stage for the first time, and the parties decided to invite comments on those draft rules.[1124] In the third meeting[1125], the delegations provided additions and modifications to the list on points of contact and requested the Secretariat to update the list.[1126] The meeting was able to finalize much work started in the earlier meetings of the signatories.[1127]

In the fourth meeting[1128], the draft report on policies and strategies promoting EIA,[1129] which was prepared by the Secretariat on the basis of information provided

---

updated version of the list of points of contact' (paragraph 12 (b)). The Secretariat was also requested to prepare a report on policies and strategies promoting EIA. The interim report (ENVWA/WG.3/R.6) of the task force on legal and administrative aspects was introduced, and the parties invited delegations to submit comments on it and entrusted 'the task force with the continuation of its work on the preparation of its final report' (paragraphs 15 and 16). The same measures were to be undertaken with respect to the report (ENVWA/WG.3/R.7) of the workshop on methodological aspects of EIA in a transboundary context (paragraphs 17 and 18).

1124 Ibid., paragraphs 22 and 23.

1125 CEP/WG.3/2 7 April 1995. The meeting was organized from 21 to 24 February 1994 in Geneva. In the meeting, it was decided that a workshop would be convened in Estonia 'with the aim of preparing proposals for strengthening the capabilities of future Parties, particularly countries in transition, to comply with the obligations under the Convention' (paragraph 16). The draft rules of procedure for the meeting of the parties were also discussed, and some amendments were made (paragraphs 22 and 23).

1126 Ibid., paragraph 11 (b).

1127 In the third meeting, it was recommended that the draft report on policies and strategies promoting environmental impact assessment (ENVWA/WG.3/R.10 and Add.1) as well as specific methodological issues of EIA in a transboundary context (paragraphs 20 and 21 of the report; the document is ENVWA/WG.3/R.13) and the final reports of the task force on legal and administrative aspects (paragraphs 17-19; the document is ENVWA/WG.3/R.12) be published.

1128 CEP/WG.3/2 7 April 1995. The fourth meeting was convened from 14 to 17 March 1995 in Geneva. In the meeting, the results (CEP/WG.3/R.3) of the workshop on strengthening the capability of future parties to comply with the obligations under the Convention were presented by the delegation of Estonia. The meeting took note of many proposals presented in the report and proposed varying policy actions to support the countries in transition. The meeting also acknowledged the reports of the two workshops on legal and administrative aspects of the Convention, i.e., the workshop on key elements for bilateral and multilateral agreements on EIA in a transboundary context (Part VII of the report; the document is in CEP/WG.3/R.4) and the workshop on public participation in EIA (CEP/WG.3/R.5). An additional development noted was the series of consultations by governmentally designated experts on the issue of specific methodological issues of EIA in a transboundary context (CEP/WG.3/R.6), especially concentrating on the question of how 'significance' is to be defined and inviting delegations to apply the criteria set forth in the document (Part VIII of the report). Of considerable importance was the progress made in preparing the draft rules of procedure for the meeting of the parties, since the Convention

by the delegations, was commended by the meeting and it was decided that the Secretariat should continuously update the report.[1130] It was also important that a great deal of development had occurred in preparing the draft rules of procedure for the upcoming first meeting of the parties. At the time of the fifth meeting[1131], the entry into force of the Convention was approaching and thus the meeting discussed the draft rules of procedure for the meeting of the parties[1132] as well as the 'possible role of the Bureau of the Meeting of the Parties'[1133]. The final session of the signatories to the Espoo Convention was designated as a special session for the preparation of the first meeting of the parties[1134] and was organized in Geneva from 11 to 13 November 1996. It was important that some projects of the meeting of the signatories were finished before the first meeting of the parties. The meeting proposed that a format for notification be adopted by the parties at their first meeting.[1135]

The Espoo Convention came into force on 10 September 1997. Thereafter, all those states which had previously formally accepted the Convention became parties to it.[1136] Those contracting states which formally accept the Convention after 10

---

required in Article 11 that the meeting of the parties must adopt the rules of procedure in their first meeting (Part IX of the report).

1129 (CEP/WG.3/R.2)

1130 The Report, Part V.

1131 CEP/WG.3/4 30 April 1996. The signatories met for the fifth time from 19 to 21 March 1996 in Geneva. In the meeting, the Secretariat was asked to update the document on policies and strategies promoting EIA (paragraph 15 of the report) and to prepare a draft report on the needs of countries in transition (paragraph 17) as well as on international efforts to strengthen EIA (paragraph 28). The Bureau, together with certain delegations and the Secretariat, were invited to prepare further proposals for a database on EIA (paragraph 22) and to prepare a revised work program for the international center on EIA (paragraph 24). The format for notification was also discussed, and some delegations were assigned the task of developing this (paragraph 25).

1132 (CEP/WG.3/R.9).

1133 (CEP/WG.3/R.10).

1134 CEP/WG.3/6 18 March 1997.

1135 Ibid., Part G.

1136 Article 17 regulates the formal acceptance of the Espoo Convention: 'This Convention shall be subject to ratification, acceptance or approval by signatory states and regional economic integration organizations'.

September 1997 become parties 'on the ninetieth day after the date of deposit'[1137] of their formal acceptance. The coming into force of the Convention made the cooperation processes of the Espoo Convention available. In addition to the amendment procedure[1138] and voting procedure[1139], an important process to develop the system of the Convention is the meeting of states parties.

The meeting of parties is convened 'in connection with the annual sessions of the Senior Advisers to ECE Governments on Environmental and Water Problems'[1140]. The strictness of this requirement is mitigated by providing that the meeting of the parties is to be arranged 'so far as possible'[1141]. In fact, the Convention stipulates that the meeting of the parties 'shall be held at such times as may be deemed necessary by a meeting of the Parties, or at the written request of any Party'[1142]. The written request must be supported by at least one-third of the parties to the Convention.

[1137] Ibid., Article 18 (3).

[1138] Any party may propose amendments to the Convention. There exist two procedures for this purpose, both ultimately requiring the consent of each contracting state. First, a consensus is sought in a meeting between all the contracting states. If this consensus is not achieved, a quorum of three-fourths of the parties present and voting is enough for an amendment to be adopted. The amendment becomes binding only when a three-fourths majority of all contracting states has ratified, approved or accepted it. The amendment enters into force for these parties on the ninetieth day after the final receipt of notification of acceptance by the depositary. After this date, the other contracting states are entitled to accept the amendment, and the amendment enters into force 'on the ninetieth day after a Party deposits its instrument of ratification, approval or acceptance of the amendments'.

[1139] The right to vote is regulated in Article 12 of the Espoo Convention. As is usual in international conventions, each party has one vote in the amendment procedure. Special voting rights are provided for in the case of the EC. Since the Community and some of its member states are parties to the Convention, and environmental protection is one of the fields of policy where the member states and the Community exercise shared jurisdiction (see Kent 1996, pp. 259-262), there is a need for special arrangements with regard to voting rights in the meeting of the parties in order to ensure the equality of contracting states. It is thus required that a decision be made before a vote is taken that either the Community will vote on behalf of all the member states or the member states parties to the Convention will exercise their individual right to vote.

[1140] Ibid., Article 11 (1).

[1141] Ibid.

[1142] Ibid.

Article 11 of the Convention also requires that the first meeting must be organized 'not later than one year after the date of the entry into force of this Convention'[1143].

Pursuant to this provision, the first meeting of parties was arranged in Oslo, Norway, 18 to 20 May 1998.[1144] This meeting of the parties also adopted the rules of procedure as required by Article 11 (d) of the Convention. These rules are very detailed in character and set out standards on how the meetings of the parties are to be organized.[1145] Another important development was that the meeting of the parties established a subsidiary body, the Working Group on Environmental Impact Assessment, 'to assist in the implementation of the Convention and the management of the work-plan'[1146]. The work plan for the period between the first meeting in Oslo to the February 2001 meeting in Sofia was defined to include various issues of importance to the future of the Espoo Convention.[1147]

The second meeting of the parties took place in Sofia, Bulgaria, from 26 to 27 February 2001.[1148] As the meeting of the parties had been organized through the rules

---

1143 Ibid.

1144 The Report of the first meeting of the parties can be found on the home page of the UN ECE at http://www.unece.org/env/eia/ [28.4.2001].

1145 Annex I to the Report, Decision I/1, Rules of Procedure. These rules consist of the following chapters: purposes, definitions, place of meetings, dates of meetings, observers, agenda, representation and credentials, officers, subsidiary bodies, secretariat, conduct of business, voting, official languages, amendments to rules of procedure, overriding authority of the Convention. This report can be found on the home page of the UN ECE at http://www.unece.org/env/eia/report.htm [28.4.2001].

1146 Annex II to the Report, Decision I/2 'Mechanisms for the implementation of the Convention', paragraph 1.

1147 Annex VI, Decision I/6, Appendix I 'Work-plan for the implementation of the Convention for the period 1998 to 2000'. It includes the following issues: reviews of strategies and policies, aspects of bilateral and multilateral cooperation, practical application of the Convention, public participation in a transboundary context, non-compliance guidelines, recent EIA developments and links with other ECE Conventions, database on environmental impact assessment, evaluation of the database, partnerships, subregional pilot workshop.

1148 Personal copy of a report from the Sofia meeting obtained by the author from the Finnish Ministry of the Environment. The report did not specify the particulars of the decisions made, and thus the preparatory documents that can be found on the home page of the UN ECE (at

of procedure and other preliminary issues, it was possible to move into the substantive issues pertaining to the Convention. Some amendments to the Convention were adopted[1149] and other important decisions were made which will be studied below. Decision II/11[1150] was the adoption of a work-plan to direct the activities of the parties until the third meeting of the parties, which will take place in the spring of 2004 in Croatia. The important activities identified include the creation of a reporting system for the implementation of the Convention, strengthening cooperation with other ECE Conventions, developing guidelines on good practice in concluding bilateral and multilateral implementation agreements, preparing a draft protocol on strategic environmental assessment, reviewing whether any amendments to the Convention are necessary, and providing draft guidance for public participation in a transboundary context. In addition, as will be studied below,[1151] an Implementation Committee was established. Significantly, the meeting of the parties decided to open the possibility for all member states of the United Nations to become parties to the Convention.[1152] There already are 37 parties to the Convention.[1153]

---

http://www.unece.org/env/eia/2p.htm [24.4.2001]) have been used. Most of the decisions taken by the meeting of parties were done in accordance with these preparatory documents.

[1149] The concept of the public in Article 1 (x) of the Espoo Convention was amended to correspond to the definition adopted in the Aarhus Convention. According to the amended version: '"The public" means one or more natural or legal persons and, in accordance with national legislation or practice, their associations, organizations or groups'. The Convention was also opened to all the member states of the United Nations. According to new Article 17 (3): 'Any other State, not referred to in paragraph 2 above, that is a Member of the United Nations may accede to the Convention upon approval by the Meeting of the Parties. The Meeting of the Parties shall not consider or approve any request for accession by such State until this paragraph has entered into force for all the States that were Parties to the Convention on February 27, 2001'. The Sofia meeting also adopted new provisions on how these amendments become binding on the states parties. First of all, new Article 17 (8) provides: 'Any State that ratifies, accepts or approves this Convention shall be deemed simultaneously to ratify, accept or approve the amendment to the Convention set out in decision II/14 taken at the second meeting of the Parties'. Finally, the meeting decided on how the amendments would enter into force: 'The Meeting decides that notwithstanding the provisions of Article 14, paragraph 4 of the Convention, this amendment shall not enter into force until the ninetieth day after it has been ratified, approved or accepted by all States and regional economic integration organizations that were Parties to the Convention at the time of the adoption of this decision'.

[1150] MP.EIA/2001/11.

[1151] See *infra* chapter V (2.6.).

[1152] New Article 17 (3) of the Espoo Convention.

The institutional development within the Espoo Convention has progressed in time. The meeting of the parties, representing the voice of states parties, has taken place twice and has adopted many important decisions pertaining to the institutional structure of the Espoo Convention. The Bureau of the Convention was established as an organ coordinating the work pertaining to the development of the system of the Convention between the meetings of the parties. Secretariat tasks are handled by the UN ECE. The Implementation Committee studies potential non-compliance with the obligations of the Convention. The system of the Espoo Convention can most appropriately be qualified as an example of a recent phenomenon in international law that has developed in the field of environmental protection in particular and which Churchill and Ulfstein refer to as an autonomous institutional arrangement (AIA).[1154]

2.2. The Espoo Convention in the Arctic

The EIA procedure in general and the transboundary EIA procedure in particular have been at the center of the Arctic-wide cooperation process from its very beginning. In fact, from a broader perspective, the Arctic-wide cooperation process as a whole can be seen as a regional EIA procedure providing a basis on which practical measures to protect the Arctic environment are drafted. In contrast to Antarctic environmental cooperation, the Arctic cooperation process started by first assessing existing and future environmental problems through a specific program, the Arctic

[1153] As of 15 May 2001, there are 37 parties to the Espoo Convention. Information obtained from the UNTS.

[1154] In a recent article, Churchill and Ulfstein argue that a special feature of recent multilateral environmental agreements has been that states parties not only conclude an agreement with binding obligations but also establish institutional frameworks, such as conferences of parties, to oversee the development of the treaty system rather than setting up traditional inter-governmental organizations for this purpose. They also perceive that such autonomous institutional arrangements (AIAs), as they call them, can be considered as inter-governmental organizations (IGO) for the purpose of applicable law even though they are of a less formal nature and less permanent than the traditional IGOs. It seems clear that the system of the Espoo Convention qualifies as such an AIA. See Churchill and Ulfstein 2000, pp. 623-659.

Monitoring and Assessment Program (AMAP), before any major environmental protection measures were concluded.[1155] One new proposal which the Arctic cooperation has yielded is the Arctic Climate Impact Assessment Programme (ACIA), which aims at studying the varying impacts ozone layer depletion and global warming may have on the Arctic environment.[1156]

The Arctic states have had different histories with regard to the Espoo Convention. Seven of the states signed the Convention on 26 February 1991, and one, the Russian Federation, on 6 June 1991. As mentioned above, the very signing of the Convention carries certain legal effects, namely the requirement not to defeat the object and purpose of the treaty prior to its entry into force. The first of the Arctic states to ratify the Convention was Sweden, which did so on 24 January 1992. Thereafter, Norway accepted the Convention as binding (23 June 1993), followed by Finland (10 August 1995), Denmark (14 March 1997), the EC (24 June 1997) and Canada (13 May 1998). Iceland is expected to ratify the Convention in the near future.

There are thus three Arctic states that are not yet parties to the Espoo Convention: the United States, the Russian Federation and Iceland. It is a bit surprising to find out that there were high expectations immediately after the signing of the Convention that the United States and the Russian Federation would ratify the Convention within a short period of time. The report from the second meeting of the signatories states: 'The delegations of...the Russian Federation and the United States of America...expected

[1155] For an analysis, see Vidas 2000, pp. 99-100.

[1156] The Arctic Council has one project now underway to study the effects of these global pollution problems on the Arctic environment. As stated by the mission statement of this programme: 'The goal of the Arctic Climate Impact Assessment (ACIA) is to evaluate and synthesize knowledge on climate variability, climate change, and increased ultraviolet radiation and their consequences...Three major volumes will be completed by 2004'. The mission statement of the Arctic Climate Impact Assessment Programme (ACIA) can be found on the home page of the Arctic Council at http://www.arctic-council.org/ [7.2.2001].

that their countries would be able to ratify the Convention in the course of 1993'[1157]. In the third report of the meeting of signatories it is stated: 'The delegations of...the Russian Federation informed the meeting that their countries expected to ratify, accede to or approve the Convention in 1994'[1158]. In addition, 'the delegations of...the United States of America reported on the progress made in their countries in order to ratify or accede to the Convention'[1159].

At the fourth meeting of the signatories, the Russian Federation was deleted from the list of those delegations that were expected to ratify the Convention 'before the end of 1995' and was placed among those states in which progress had been made towards ratification, together with the United States.[1160] This situation persisted in the fifth meeting of the signatories.[1161] In the final meeting of the signatories, it was only reported that the Russian Federation, as well as some other delegations, expected to ratify the Convention at a later stage.[1162] The United States' delegation did not even take part in this final, special meeting of the signatories.[1163]

From the very beginning of the Arctic cooperation process, the Espoo Convention has enjoyed a privileged position. This is only natural given that all eight Arctic states that participate in the Arctic cooperation process are signatories to the Convention.

[1157] Report of the second meeting of the signatories to the Espoo Convention, ENVWA/WG.3/4 17 December 1992.
[1158] Report of the third meeting of the signatories to the Convention on Environmental Impact Assessment in a Transboundary Context, Part IV, paragraph 9. ENVWA/WG.3/6, 9 March 1994.
[1159] Ibid.
[1160] Report of the fourth meeting of the signatories to the Espoo Convention, CEP/WG/.3/2, 7 April 1995, paragraph 10.
[1161] Report of the fifth meeting of the signatories to the Espoo Convention, CEP/WG.3/4, 30 April 1996, paragraph 8.
[1162] Report of the special session of the signatories, CEP/WG.3/6, 18 March 1997, paragraph 10.
[1163] Ibid., paragraph 2.

The Convention was expressly promoted as early as 1993, in the Nuuk Declaration. According to paragraph 8:

> Therefore, we shall maintain, as appropriate, or put into place as quickly as possible, an internationally transparent domestic process for the environmental impact assessment of proposed activities that are likely to have a significant adverse impact on the Arctic environment and are subject to decisions by competent national authorities. To this end we support the implementation of the provisions of the Convention on Environmental Impact Assessment in a Transboundary Context.

It can be asked whether the Arctic states have committed themselves to maintain or put into place an internationally transparent EIA procedure in this Article. Although the modal 'shall' is used, it is qualified by the expressions 'as appropriate' and 'as quickly as possible', which convey that the obligation is only one of attempt. Paragraph 9 of the Nuuk Declaration is also relevant: 'We underline the importance of prior and timely notification and consultation regarding activities that may have significant adverse transboundary environmental effects'. Although not establishing a legal obligation, 'underline' in this declaration conveys the importance placed on transboundary EIA.

The 1996 Inuvik Declaration considered the Espoo Convention much more briefly: 'reaffirming the support of our countries...on Environmental Impact Assessment in a Transboundary Context'[1164]. In the 1997 Alta Declaration, the Arctic states agreed to apply the 1997 EIA Guidelines and the 1997 Offshore Guidelines. The EIA

[1164] Preface to the 1996 Inuvik Declaration.

Guidelines contain a Section on transboundary impact, to be studied below, that specifically mentions the Espoo Convention.[1165]

2.3. Preliminary Issue of Reservations

When Canada ratified the Espoo Convention on 13 May 1998, it made the following reservation:

> Inasmuch as under the Canadian constitutional system legislative jurisdiction in respect of environmental assessment is divided between the provinces and the federal government, the Government of Canada in ratifying this Convention, makes a reservation in respect of proposed activities (as defined in this Convention) that fall outside of federal legislative jurisdiction exercised in respect of environmental assessment.[1166]

This reservation provoked objections from many contracting states, including the three Nordic Arctic states, Sweden, Finland and Norway.[1167] All three are of the opinion that such a reservation is contrary to the well-established rule of customary international law and the Vienna Convention on the Law of Treaties, which stipulates that reservations contrary to the object and purpose of the treaty are not permitted.

[1165] Section 11, 'Transboundary impacts' contains the following reference (pp. 40-41): 'The UN ECE Convention on EIA in a Transboundary Context, the Espoo Convention (1991, entered into force in 1997), provides a comprehensive framework for dealing with activities likely to have significant adverse transboundary impacts'.

[1166] Information obtained from the UNTS.

The Finnish objection reads as follows:

> In the view of the Government of Finland the general reservation made by the Government of Canada does not adequately clarify to which extent Canada considers itself bound by the Convention. It is of fundamental importance that States are prepared to undertake legislative changes necessary to comply with their obligations under their treaties. Furthermore, according to Article 19 of the Vienna Convention on the Law of Treaties of 23 May 1969 as well as customary international law a reservation incompatible with the object and purpose of a treaty shall not be permitted. Accordingly, Finland objects to the general reservation of Canada as not compatible with the object and purpose of the [Convention].[1168]

All three Nordic Arctic states consider the reservation inadmissible. Indeed, there are factors which speak against Canada's reservation as well. First of all, as will be shown below, the Espoo Convention suffers from an inherent lack of reciprocity of treaty obligations between the states parties because it only harmonizes the national EIA procedures to a modest degree. This lack of reciprocity is brought into sharp relief by Canada through its making the application of the Convention dependent on whether the federal government has legislative jurisdiction over the case. Second, it is hard to define the extent of the reservation. As this is connected to the internal law of Canada, it is the development of internal constitutional law in general and EIA procedure in particular that will shape the content of Canada's obligations towards other contracting states. On the other hand, it can, of course, be argued that Canada limited the applicability of the principle of reciprocity of treaty obligations in this specific case by providing for an express reservation to this effect.

[1167] Other states also objected to the reservation made by Canada: Spain, Italy, France and Luxembourg. Information obtained from the UNTS.

Canada replied to the objections made by the other contracting states. On 21 January 2000, the Secretary-General of the United Nations received from the Government of Canada the following communication:

> The Government of Canada notes that some States have formulated objections to the reservation of the Government of Canada to the Espoo Convention. The Government of Canada wishes to reaffirm its view that a reservation in respect of proposed activities (as defined in the Convention) that fall outside federal legislative jurisdiction exercised in respect of environmental assessment is compatible with the object and purpose of the Convention and is thus admissible. In reaffirming its position on this matter, the Government of Canada refers to the negotiating history of the Convention and specifically to the sixth and final meeting of the Working Group to elaborate a draft Convention. At that meeting, the states present agreed to delete a draft article that would have prohibited all reservations to the Convention. It was and remains Canada's understanding that the agreement to delete the prohibition on reservations was linked directly with a further decision not to include a "federal clause" within the Convention.[1169]

The reasons for Canada's reservation are quite understandable. The confederal structure of its government makes it hard for the state to act towards other states since decision-making in matters of environmental protection, including EIA, is also shared between federal, provincial and territorial governments.[1170] Considering the complex

[1168] Information obtained from the UNTS. The Finnish, Swedish and Norwegian objections are almost identical in content.

[1169] It is also provided: 'Canada further wishes to state that Canada's reservation to the Espoo Convention is an integral part of Canada's ratification of the Convention and is not severable therefrom. Canada can only accept treaty relations with other states on the basis of the reservation as formulated and in conformity with Article 21 of the Vienna Convention on the Law of Treaties'. Information obtained from the UNTS.

[1170] The joint organ of federal, provincial and territorial governments in environmental protection affairs, the Canadian Council of Ministers of the Environment (CCME), has tried to solve these problems of shared responsibility. A Canada Wide Accord on Environmental Harmonization was concluded in January 1998 to provide a basic framework for cooperative action. Important from the viewpoint of this study is the Sub-Agreement on Environmental Assessment signed between the federal, provincial and territorial governments. The Sub-Agreement aims for a single assessment with only one lead party, either the federal, provincial or territorial government. Article 5.6 provides

events surrounding the reservation made by Canada, it is necessary to examine what the legal effects of the reservation made by Canada in international law are, especially in light of the objections by the Nordic states.

The development of the customary rules on reservations culminated in the conclusion of the 1969 Vienna Convention.[1171] The system of the Convention in regard to reservations is built in such a way that the contracting states are permitted to decide this matter by their explicit agreement on the issue, by allowing for the making of reservations in general, allowing only certain kinds of reservations or prohibiting them altogether.[1172] In other words, states are free to decide the issue of reservations, and if they do not, the presumption exists that they have the right to make reservations. However, if a reservation goes against the object and purpose of a treaty, it is prohibited.[1173] This is specified in Article 20 (2) of the Convention.[1174]

---

the basic rules for determining which is to act as the lead party. One of the criteria enumerated in Article 5.6.4 is: 'interprovincial, interterritorial or international considerations', providing some room for the federal government to act as the lead party if transboundary impact on the territory of other states is likely. The Canada Wide Accord on Harmonization, as well as the Sub-Agreement on Environmental Assessment, can be found on the home page of the Canadian Council of Ministers of the Environment at http://www.mbnet.mb.ca/ccme/3e_priorities/3e.html [16.3.2001].

1171 The legal term 'reservation' means 'a unilateral statement, however phrased or named, made by a State, when signing, ratifying, accepting, approving or acceding to a treaty, whereby it purports to exclude or modify the legal effect of certain provisions of a treaty in their application to that State' (Article 21 (1d) of the Vienna Convention). If the parties have not agreed on reservations, the Vienna Convention contains rules for these situations. The difficulties relate to multilateral treaties; for instance, must there be approval by the whole treaty community of a reservation made by one member if no agreement has been reached by the parties? The ICJ answered this in the negative in its Advisory Opinion on whether it was legal to make reservations to the Genocide Convention, a multilateral treaty with extensive participation (Reservations to the Convention on Genocide, Advisory Opinion: I.C.J. Reports 1951, p. 15). The Court provided that if a reservation does not go against the object and purpose of a treaty, it is legal and thus admissible. By the time the 1969 Vienna Convention was negotiated, the ICJ's liberal approach had been widely accepted as representing the customary law on the subject (Brownlie 1998, pp. 613-614). For a general treatment, see Horn 1988.

1172 Article 19 of the Vienna Convention.

1173 Ibid.

1174 Ibid. Article 20 (2) reads: 'When it appears from the limited number of the negotiating States and the object and purpose of a treaty that the application of the treaty in its entirety between all the parties is an essential condition of the consent of each one to be bound by the treaty, a reservation requires acceptance by all the parties'.

Since it is normally left for each state party to decide whether a reservation goes against the object and purpose of the treaty, the other contracting parties must make a choice.[1175] If a reservation is of great importance to a state party, and that party does not want to establish a legal relation with the reserving state, the party must clearly express this intention in its objection to the reservation.[1176] Furthermore, it must make this reaction known within the time period stipulated by Article 20 (5) of the Vienna Convention; should this time lapse, the treaty is in force for the parties as modified by the reservation.[1177] According to Article 21 (3), the presumption is that if a state party to the treaty objects to a reservation, not expressly declaring that it views the reservation as terminating the legal relationship between it and the reserving state, the treaty is in force between the parties as modified by the reservation. The provisions to which the reservation relates are not applicable in the legal relationship between the reserving state and the objecting state.[1178]

The three Nordic Arctic states that objected to the Canadian reservation perceived it as inadmissible. This is in line with the recent practice of the Nordic states that the reserving state should not benefit from its reservation to the treaty with respect to the objecting states.[1179] Yet the law of reservations still seems to be based on the Vienna Convention in that the reservation modifies the treaty relation if the objecting states have not terminated their treaty relationship with the reserving state within a certain period of time. None of the three Nordic Arctic states that have objected to the Canadian reservation have declared that they consider their treaty relationship with Canada terminated. According to the customary law of treaties, then, the Espoo

[1175] Bowett holds that objections should not and cannot be made against impermissible reservations because such reservations are invalid as such. However, the system of the Vienna Convention clearly entitles the other states to decide for themselves since no other system is available. Bowett 1976-1977, pp. 67-92.

[1176] Article 20 (4b) of the Vienna Convention.

[1177] Ibid., Article 20 (5).

[1178] For arguments in a similar vein, see Sinclair 1984, pp. 76-77.

Convention is applicable between Canada and these three Arctic states as modified by the reservation.[1180]

Thus, in the legal relationships between Canada and the three Nordic states, the Espoo Convention is applicable to the extent of the reservation. Since the Canadian reservation is of such a wide character, the Nordic Arctic states are entitled not to apply the Convention with respect to Canada if, for instance, a certain activity is governed by provincial jurisdiction rather than the central government. However, Canada's reservation is not as problematic in the Arctic as in other parts of Canada since the reservation concerns only those cases where a province has jurisdiction over a certain activity. There are no provinces in the Arctic, only two territories and the settlement areas of indigenous peoples, all of which have a different constitutional status than that of a province. Hence, the Canadian Arctic is always under federal legislative jurisdiction and there is no obstacle to Canada's applying the Espoo Convention fully in its Arctic areas. It can thus be recommended that if the Espoo Convention applies between Canada and one or some of the Nordic Arctic states, it should be applied in its entirety. A number of other reservations have been made to the Espoo Convention which have an effect on the Arctic.[1181]

[1179] See Klabbers 2000, pp. 179-193.

[1180] Article 21 (3) of the Vienna Convention.

[1181] See *supra* note 836 on the reservation made by Denmark. With respect to the EC, it is clear that it is the member states which are responsible for the application of the Convention in their territory. The EC also has some responsibilities in the field of implementation as a contracting party to the Espoo Convention. Since the competence in the field of environmental protection is shared between the Community and the member states, the Community made a 'declaration upon signature and confirmed upon ratification' that: 'It is understood, that the Community member States, in their mutual relations, will apply the Convention in accordance with the Community's internal rules, including those of the EURATOM Treaty, and without prejudice to appropriate amendments being made to those rules'. The basis for the Community's competence in this field is evidently Council Directive 85/337/EEC of 27 June 1985 which does not, however, provide as extensive a transboundary EIA procedure as that required by the Espoo Convention. Hence, the EC also declared upon approving the Convention that: 'It (the Directive) enables the Community to comply with most of the obligations under the Espoo Convention. Member States are responsible for the performance of those obligations resulting from the Espoo Convention not currently covered by Community law and more specifically by Directive 85/337/EEC'. The obligation of the Community is thus to ensure that the member states provide for an internationally transparent EIA procedure

2.4. Scope of Application of the Espoo Convention

Generally speaking, the Espoo Convention is applicable to all stationary activities likely to cause transboundary impact. However, upon closer scrutiny, the scope of the Convention has been refined to apply to certain activities having clearly defined potential transboundary impact. Before elaborating upon the transboundary pollution situations that come within the scope of the Convention, it will be appropriate to review its general scope.

The Espoo Convention is a multilateral international treaty that has been concluded among several states. It is the states parties, which determine the geographical scope of a treaty within the limits of general international law and the rights of third states. Since it is an international treaty, the presumption is that the Espoo Convention applies to the whole of each contracting state's territory unless the contracting states decide otherwise. As provided by Article 29 of the Vienna Convention: 'Unless a different intention appears from the treaty or is otherwise established, a treaty is binding upon each Party in respect of its entire territory'.

As mentioned above, 'territory' denotes those physical areas over which a state exercises territorial sovereignty.[1182] The contracting states of the Espoo Convention

---

towards each other through Directives. The Community is not obligated to legislate an internationally transparent EIA procedure towards third states, i.e., those outside the EC. However, it can be asked whether the Community is responsible for ensuring that the member states provide for an internationally transparent EIA procedure towards third states that are parties to the Espoo Convention. This would mean that even if a member state was not a party to the Espoo Convention, it would be obligated to apply the Convention when dealing with a non-member state which was a party to the Espoo Convention. This issue was clarified by the EC in the Declaration it made upon signature and confirmed upon ratification: 'The Community underlines that Directive 85/337/EEC does not cover the application of the Espoo Convention between the Community on the one hand and non-member States party to the Espoo Convention on the other hand' Information obtained from the UNTS.

[1182] See *supra* chapter IV (3.1.).

have defined the geographical scope of the Convention as being larger than sovereign areas through their use of the term 'jurisdiction'[1183]. The concept covers areas of jurisdictional competence such as the Exclusive Economic Zone (EEZ) and continental shelf. As was examined above, there are areas in the Arctic whose status is under dispute by the Arctic states,[1184] these disputes stemming from the fact that two states have an overlapping claim to the same area. How does the Espoo Convention apply to these areas?

If one of the disputant states authorizes an offshore hydrocarbon exploitation project to operate in these disputed areas, it is, if both states are parties to the Convention, required to apply the Convention normally since the activity takes place in its area of jurisdictional competence, although the other state may also claim the same area.[1185] This is confirmed by the definition of the term 'transboundary impact' as 'caused by a proposed activity the physical origin of which is situated wholly or in part within the area under the jurisdiction of another Party'[1186]. In addition, the Convention defines the term 'party of origin' to mean the contracting party or parties, implying

---

[1183] Article 1 (ii) of the Espoo Convention reads: '"Party of origin means" the Contracting Party or Parties to this Convention under whose jurisdiction a proposed activity is envisaged to take place'. Article 1 (viii) reads: '"Transboundary impact" means any impact...which is situated wholly or in part within the area under the jurisdiction of another Party'.

[1184] Such controversies exist between the United States and Canada over their respective maritime zones in the Beaufort Sea and between Norway and the Russian Federation in the Barents Sea; See supra chapter II (3). In addition, there is a dispute between the contracting parties to the Svalbard Treaty as to whether Norway is entitled to claim exclusive EEZ and continental shelf zones around the Svalbard Islands, thus excluding the economic rights guaranteed by the Treaty to the other contracting states. See *supra* chapter II (2).

[1185] The situation is a bit different in the context of the Svalbard Islands since no other contracting state to the Svalbard Treaty disputes Norway's right to extend its jurisdiction over the water-column and sea-bed around Svalbard, but they do dispute Norway's right to exclude the economic rights of the other contracting states in these areas. If one of the contracting states to the Svalbard Treaty and the Espoo Convention were to authorize hydrocarbon exploitation of the Svalbard shelf, it would not be obligated to apply the Espoo Convention since it is only the territorial state that is obligated under the Convention. Yet this state would be obligated to observe the due diligence principle of general international law, which requires measures from states in respect of activities under their control.

that two or more contracting states can be origin states. This is evident when two or more states have a joint project in their areas of jurisdiction which is likely to cause significant adverse transboundary impact to the environment of other contracting states.

Since the question of the entry into force of the Convention has already been dealt with above, an important question related to the temporal scope of the Convention is whether an activity should be assessed in accordance with the Espoo Convention if the planning process started before the entry into force of the Convention. This question has an ongoing relevance because three of the Arctic states are still to become parties to the Convention. If a treaty contains no provision on the retroactivity of treaty obligations, the following presumption applies in the Vienna Convention:

> Unless a different intention appears from the treaty or is otherwise established, its provisions do not bind a Party in relation to any act or fact which took place or any situation which ceased to exist before the date of the entry into force of the Treaty with respect to that party.[1187]

The planning process of an activity can not be described as an 'act or fact', since it is a long-term process, not a single event. Thus, in accordance with Article 29 of the Vienna Convention, an activity can be seen as a 'situation' and the pivotal question becomes whether the situation has ceased to exist before the entry into force of the Espoo Convention in respect of that activity's state of location.

When does the planning process of an activity cease to exist? The planning process can be seen as terminated when the licencing authority makes its decision on the

---

1186 Article 1 (viii) of the Espoo Convention. From a practical point of view, it is more likely that such a move would result in the parties' being forced to re-open negotiations over the location of their maritime boundaries.

1187 Article 29 of the Vienna Convention.

future of the activity. Thereafter, the plan for an activity ceases to be of interest to the Espoo Convention because it has already been given governmental approval or refusal. Moreover, from a practical viewpoint, it would be difficult to initiate a transboundary EIA procedure after a licencing decision has been made. Consequently, all those activities which have not yet been given a licence to operate potentially come under the scope of the Convention, even if the planning process associated with them started before the entry into force of the Convention for the origin state.

The Vienna Convention regulates the general termination of treaty obligations. A treaty is in force until it is terminated by the provisions of that specific treaty or 'by consent of all the Parties after consultation with the other contracting States'[1188]. The Espoo Convention includes no provisions on the general termination of the Convention and hence can be terminated by the consent of all the contracting states. The Convention does stipulate that the contracting states are entitled to withdraw from the Convention four years after the 'Convention has come into force with respect to a Party'[1189]. However, a withdrawal notice does not absolve a state from its obligations as of the moment it has submitted this written notification to the depositary; rather, as the Convention provides, withdrawal takes effect on the ninetieth day after the date of its receipt by the depositary.[1190]

All activities which come under any national licencing system are included in the subject-matter scope of the Convention. 'Activity' is defined in Article 1(v): '"Proposed activity" means any activity or any major change to an activity subject to a decision of a competent authority in accordance with an applicable national procedure'. It must be noted, however, that the national environmental protection systems of the contracting states define which activities come within the scope of the

[1188] Ibid., Article 54 (b).
[1189] Article 19 of the Espoo Convention.

Convention, as evidenced by the phrase 'in accordance with an applicable national procedure'[1191]. On the other hand, as will be shown below,[1192] the contracting states must establish a national licencing procedure, at least in regard to the activities listed in Appendix I of the Convention.

Proposed activities include all activities which are subject to a national licencing procedure. This applies to both mobile and stationary activities, but in practice only stationary activities come within the scope of the Convention since all national EIA procedures are primarily based on such activities. This is also apparent from the list in Appendix I. On the other hand, most pollution from certain stationary activities to which EIA and transboundary EIA procedure apply, for instance, motorways,[1193] comes from mobile activities.

The use of the term 'activity' confines the scope of the Convention to individual activities only, thus excluding plans or programs. This is specified in Article 2(7):

> Environmental impact assessments as required by this Convention shall, as a minimum requirement, be undertaken at the project level of the proposed activity. To the extent appropriate, the Parties shall endeavour to apply the principles of environmental impact assessment to policies, plans and programmes.

As indicated by the use of the phrase 'to the extent appropriate' and 'endeavour', the Convention imposes only an obligation of attempt to carry out strategic-level assessment, as the assessment of policies is generally called. This requirement is being re-assessed at the moment since the second meeting of the parties established a

---

1190 Ibid.
1191 Ibid., Article 1 (v).
1192 See *infra* chapter V (2.5.1.).
1193 See paragraph 7 of Appendix I of the Espoo Convention.

'subsidiary body, called the open-ended *ad hoc* Working Group on the Protocol, with a mandate to prepare a legally binding instrument in the form of a Protocol'[1194]

A proposed activity must cause 'significant adverse transboundary impact'[1195] to fall within the scope of the Convention. There are two main sub-elements which specify the consequences an activity must have in order to be covered by the Convention. The concept of impact defines on a general level the kind of consequences which the proposed project must have,[1196] and is defined very broadly in the Convention. Article 1 (vii) defines the concept of environment in a comprehensive manner but it does not define what kind of effects an activity must have on the environment, and thus fails to exclude any effects, for instance, alteration of the environment by noise. The concept of the environment covers living beings, humans, flora and fauna; environmental media, i.e., soil, air and water; and the cultural environment, for instance, the landscape and historical monuments. The concept of the environment provided here also includes interaction among living beings, environmental media and the cultural environment, i.e., ecosystems.[1197] Moreover, human social relations, 'cultural heritage or socio-economic conditions' are included, albeit only indirectly.

What kinds of effects on the environment must result before an EIA is invoked? Here again the broad definition of 'impact' figures significantly. 'Impact' means 'any

[1194] Paragraph 1 of Decision II/9 'Strategic Environmental Assessment' of the second meeting of the parties (MP.EIA/2000/9, 7 December 2000). The objective, also in paragraph 1, is 'finalizing [the Protocol] so that it can possibly be adopted at an extraordinary meeting of the Parties to the Convention to be convened on the occasion of the Fifth Ministerial Conference "Environment for Europe" to be held in Kiev, Ukraine'. The *ad hoc* group had its first meeting on 14 to 16 May 2001, the agenda of which can be found on the home page of the UN ECE at http://www.unece.org/env/eia/ad-hocwg.htm [7.3.2001].

[1195] Article 3 (1) of the Espoo Convention.

[1196] Ibid. Article 1 (vii) reads: '"Impact" means any effect caused by a proposed activity on the environment including human health and safety, flora, fauna, soil, air, water, climate, landscape and historical monuments or other physical structures or the interaction among these factors; it also includes effects on cultural heritage or socio-economic conditions resulting from alterations to those factors'.

[1197] For a similar interpretation of the EIA directive of 1985, see Peters 1996, pp. 210-213.

effect' on the above-mentioned environmental elements, and an activity must only be able to cause some kind of alteration to the environment in order to be covered by the Convention. If an activity alters, for instance, socio-economic conditions, it does not fall within the scope of the Convention. Such a situation might arise, for instance, if an activity cannot alter physical factors but can have a psychological effect on socio-economic life on the other side of a national border. The 'impact' of an activity must thus entail adverse physical effects. It must also exceed a certain threshold of importance, called 'significant' in the Convention. An activity must also have an impact on physical areas under the jurisdiction of another contracting state, meaning that impacts on international areas are excluded from the scope of the Convention.[1198] As was shown above, in these cases the EEZ can be considered part of a state's jurisdiction, and not part of the high seas.

2.5. The Transboundary EIA Procedure

The goal of the Espoo Convention is outlined in Article 2 (1): 'The Parties shall, either individually or jointly, take all appropriate and effective measures to prevent, reduce and control significant adverse transboundary environmental impact from proposed activities'[1199]. In order to enforce this goal, the Convention creates two kinds of procedural rules, those which obligate the contracting states to reform their national EIA procedures and prior licencing procedures in order to provide a basis for the transboundary EIA procedure and those which obligate them to ensure that the potentially affected state and its public can participate in the transboundary EIA procedure.[1200]

[1198] See Article 1 (viii) of the Espoo Convention.
[1199] Ibid., Article 2 (1).
[1200] This part of the thesis relies heavily on the present author's article published in the Finnish Yearbook of International Law (1997), Vol. VIII, pp. 161-199 (Koivurova 1997 II). An official interpretation of the provisions of the Convention can be found on the home page of the UN ECE at http://www.unece.org/env/eia/procstep.htm [12.5.2001].

2.5.1. Requirements of the National Environmental Protection System of the Contracting States

The transboundary EIA procedure of the Convention is based on the functioning of the national EIA system of the origin state. As will be shown, the Convention has been organized in a flexible manner so that even a variety of EIA procedures can act as the basis for a transboundary EIA procedure. It is thus very important to study what kind of requirements the Convention sets out for the national EIA procedures of the contracting states.

The Espoo Convention focuses on the activities listed in Appendix I of the Convention.[1201] The aim is to make sure that the EIA procedures of the contracting states are capable of providing the basis for an efficient transboundary EIA procedure, at least with respect to these activities. On the other hand, as provided in Article 2 (2), the Convention does not set unduly high requirements for the contracting states' national EIA procedures:

> Each party shall take the necessary legal, administrative or other measures to implement the provisions of this Convention, including, with respect to proposed activities listed in Appendix I that are likely to cause significant adverse transboundary impact, the establishment of an environmental impact assessment that permits public participation and preparation of the environmental impact assessment documentation described in Appendix II.

The Convention seeks to ensure that the national EIA procedures of the contracting states are capable of initiating the transboundary EIA procedure in regard to the activities listed in Appendix I. However, there is no requirement that each category of activities listed in Appendix I must be subjected to a separate EIA procedure, since another criterion also applies: the potential consequences of the activity must also be

---

[1201] The seven appendices of the Espoo Convention 'form an integral part of the Convention'. Article 10 of the Espoo Convention.

likely to cause significant adverse transboundary impact. Thus, the contracting states are not obligated to establish an EIA procedure which always assesses the environmental impact of the activities listed in Appendix I.

The consequence of this flexible approach to harmonization is that the contracting states are not even required to establish a separate EIA procedure, but, what is known as the sectoral approach, can maintain their EIA procedures where multiple laws and officials are in charge of certain activities listed in Appendix I. It is evident that the contracting states, by not harmonizing the national EIA procedures more, chose to provide flexibility for the content of the national EIA procedure. It is, of course, possible that a sectoral system might be arranged so as to fulfill the obligations of the Convention, but in practice such a system encounters many difficulties. If multiple laws regulate the EIA procedure in regard to the activities listed in Appendix I, and multiple authorities are in charge of implementing these laws, it is possible that the obligations of the Convention cannot be efficiently observed simply because it is too difficult to ensure that each national body is adequately aware of its particular obligations.[1202]

The Espoo Convention imposes only two requirements on the national EIA procedures. Article 2(2) requires that the national EIA procedures of the contracting states contain 'public participation and preparation of the environmental impact assessment documentation described in Appendix II'. The Convention does not require any specific type of EIA procedure in preparing the Environmental Impact Statement (EIS) document, for instance, one operating through the developer or

---

[1202] Ebbesson conducted a study on whether, and to what extent, the Swedish licencing authorities were aware of the obligations of the NEPC. According to him, only a few authorities were aware of the NEPC and its requirements. He concludes that the role of the National Environmental Protection Agency (Naturvårdsverket) should be strengthened. This is only natural since if there exist multiple organs carrying out the duties of the NEPC, it is also much more difficult to educate these organs their responsibilities in comparison to a situation where a co-ordinating organ has a role of overseeing the implementation. Ebbesson 1992.

through an independent government panel; it only set out the minimum conditions for an EIS. These minimum requirements for the EIS are contained in Appendix II of the Convention. No independent scoping procedure is required.

The Convention organizes the participation of the public of the affected state through the non-discrimination principle: the public of the affected state enjoys the same rights of participation as the public of the origin state.[1203] The content of the national EIA procedure with regard to the participation of the public is very important, for it determines the participation rights of the public of the origin and the affected states. Unfortunately, public participation is only mentioned, with no details provided on how this participation should be arranged or at what stage of an EIA procedure it should occur.

Crucial to a transboundary EIA procedure is the time at which the public of the origin state is informed of the pending national EIA procedure. The Espoo Convention connects two important events of a transboundary EIA procedure to this moment. First, the transboundary EIA procedure must be initiated no later than the date on which the origin state's public is informed. Second, since the public of the affected state enjoys the same rights of participation as the public of the origin state, the initial information about the upcoming EIA procedure marks the moment when the public of the affected state must also be involved in a transboundary EIA procedure.

The Convention does not require any specific type of national EIA procedure to be established in this respect. It only stipulates that the initiation of a transboundary EIA procedure must occur at the latest when the public of the origin state is informed. If the national EIA legislation provides that the public need be informed only after the completion of the EIS, the efficiency of the transboundary EIA procedure is considerably weakened. Since the EIS is prepared in accordance with national EIA legislation by either a governmental agency or a developer, the agency or developer must also study the likely transboundary impacts of the proposed activity. However, if the transboundary EIA procedure starts only after the governmental agency or developer has completed the EIS, the document will contain no evaluation of potential transboundary impacts; if it does, these will be based only on the speculative knowledge of possible impacts since no input from the affected state was available before the initiation of the transboundary EIA procedure.

It is, of course, possible that the origin state will start the transboundary EIA procedure independently of the functioning of its national EIA procedure. If this occurs ad hoc, without any legislative background, it does not provide a sufficient basis for the initiation of a transboundary EIA procedure. It is highly desirable that if the national EIA procedure initiates the transboundary EIA procedure only after the completion of the EIS document, the early initiation of the transboundary EIA procedure should be guaranteed through general legislation, or at least through administrative guidelines.

An obligation also exists to establish prior licencing procedures for at least the activities listed in Appendix I of the Convention. This is set forth in Article 2(3):

> The Party of origin shall ensure that in accordance with the provisions of this Convention an environmental impact assessment is undertaken prior to a decision to authorize or undertake a proposed activity listed in Appendix I that is likely to cause a significant adverse transboundary impact.

The contracting states are obligated to establish not only a minimum national EIA procedure but also licencing procedures in regard to the activities listed in Appendix I. As mentioned above, the contracting states retain much discretion as to the kind of EIA procedure they adopt since the Convention does not require any compulsory

---

[1203] Article 2 (6) of the Espoo Convention.

initiation of the transboundary EIA procedure in regard to the listed activities. The situation is different with respect to the establishment of national licencing procedures: the contracting states must establish prior licencing procedures for all the listed activities. If a certain category of the listed activities should fail to fall within the scope of a national licencing procedure, the transboundary EIA procedure of the Convention could not work properly. Since the transboundary EIA procedure is supposed to conclude with the decision by the licencing authority of the origin state, this part of the transboundary EIA procedure would fail to function if no licencing procedure were established.

2.5.2. Transboundary Screening Procedure

As mentioned above, the Espoo Convention does not require the contracting states to establish or reform their national EIA procedures if these procedures already make some provision for public participation and provide the minimum requirements for the EIS set out in Appendix II. Therefore, the contracting states are left considerable discretion as to how they will ensure that the transboundary EIA procedure is initiated with regard to those activities which are likely to cause a significant adverse transboundary impact.

The Convention establishes two categories of activities for which different screening procedures are available: those listed in Appendix I, and those not listed. The Convention contains a presumption that a transboundary EIA procedure must be initiated in regard to the activities listed in Appendix I of the Convention.[1204] These activities are not, however, automatically included in the scope of the Convention;

1204 Ibid. Article 3 (1) reads: 'For a proposed activity listed in Appendix I that is likely to cause a significant adverse transboundary impact, the Party of origin shall, for the purposes of ensuring an adequate and effective consultations under Article 5, notify any Party which it considers may be an affected Party as early as possible and no later than when informing its own public about that proposed activity'.

rather, it is presumed that in normal circumstances they are 'likely to cause a significant adverse transboundary impact'[1205]. This presumption can be rebutted, for instance, by showing that substantial distance from an international border constitutes special circumstances and excludes the listed activity from the scope of the Convention.

The activities chosen for inclusion in Appendix I are known to be environmentally harmful.[1206] Some of these activities are defined in such a manner that all the activities described, irrespective of their scope, come under the definition, for instance, 'Offshore hydrocarbon production'[1207].

[1205] Ibid., Article 3 (1).

[1206] Ibid. The categories of activities listed in Appendix I are the following: '1. Crude oil refineries (excluding undertakings manufacturing only lubricants from crude oil) and installations for the gasification and liquefaction of 500 tonnes or more of coal or bituminous shale per day. 2. Thermal power stations and other combustion installations with a heat output of 300 megawatts or more and nuclear power stations and other nuclear reactors (except research installations for the production and conversion of fissionable and fertile materials, whose maximum power does not exceed 1 kilowatt continuous thermal load). 3. Installations solely designed for the production or enrichment of nuclear fuels, for the reprocessing of irradiated nuclear fuels or for the storage, disposal and processing of radioactive waste. 4. Major installations for the initial smelting of cast-iron and steel and for the production of non-ferrous metals. 5. Installations for the extraction of asbestos and for the processing and transformation of asbestos and products containing asbestos: for asbestos-cement products, with an annual production of more than 20.000 tonnes finished product; for friction material, with an annual production of more than 50 tonnes finished product; and for other asbestos utilization of more than 200 tonnes per year. 6. Integrated chemical installations. 7. Construction of motorways, express roads and lines for long-distance railway traffic and of airports with a basic runway length of 2.100 metres or more. 8. Large-diameter oil and gas pipelines. 9. Trading ports and also inland waterways and ports for inland-waterway traffic which permit the passage of vessels of over 1.350 tonnes. 10. Waste-disposal installations for the incineration, chemical treatment or land-fill of toxic and dangerous wastes. 11. Large dams and reservoirs. 12. Groundwater abstraction activities in cases where the annual volume of water to be abstracted amounts to 10 million cubic metres or more. 13. Pulp and paper manufacturing of 200 air-dried metric tonnes or more per day. 14. Major mining, on-site extraction and processing of metal ores or coal. 15. Offshore hydrocarbon production. 16. Major storage facilities for petroleum, petrochemical and chemical products. 17. Deforestation of large areas'.

[1207] Ibid., paragraph 15 of Appendix I.

Some categories of activities provide for threshold levels, e.g.,:

> Crude oil refineries (excluding undertakings manufacturing only lubricants from crude oil) and installations for the gasification and liquefaction of 500 tonnes or more of coal or bituminous shale per day.[1208]

In some, the threshold is described in very general terms, e.g., 'major', providing for more discretion. Paragraph 4 reads: 'Major installations for the initial smelting of cast-iron and steel and for the production of non-ferrous metals'[1209]. It is also specified that the potential consequences must be 'likely' to cause a significant adverse transboundary impact.[1210]

In order to enforce this presumption with regard to the listed activities, the Convention contains a special mechanism for evaluating whether the criteria provided in Article 3(1) are fulfilled. If the origin state decides not to initiate a procedure with regard to an activity listed in Appendix I, the concerned states are, according to Article 2 (8), obligated to discuss whether the transboundary EIA procedure should be applied. If discussions between the states fail, either party, normally the affected state, has the right to initiate an inquiry commission procedure in which an expert panel gives its opinion on whether the activity is likely to cause a significant adverse transboundary impact.[1211] Crucially, the affected state is entitled to initiate the inquiry commission procedure without the consent of the origin state and may even obtain the opinion from the commission without that state's participation.[1212] As the decision of the inquiry commission is only an opinion, it is not directly legally binding. From a practical point of view, the inquiry commission

---

1208 Ibid., paragraph 1.
1209 Ibid., paragraph 4.
1210 Ibid., Article 3 (1).
1211 Ibid.
1212 Ibid., Appendix IV of the Espoo Convention.

procedure puts pressure on the decision of the origin state whether or not to initiate the transboundary EIA procedure.

The Convention contains a reverse presumption with respect to those activities not listed in Appendix I: it is presumed that they do not fulfill the criteria laid down in Article 3(1) in normal circumstances. This presumption can be rebutted by showing that exceptional circumstances exist; for instance, the activity may be located very near an international border. The criteria for these exceptional circumstances are enumerated in Appendix III of the Convention:

> (a) Size: proposed activities which are large for the type of the activity;
> (b) Location: proposed activities which are located in or close to an area of special environmental sensitivity or importance (such as wetlands designated under the Ramsar Convention, national parks, nature reserves, sites of special scientific interest, or sites of archeological, cultural or historical importance); also, proposed activities in locations where the characteristics of proposed development would be likely to have significant effects on the population;
> (c) Effects: proposed activities with particularly complex and potentially adverse effects, including those giving rise to serious effects on humans or on valued species or organisms, those which threaten the existing or potential use of an affected area and those causing additional loading which cannot be sustained by the carrying capacity of the environment.

In normal circumstances, activities not listed in Appendix I are seen as falling outside the scope of the Convention. However, a mechanism is provided for unlisted activities whereby the affected state is entitled to enter into discussions with the origin state.[1213] If exceptional circumstances exist, for instance, if the activity is located very near an international border, the Convention will apply.[1214] In the Arctic, it would seem that criterion (b) in particular may require the initiation of the transboundary EIA procedure even if the activity is not on the list in Appendix I,

[1213] Ibid., Article 2 (5).
[1214] Ibid., Appendix III (2).

because there are many varying kinds of nature parks and reserves in the region. The Arctic contains a vast number of protected areas of varying legal status that have been mapped by the CAFF under its protected areas program (see figure 8).[1215] One of the important aspects of the Espoo Convention is that military activities, although not included in Appendix I, may come within its scope if, for instance, they are located very near a border.

A transboundary EIA procedure applies only if both of the states concerned so agree. This provision pertains to both listed and non-listed activities. Consent to apply the procedure is expressed in accordance with the provisions of the Convention, the origin state providing notification and the affected state furnishing a response that it will participate in the procedure.[1216] Even if the affected state resorts to the inquiry commission procedure, and the Commission determines that the procedure of the Convention should be applied, it is the task of the origin state to initiate the procedure through notification.

1215 See the home page of CAFF at http://www.grida.no/caff/cpanstratplan.htm [23.4.2001].
1216 The consent of the origin state is expressed through the notification stipulated in Article 3 (1) of the Espoo Convention and the affected state's consent through the participation response provided for in Article 3 (3-4) of the Convention.

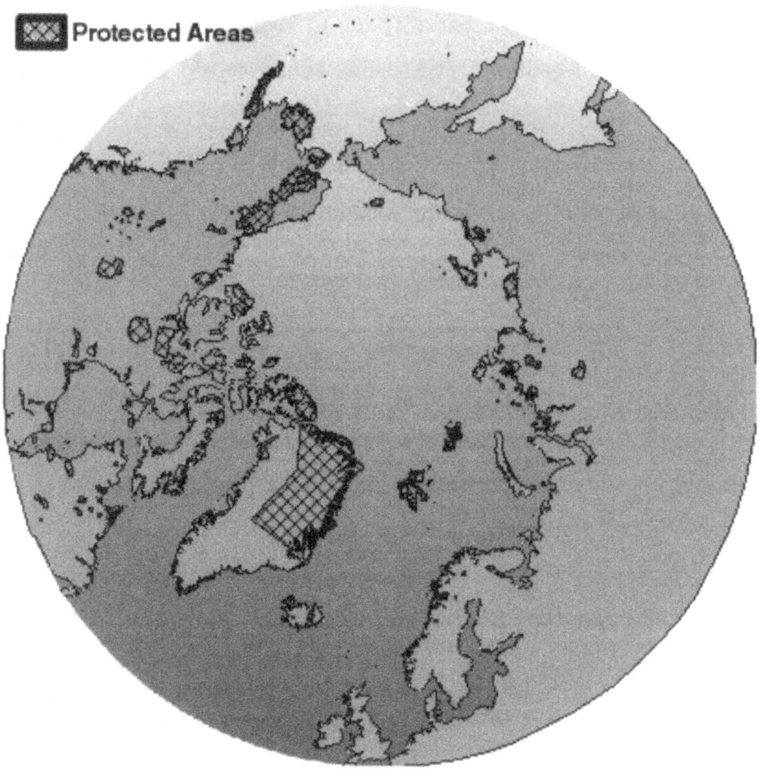

Figure 8 (source: home page of the UNEP Grid Arendal at http://maps.grida.no/arctic/)

It must be determined separately whether a refusal by the origin state to initiate a transboundary EIA procedure amounts to a breach of its treaty obligation. As shown above, the origin state is obligated to initiate the procedure if a significant adverse transboundary impact is likely to follow from a proposed activity. With regard to the activities listed in Appendix I, there is a strong presumption that, in normal circumstances, the procedure must be initiated if the origin state cannot show the existence of special circumstances. Hence, the legality of the state's failure to initiate the transboundary EIA procedure in these cases entails a burden of proof to show the existence of special circumstances. If the origin state declines to initiate the procedure after an affirmative finding of the Inquiry Commission, it is clearly in violation of its treaty obligations under the Convention.

The same logic does not apply to proposed activities that are not listed in Appendix I of the Convention. Here it is the affected state that has to show the existence of special circumstances in order for the Convention to be applicable. Breach of the treaty by the origin state due to failure to initiate the procedure is much more difficult to show in these cases. However, if, for instance, the non-listed activity is planned in a location very near the border and is likely to pollute a natural park on the other side, non-initiation on the part of the origin state would clearly amount to a breach of treaty.

The timing of the initiation of the transboundary EIA procedure is specified in Article 3(1):

> the Party of origin shall, for the purposes of ensuring adequate and effective consultations under Article 5, notify any Party which it considers may be an affected Party as early as possible and no later than when informing its own public about that proposed activity.

The provision sets a minimum obligation, and a possible one. The deadline for notification is the time at which the public of the origin state is informed. If the origin state is in a position to carry out notification earlier than the deadline outlined above, even though its national EIA procedure does not provide for the early notification of the public, it must do so. Although the obligation to notify 'as early as possible' is a bit vague, it still establishes an obligation of attempt on the part of contracting states.

Article 3 (2, 3, 4, 5) regulates the first round of information exchange between the concerned states. The applicability of these provisions is dependent on the moment at which the origin state is obligated to initiate the transboundary EIA procedure. If the notification takes place only after the EIS has already been completed, it is only required that the potentially affected state be notified of the EIS; there is no need for preliminary information exchange in these cases.

However, in a typical case, the origin state is in a position to initiate the transboundary EIA procedure before the EIS is conducted. First, the origin state must notify the potentially affected state. The minimum content of this notification is outlined in Article 3 (2).[1217] The affected state is obligated to respond to the notification 'within the time specified in the notification'[1218] and must indicate whether it wants to participate in the procedure. According to Article 3 (4), it remains for the potentially affected state to ensure that it maintains its rights to participate in the transboundary EIA procedure by an express notification. If it does not respond within the time specified in the notification, or indicates that it does not want to participate, the origin state is not obligated to involve the potentially affected state in

[1217] Article 3 (2) of the Espoo Convention reads: '(a) Information on the proposed activity, including any available information on its possible transboundary impact; (b) The nature of the possible decision; and (c) An indication of a reasonable time within which a response under paragraph 3 of this Article is required, taking into account the nature of the proposed activity'.
[1218] Ibid., Article 3 (3).

the transboundary EIA procedure.[1219] If the origin state receives a reply from the potentially affected state expressing its willingness to participate in the procedure, the former is obligated to supplement the information that was delivered in the notification.[1220] If such information was delivered already at the stage of notification, it need not be delivered a second time.

Since the Espoo Convention sets out only general rights and obligations for the contracting states, it leaves much discretion to the parties in implementing and applying obligations. This is the reason why the first meeting of the parties adopted two decisions related to the notification procedure between the concerned states. The first decision constituted an agreement as to the points of contact stipulating that notifications must be delivered to the points of contact of the contracting states 'unless otherwise provided in bilateral or multilateral agreements or other arrangements'[1221]. If no point of contact has been nominated, the notification must be transmitted to the ministry of foreign affairs of the potentially affected state. The second decision at the meeting adopted a format for notification.[1222]

2.5.3. Preparation of the Transboundary EIS Document

The Espoo Convention lays down the minimum requirements for the EIS document to be produced by the national EIA procedure. This is of great importance for the functioning of the transboundary EIA procedure since the potentially affected state is

---

1219 Ibid. Article 3 (4) stipulates that if the potentially affected state does not respond in time, or indicates that it does not want to participate, 'the provisions in paragraphs 5, 6, 7 and 8 of this Article and in Articles 4 to 7 will not apply'.

1220 Ibid. Article 3 (5) reads: '(a) Relevant information regarding the environmental impact assessment procedure, including an indication of the time schedule for transmittal of comments; and (b) Relevant information on the proposed activity and its possible significant adverse transboundary impact'.

1221 Report of the first meeting of the parties, Annex III, Decision I/3.

1222 Ibid., Annex IV, Decision I/4, and the appendix. The following issues are dealt with: background, specific information requirements and form of notification.

provided with all the information produced by the national EIA procedure.[1223] According to Appendix II, the main requirements of the national EIA procedures of the contracting states are the following.

The EIS must contain reasonable alternatives to the proposed activity, including a no-action alternative. If the proposed activity is site-specific, meaning that it can operate only in a certain location, the alternatives are technological solutions, different size-alternatives, and a no-action alternative. If the proposed activity is not site-specific, the alternative plans for location are important.[1224] The EIS must include the 'potential environmental impact of the proposed activity and its alternatives and an estimation of its significance'[1225]. Thus, not only must the impact of the main proposed plan for an activity and its possible alternatives be assessed but also an estimation of the significance of all these plans on the environment must be given.

The determination of the objects of an environmental study is defined in Article 1(vii). The EIS must also contain proposals for measures to mitigate the effects of the proposed activity and of each alternative.[1226] It is also important that the EIS clarify

---

[1223] The concerned states may exclude certain information from the EIS document, or from the notification. Article 2(8) of the Convention states: 'The provisions of this Convention shall not affect the right of Parties to implement national laws, regulations, administrative provisions or accepted legal practices protecting information the supply of which would be prejudicial to industrial and commercial secrecy or national security'. This rule gives rather broad discretion to the origin state to exclude information from the EIS. However, like all rules of exception, it has to be interpreted narrowly.

[1224] Paragraphs a and b, Appendix II of the Espoo Convention.

[1225] Ibid., paragraph d.

[1226] Ibid. The minimum content stipulated in Appendix II of the Convention includes the following elements: '(a) a description of the proposed activity and its purpose; (b) a description, where appropriate, of reasonable alternatives (for example locational or technological) to the proposed activity and also the no-action alternative; (c) a description of the environment likely to be significantly affected by the proposed activity and its alternatives; (d) a description of the potential environmental impact of the proposed activity and its alternatives and an estimation of its significance; (e) a description of mitigation measures to keep adverse environmental impact to a minimum; (f) an explicit indication of predictive methods and underlying assumptions as well as the relevant environmental data used; (g) an identification of gaps in knowledge and uncertainties encountered in compiling the required information; (h) where appropriate, an outline for monitoring

the uncertainties surrounding the collection of information, thereby implementing the precautionary principle. Finally, the results of the EIS must be presented in an understandable manner. In the Arctic context, the Arctic Monitoring and Assessment Program (AMAP) has been very important since it has provided baseline information to be used in making these Arctic impact assessments. In 1997, the AMAP published its comprehensive Assessment Report, which contained information on the characteristics of the Arctic environment from different perspectives.[1227]

Oddly enough, Appendix II does not contain any reference to potential transboundary impact even though that is precisely what it deals with. The Convention only requires the contracting states to ensure that their national EIA procedures study the above-mentioned issues and present them in a certain manner; it does not require any separate focus on transboundary impacts. This is due to the fact that the contracting parties were not able to harmonize their national EIA procedures any more than they already had. As the Convention now accepts a national EIA procedure which involves the public only after the EIS has been completed, it also accepts that the transboundary EIA procedure starts after the EIS has been done. There was thus no possibility to include transboundary impacts in the list of Appendix II since this would have excluded from the Convention states that did not have an appropriate national EIA procedure.

However, if the national EIA procedure involves the public before the EIS is prepared, for instance through a scoping procedure, the Convention requires that the potentially affected state provide information to the origin state and that the public of

---

and management programmes and any plans for post-project analysis; and (i) a non-technical summary including a visual presentation as appropriate (maps, graphs, etc.)'.

[1227] AMAP Assessment Report 1998. There are also other cooperation processes underway to provide for baseline information from the Arctic environment. For instance, the 1994 Agreement on Cooperation in the Prevention of Pollution of the Environment in the Arctic Between the United States and the Soviet Union contains obligations for the parties to exchange information on the state

the potentially affected state be able to participate in this procedure. In Article 3 (6) it is stipulated that:

> An affected Party shall, at the request of the Party of origin, provide the latter with reasonably obtainable information relating to the potentially affected environment under the jurisdiction of the affected Party, **where such information is necessary** for the preparation of the environmental impact assessment documentation. The information shall be furnished promptly and, as appropriate, through a joint body where one exists.[1228] (emphasis added)

The condition for the delivery of information from the affected state to the origin state is that the information be 'necessary for the preparation of the environmental impact assessment documentation'[1229]. In normal circumstances, there would always seem to exist a need for information about the environmental conditions of the area likely to be affected. Since the initiation of a transboundary EIA procedure means that both states have agreed that a significant adverse transboundary environmental impact is likely, and the origin state is not competent to study the potential environmental impact on the area of the affected state, there presumptively exists the need for such information. The EIS would definitely remain inadequate without details of environmental conditions in the area likely to be affected. If and when a need exists to obtain information from the affected state, the affected state must provide that information to the origin state. This may happen through a request by the origin state, but if no such request is forthcoming, and the need exists, the affected state may provide this information. Article 3 (6) provides that such information be delivered 'through a joint body where one exists'[1230].

---

of the Arctic environment as well as in what way the anthropogenic pollution effects the indigenous peoples and the environment in general.

1228 Article 3 (6) of the Espoo Convention.

1229 Ibid.

1230 The 1992 Transboundary Watercourses Convention contains provisions which are relevant here. First of all, the so-called riparian parties of transboundary watercourses are obligated to establish joint bodies. One of the tasks of these joint bodies is 'to participate in the implementation of environmental impact assessments relating to transboundary waters, in accordance with appropriate international regulations'. Hence, the riparian parties are obligated to make use of these joint bodies

The Convention does not require that the public of the affected state be given an opportunity to participate in the preparation of the EIS document. However, if the national EIA procedure of the origin state provides this opportunity to the public of the origin state, the non-discrimination principle of Article 2(6) accords the same right to the public of the affected state. In such cases, the states concerned must ensure that the public of the affected state is given the same possibility to influence the scope of the EIS as the public of the origin state. The Convention contains a separate provision for these situations. Article 3(8) provides general guidelines for the practical organization of the participation of the public of the affected state in the preparation of an EIS. The states concerned must inform the public 'likely to be affected'[1231] and provide them with an opportunity to submit comments either directly to the competent authority - depending on the EIA procedure of the origin state - or indirectly via the origin state's contact authority. In accordance with the non-discrimination principle, the manner and form of participation depend on the EIA legislation of the origin state.

2.5.4. Reaction to the Completed EIS by the Affected State and its Public

After completion of the EIS, the states concerned are obligated to agree upon the time frame for consultations. They are to enter into consultations 'without undue delay'[1232] concerning, e.g., the potential transboundary impact and measures to reduce or eliminate the impact. The Convention provides certain guidelines for the possible subjects of discussion.

---

in implementing the transboundary EIA procedures provided by the two conventions. The Transboundary Watercourses Convention emphasizes that especially in the exchange of information the joint bodies must be used by the riparian parties.

[1231] Article 3 (8) of the Espoo Convention.

[1232] Ibid., Article 5.

Article 5 provides the following suggestions for the topics of consultation between the concerned states:

> (a) Possible alternatives to the proposed activity, including the no-action alternative and possible measures to mitigate significant adverse transboundary impact and to monitor the effects of such measures at the expense of the Party of origin;
> (b) Other forms of possible mutual assistance in reducing any significant adverse transboundary impact of the proposed activity; and
> (c) Any other appropriate matters relating to the proposed activity.

The Convention again provides a flexible approach since it is suggested that the concerned states discuss, for instance, forms of mutual assistance. In the Arctic, such issues should be very important between the Russian Federation and its neighboring states because of the gap between the standards of living. Article 5 also provides that 'consultations may be conducted through an appropriate joint body, where one exists'. Joint organs available for use in the consultation phase also exist for purposes of Arctic transboundary EIA.[1233]

[1233] The following joint bodies are available also for the execution of transboundary EIA procedures taking place in the Arctic. The 1971 Agreement Between the Government of Canada and the Government of Norway on Sealing and the Conservation of the Seal Stocks in the North-west Atlantic establishes a Commission, for instance, to submit proposals to Canada and Norway concerning scientific research to be undertaken jointly or separately with respect to sealing and the conservation of the seal stocks. The Finnish Russian Borderwater Commission has power to make necessary investigations, and thus also impact assessments, and obtain assistance from the officials of both of the parties to this task (for more details, see *supra* chapter IV (5.1.)). The Finnish Norwegian Borderwater Commission is entitled to carry out preventive investigations, and thus also impact assessment studies, and receive assistance from both of the parties (information of the Commission can be found at http://www.vyh.fi/kvasiat/rajavesi/snrajave.htm [29.4.2001]). The Porcupine Caribou Herd Agreement between the United States and Canada also provides some powers for a joint organ. First of all, if an activity in the area of either party 'could significantly affect the conservation of Porcupine Caribou Herd or its habitat', the origin state must notify a joint organ of the parties. The so-called Board has a mandate to make recommendations, which are expressly non-binding in law (see more closely *supra* chapter IV (4.1.2.)). In the 1992 Paris II Convention, the Commission, the joint organ of all parties to this multilateral treaty, has some tasks in the situations studied here. If the potentially affected state so requests, the origin state must

By agreeing to initiate the procedure set out in the Convention, the states concerned accept that there is the likelihood of a significant adverse environmental impact from the proposed activity. And, as the goal of the Convention is to 'take all appropriate and effective measures to prevent, reduce and control' such impacts, it is recommendable, but not obligatory, to come up with plans to reduce the likely transboundary impact of the proposed project. It is also important to have regard to those international treaties providing for procedures in cases of potential accidents at this phase. There are multiple treaties applicable in the Arctic that provide for different procedures in case of accident.[1234] It is, however, possible that the EIS will

---

arrange for consultations 'with a view to negotiating a co-operation agreement'. At the request of any party concerned, the Commission is entitled to 'consider the question' and make recommendations 'with a view to reaching a satisfactory solution'. The 1992 Agreement on Cooperation in the Field of Environmental Protection Between the Finnish Republic and the Russian Federation (author's own translation) establishes a joint organ called the Environmental Cooperation Commission. It is important that one of the specific fields of this environmental protection cooperation between the parties is, according to Article 3 of the Agreement, environmental impact assessment. The 1992 Agreement between the Government of Canada and the Government of the Russian Federation on Cooperation in the Arctic and the North establishes a Mixed Commission on Cooperation in the Arctic and the North under Article 4. The Annex to the agreement lists, e.g., effects and transport of contaminants as one possible area of cooperation. Through the 1993 Environmental Cooperation Agreement, parties agreed to establish a Canadian-Russian Mixed Environmental Commission (Article V). One part of its mandate is environmental impact assessments (Article II). The 1994 Agreement on Cooperation in the Field of Protection of the Environment and Natural Resources between the United States and the Russian Federation establishes a joint organ called the Joint Committee. In this context it is important that the fields of cooperation enumerated in Article 3 include: '5. arctic and subarctic resources; 6. environmental impact assessment'.

1234 General accident treaties that apply also in the Arctic include: The 1986 Convention on Early Notification of a Nuclear Accident of 1986. The 1986 Convention on Assistance in the Case of a Nuclear Accident or Radiological Emergency. The 1993 Convention Concerning the Prevention of Major Industrial Accidents. The 1992 Convention on the Transboundary Effects of Industrial Accidents. The 1989 Agreement Between Denmark, Finland, Norway and Sweden on Cross-Border Cooperation in Preventing or Limiting the Damage in Emergency Situations Having Effect on Humans, Property or the Environment (author's own translation). Accident treaties focusing on oil-related accidents and applying in the Arctic include: The 1990 International Convention on Oil Pollution Preparedness, Response and Co-Operation. The 1971 Agreement Concerning Co-Operation in Measures to Deal with Pollution of the Sea by Oil. The 1974 Agreement Relating to the Establishment of Joint Pollution Contingency Plans for Spills of Oil and Other Noxious Substances (Canada and the United States). General treaties that contain regulation of accidents are: The 1982 LOSC (Article 198) and the 1992 Convention on Biological Diversity (Article 14 (d)). The following Arctic-specific accident treaties are: The 1989 Agreement Between the United States

show that no significant adverse transboundary environmental impact from the proposed activity is likely, making it also less likely that agreements between the concerned states are necessary.

The mitigation of a likely significant adverse transboundary impact may be achieved, for instance, by selecting less polluting alternatives to the main proposed idea or by executing measures reducing the impact of the main proposed activity. Since the alternatives, their environmental impacts, and possible mitigation measures will have already been recorded in the EIS, the states have a good basis for finding measures to reduce the likelihood of significant adverse transboundary damage.

The Espoo Convention requires that, at a minimum, the national EIA procedures of the contracting states should provide for participation of the public after the completion of the EIS. Hence, after the completion of the EIS, both the public of the affected and of the origin state are entitled to participate in the transboundary EIA procedure. The specific manner and form of this participation is determined by the legislation of the origin state. Suggestions regarding the practical organization of the participation of the public of the affected state are given in Article 4(2).[1235] The concerned states should agree upon how to inform the public of the affected state. The active participation of the public of the affected state should occur either directly, through comments and objections delivered to the competent authority, or indirectly

---

and the Soviet Union Concerning Cooperation in Combatting Pollution in the Bering and Chukchi Seas in Emergency Situations. The 1983 Agreement Between the Government of the Kingdom of Denmark and the Government of Canada for Cooperation relating to the Marine Environment, Annexes A and B. The 1994 Agreement Between the Government of the Kingdom of Norway and the Government of the Russian Federation on Cooperation and Response to Oil Pollution in the Barents Sea. The 2000 Memorandum of Understanding Between the Government of Norway and the Government of the Russian Federation on Cooperation on search and rescue. See also the Field Guide for Oil Spill Response in Arctic Waters made by the EPPR working group of the Arctic Council at http://www.arctic-council.org/fldguide/index.asp [12.3.2001].

1235 Ebbesson (1996, p. 189) criticizes the Espoo Convention's system of participation of the public of the potential affected state for its failure to specify the 'form nor intensity of public participation'. See also Ebbesson 1999, pp. 47-55.

via the contact authority of the origin state. Since the participation rights of the public of the affected state are the same as those of the public of the origin state, the former may always try to participate directly without mediation by the concerned states if the public of the origin state enjoys such rights.

2.5.5. The Licencing Decision

The Espoo Convention requires licencing procedures to be established for the activities listed in Appendix I of the Convention. It is these national licencing procedures that produce the final decision on the permissibility of an activity and, if it is permitted, the terms on which it must operate. In this sense, the Convention implements the 'polluter pays' principle, since the economic enterprise internalizes, in principle at least, the costs of its pollution.

The transboundary EIA procedure produces information about likely transboundary impact, as well as proposals for its mitigation, for the licencing authority. Neither the outcome of the consultations between the states concerned nor the comments of the public of the affected state dictate the decision of the transnational permit authority. Nollkaemper is of the opinion that the transboundary EIA procedure of the Espoo Convention: 'inevitably [does] not guarantee that the state of origin should follow a particular decision'[1236]. Article 2 (1) of the Convention can most adequately be seen as fixing the object and purpose of the treaty rather than requiring material measures of its own. Still, as was shown above, the customary law principle of due diligence and the goal of sustainable development govern these situations as well.[1237]

The main aim of the transboundary EIA procedure of the Espoo Convention is to provide more information for the national EIA procedure, which normally only

[1236] Nollkaemper 1992, pp. 168-169.
[1237] See *supra* chapter IV (7.).

collects information on a national basis. All this information, both nationally produced and internationally, goes to the licencing authority of the origin state, which makes the final decision. With respect to the listed activities of Appendix I, the national licencing procedures normally operate on the basis of balancing norms, i.e., material norms that require the weighing of positive and negative aspects of the activity. In normal circumstances, the transboundary EIA procedure identifies precisely these negative aspects of the activity. The Convention ensures that the results of the transboundary EIA procedure have relevance in the licencing decision-making in Article 6(1):

> The Parties shall ensure that, in the final decision on the proposed activity, due account is taken of the outcome of the environmental impact assessment, including the environmental impact assessment documentation, as well as the comments received pursuant to Article 3, paragraph 8 and Article 4, paragraph 2, and the outcome of the consultations as referred to in Article 5.

The Convention also requires an account of the decision-making process related to the transboundary EIA procedure. The origin state must provide to the affected Party the final decision on the proposed activity along with the reasons and considerations on which the decision was based.[1238] If one of the states concerned acquires additional information after the final decision has been made but before the activity has begun, either of the states concerned has the right to initiate consultations 'as to whether the decision needs to be revised'[1239]. Additional requirements for the exercising of this right are that the information was not available at the time a decision was made and that it could have materially affected the decision.[1240]

[1238] Article 6 (2) of the Espoo Convention.
[1239] Ibid., Article 6 (3).
[1240] Ibid.

The final decision must also take into account the comments and objections of the public of the affected state.[1241] Curiously enough, according to the wording of Article 6 (2), the final decision must only be delivered to the affected state, not to the public of that state. The right of the public of the affected state to non-discrimination pertains only to the transboundary EIA procedure since this is the only procedure for which the Convention ensures public participation. Thus, there exists no clear-cut obligation to inform the public of the affected state of the final decision. If the public of the affected state has participated in the transboundary EIA procedure, it is of course highly desirable that the affected state should disseminate the decision it receives from the origin state to them.

2.5.6. Post-Project Analysis

Only a few national EIA procedures contain compulsory post-project analysis.[1242] Although the Convention contains provisions on such a procedure, these only obligate states to discuss such measures. The Convention suggests that the EIS should already contain a plan for post-project analysis.[1243] Transboundary post-project analysis is voluntary for the states concerned; the EIS may contain this outline at the states' discretion.

If either of the states concerned so requests - which is most likely in the consultation phase - they must together determine whether a post-project analysis is to be carried out. Where the states agree on such an analysis, they are advised by Appendix V of the Convention. If either state concludes, through the post-project analysis, that a significant adverse transboundary impact has occurred, it must inform the other state.

[1241] Ibid., Article 6 (1).
[1242] See, e.g., Gilpin 1995, pp. 26-28.
[1243] Appendix II (h) of the Espoo Convention reads: 'Where appropriate, an outline for monitoring and management programmes and any plans for post-project analysis'.

The states are then obligated to 'consult on necessary measures to reduce or eliminate the impact'[1244].

2.6. Dispute-resolution and Compliance Mechanism

The Convention contains two sets of dispute-resolution provisions: Articles 2 (7) on the inquiry commission procedure and Article 15 on general settlement of disputes. As described above, the inquiry commission procedure is a special dispute-resolution procedure relating only to the phase when the decision is made as to the initial applicability of the transboundary EIA procedure to a particular case. This procedure overrides the general dispute settlement between the parties as regulated by Article 15 since, as already mentioned, the affected state is entitled to initiate the inquiry commission procedure even without the participation of the origin state; the use of the forms of dispute-resolution outlined in Article 15 require the consent of both states.

The general dispute-resolution provision of the Espoo Convention is dependent on the consent of both concerned states.[1245] The parties are also given a possibility to declare that they will accept a third-party dispute-resolution mechanism, either the arbitration procedure set out in Appendix VII of the Convention or submission to the

[1244] Ibid., Article 7.

[1245] Ibid. According to paragraph 1 of Article 15: 'If a dispute arises between two or more Parties about the interpretation or application of this Convention, they shall seek a solution by negotiation or by any other method of dispute settlement acceptable to the parties to the dispute'.

ICJ.[1246] If both of the concerned states make such a declaration, the dispute automatically goes to the ICJ unless the states decide otherwise.[1247]

The second meeting of the contracting parties adopted a third means of resolving controversies, aimed at complementing the two mentioned above: Decision II/4, the 'review of compliance'.[1248] In paragraph 1, what is known as the Implementation Committee was established 'for the review of compliance by the Parties with their obligations under the Convention with a view to assisting them fully to meet their commitments'[1249]. The structure of the Implementation Committee and its objectives and functions are laid down in paragraphs 1-3 of the Annex to the Decision.[1250] Here it will be useful to study the review procedure in relation to situations in which a dispute between parties arises over a transboundary EIA procedure. The procedure, of course, applies to all situations where a party is not in compliance with the obligations of the Espoo Convention.

If the origin state does not fulfill its duties related to the transboundary EIA procedure, the affected state may notify the Secretariat of the Convention with corroborating information. The Secretariat will then inform the origin state and allow it to deliver information of its own. Thereafter, the Secretariat will transmit these sets

1246 Ibid. Paragraph 2 reads: 'When signing, ratifying, accepting, approving or acceding to this Convention, or at any time thereafter, a Party may declare in writing to the Depositary that for a dispute not resolved in accordance with paragraph 1 of this Article, it accepts one or both of the following means of dispute settlement as compulsory in relation to any Party accepting the same obligation: (a) Submission of the dispute to the International Court of Justice; (b) Arbitration in accordance with the procedure set out in Appendix VII'.

1247 Ibid. Paragraph 3 reads: 'If the parties to the dispute have accepted both means of dispute settlement referred to in paragraph 2 of this Article, the dispute may be submitted only to the International Court of Justice, unless the parties agree otherwise'.

1248 Personal copy of a report from the Sofia meeting obtained by the author from the Finnish Ministry of the Environment. The report did not specify the particulars of the decisions, and thus the preparatory documents that can be found on the home page of the UN ECE (at http://www.unece.org/env/eia/2p.htm [24.4.2001]) have been used (MP.EIA/2001/4). The decisions were taken mainly on the basis of these preparatory documents.

1249 Paragraph 1 of Decision II/4.

of information to the Committee.[1251] It is also possible that the origin state will notify the Secretariat that it is unable to comply with the obligations of the Convention, either after the notification by the affected state or on its own initiative. In these cases, the origin state is obligated to inform the Secretariat of its reasons for non-compliance, and this information is delivered to the Committee.[1252] The Implementation Committee may also, on its own initiative, commence the procedure with the party in breach of its obligations.[1253]

The Implementation Committee has several options in proceeding with the dispute: it may request further information on the dispute through the Secretariat or even gather information in the territory of the concerned states, but only at their invitation.[1254] The Committee must ensure the confidentiality of the information delivered to it.[1255] The concerned states are entitled to participate in the proceedings but not in the preparation and adoption of any report or recommendations of the Committee.[1256] The Committee is required to aim for a consensus, but if such cannot be obtained the report must reflect the views of all the Committee members.[1257] The report and recommendations of the Committee are then delivered to a meeting of the contracting parties, which decides on the measures to bring about compliance with the Convention as well as measures to assist the party in doing so. If consensus cannot be reached, the meeting of the parties decides the content of measures 'by a three-fourths majority vote of the Parties present and voting'[1258].

[1250] According to paragraph 1 of the Annex to Decision II/4, the Implementation Committee consists of eight parties to the Convention which all shall appoint a member to the Committee.
[1251] Ibid., paragraph 4 (a).
[1252] Ibid., paragraph 4 (b).
[1253] Ibid., paragraph 5.
[1254] Ibid., paragraph 6.
[1255] Ibid., paragraph 7.
[1256] Ibid., paragraph 8.
[1257] Ibid.
[1258] Ibid., paragraph 11.

According to paragraphs 12 and 13, the general dispute-resolution and the inquiry commission procedure override the review procedure if they are applicable. The review procedure is mainly intended to complement these basic dispute-resolution procedures in cases of dispute over potential transboundary pollution. As mentioned above, a party is entitled to initiate the inquiry commission procedure without the consent of the origin state, and even obtain a decision by the Inquiry Commission without the participation of the origin state. The review procedure can, in these situations, complement the inquiry commission procedure since if the origin state has not participated in the inquiry commission procedure, it is likely that it will refuse to initiate such a procedure even after a finding of the Inquiry Commission to the effect that significant adverse transboundary impacts are likely, or to start any other means of dispute settlement on the basis of Article 15. If the origin state does not want to engage in diplomatic negotiations or submit the matter for third-party settlement, there is room for a review procedure by which the affected state can address the problem.

The three means of resolving controversies are different from the viewpoint of method of trying to solve the pending dispute over transboundary EIA procedure. The inquiry commission procedure relies on factual information as a way to overcome the differences of opinion, whereas the general dispute settlement and the review procedure rely on political pressure. The structural difference between the review procedure and the general dispute settlement means of Article 15 is clear in that the former places the origin state against the treaty community, whereas the latter is mainly a bilateral or trilateral matter between the concerned states of a transboundary EIA procedure.[1259] The review procedure provides the last resort in which the

[1259] For an examination of the problems of having a community-based non-compliance procedure and bilateral means of dispute settlement existing side by side, see Koskenniemi 1992, pp. 123-162. For a review of compliance control in international environmental agreements, see Marauhn 1996, pp. 696-731. Interestingly, the meeting of parties decided in the Sofia meeting to develop a reporting system 'on how the obligations of the Convention have been complied with, both at the general level and by particular Parties'(Chapter 2 of Decision II/11, 'adoption of the work-plan').

potentially affected state can address its concerns over the possible transboundary impact of the proposed activity.

2.7. Adaptation of the Espoo Convention to the Arctic

International cooperation having relevance in the Arctic has increased markedly since the end of the Cold War.[1260] Some of these processes, such as Arctic-wide cooperation, involve governments and indigenous peoples; others, such as the Barents Euro-Arctic Council, provide cooperation between central governments on the one hand, and provincial governments on the other;[1261] yet another group, such as the Northern Forum, focuses on cooperation between the northern cities.[1262] Furthermore, there are many bodies, such as the International Arctic Science Committee (IASC), which promote cooperation among non-governmental entities.[1263] Some of these cooperation processes have sought to contribute to how the EIA procedure can best be conducted in Arctic conditions.

The Arctic EIA has been advocated in the conferences of the Arctic Parliamentarians.[1264] The Nordic Council's parliamentary conference in 1993[1265]

The implementation committee that is in charge of the review procedure is to develop recommendations for revision of a questionnaire 'for reporting for future reviews of the implementation of the Convention'.

[1260] For an analysis of the present state of Arctic cooperation processes, see Young 2000, especially chapter 2 entitled 'A Cacophony of Arctic Initiatives'.

[1261] For an article reviewing briefly all the main international organizations operating in the Arctic, including the Barents Euro-Arctic Council, see Langlais 2000 (chapter 2). For more information about the Barents Council at http://www.barents.no/engelsk/index2.htm [8.2.2001]. For general reading, see Stokke and Tunander 1994.

[1262] Langlais 2000 (chapter 6). For more information about the Northern Forum at http://www.northernforum.org/ [8.2.2001].

[1263] Ibid., chapter 4. For more information about the IASC at http://www.iasc.no/ [8.2.2001].

[1264] Ibid., chapter 7. For more information about the Standing Committee of Parliamentarians of the Arctic Region at http://www.grida.no/parl/ [8.2.2001].

[1265] This Conference, which took place in Reykjavik, Iceland, was titled 'Nordic Council's International Conference for Parliamentarians on Development and Protection of the Arctic Region'. All the declarations of the Conferences can be found on the home page of the Standing Committee. See the previous footnote for address.

'decided to establish a Standing Committee of Parliamentarians of the Arctic Region'[1266], and at the 1996 Yellowknife Conference the Parliamentarians requested their governments to pay attention to the Arctic EIA.[1267] However, it was the 1998 Salekhard Conference where the Arctic EIA was one of the main themes. A reference to EIA Guidelines was provided in the opening paragraphs:

> Taking into account...the agreement from the Fourth Ministerial Meeting of the Arctic Environmental Protection Strategy that all Arctic countries should endeavour to apply the AEPS guidelines for Environmental Impact Assessment in the Arctic[1268].

The Parliamentarians ask, in very strong terms, their governments to:

> implement within the respective national structures the guidelines for Environmental Impact Assessment in the Arctic, as developed by the Arctic Environmental Protection Strategy, by including them in all proposals, plans and decisions for activities that could have impact on the Arctic environment[1269].

[1266] Reykjavik Declaration of 1993. The Standing Committee of Parliamentarians of the Arctic Region was established in September 1994. Already in this conference the concept of EIA was taken up: 'ask the governments to...ensure that plans to make use of Arctic living and non-living resources also include Environment Impact Assessment (EIA), at a sufficiently early stage of the process'.

[1267] According to the Yellowknife declaration, the Parliamentarians 'ask our respective governments to...ensure that at an early stage, formal environmental assessment, with adequate input from resident indigenous peoples, is included in all plans and decisions that could have impact on Arctic landscapes or waters".

[1268] Paragraph 13 of the Salekhard Declaration.

[1269] Ibid., paragraph 50. In the Salekhard Declaration it has also been stated: '44. ensure that national authorities support and work actively toward the implementation of the recommendations and activities of the Arctic Council, including the programmes of the Arctic Environmental Protection Strategy, and the principles and guidelines of the Alta Declaration' and 'develop and implement policies that give full participation to indigenous Arctic peoples in development decisions in each northern country, including assessment of environmental impacts, which ensure that northern residents receive a fair and equitable share of the benefits of economic development in the Arctic, and which foster Arctic-based economic enterprises...59. consider carefully the recommendations and requests made by Parliamentarians at this Conference, and to report on actions taken, to the Fourth Conference of Parliamentarians of the Arctic Region...Furthermore...62. the Conference asks the Standing Committee to continue its work, in which the major tasks will be; 1) to follow up the implementation of the recommendations and requests from this conference to governments and international bodies, including requesting the

The next conference, organized in summer 2000 in Rovaniemi, Finland, also acknowledged the Arctic EIA but no longer in such striking terms.[1270]

The Barents Euro-Arctic Council and the Nordic Council of Ministers have also dealt with the issue of EIA in the Arctic. The Barents Council established a project to study EIA in the Barents region. The project was financed by the Nordic Council of Ministers. The goal of the project was to harmonize EIA practices in the Barents region and to increase efficiency in the application of EIA procedures in the region. However, the proposal did not lead to any concrete actions.[1271]

In 1989, the Nordic Council of Ministers' Committee for the Environment decided to convene an *ad hoc* working group on EIA, which gradually developed into an organ

governments of member countries to produce, six months before the next Conference of Parliamentarians of the Arctic Region, information for public discussion about actions taken or not taken in response to those recommendations and requests, and the results of such actions'. The Arctic Council was asked 'to increase and ensure the effective contribution of indigenous people and their societies in all policies and development decisions in the Arctic regions including assessment of environmental and social impacts, ensuring that the rights of Arctic residents are adequately protected, and that the benefits from development and utilization of Arctic resources are equitably shared' (paragraph 54).

[1270] Paragraph 4 of the Rovaniemi Declaration notes 'the increasing international and inter-country character of activities in the Arctic related to science, environmental monitoring, environmental impact assessment, and health and cultural issues'. The Rovaniemi Declaration (paragraph 21) also states: 'Ask the governments in the Arctic Region and the institutions of the European Union to...commit to continued support of the Arctic Council, with adequate financing of its activities including efficient secretarial support, involvement of the Permanent Participants, and integration or liaison of its respective monitoring, research, and socio-economic activities with related national activities, including implementation of compatible environmental impact assessment guidelines and procedures; and strengthening in appropriate areas its regional components, including involvement of regional authorities and organizations, and consideration of means to enhance Arctic regional and circumpolar transportation and communication systems'.

[1271] Author's personal copy obtained from the Ministry of the Environment. The Barents Euro-Arctic Council, and more specifically its regional council, has three permanent working groups of which one is the Task Force on the Environment.

with its own secretariat.[1272] One interesting development from the viewpoint of this study was a project initiated in spring 1997 called 'Coordinated Application of the Espoo Convention' that was financed by the Nordic Council of Ministers and the *ad hoc* Group for EIA.[1273] The *ad hoc* Group for EIA was terminated in 1998,[1274] and the task of maintaining links between the Nordic countries in the area of EIA was transferred to Nordregio, an institution established by the Nordic Council of Ministers for research, education and documentation on spatial development.[1275] Nordregio maintains a Nordic EIA network,[1276] which has recently published an in-depth study of the EIA procedures in force in each of the five Nordic states.[1277] Significantly for the present study, the publication reviews the EIA systems of Iceland[1278], Denmark[1279], Norway[1280], Sweden[1281] and Finland[1282] from the

---

1272 The history of the *ad hoc* group is reviewed in the final report, chapter 1. See *infra* footnote 1274.

1273 Husby and Tesli 1999, p. 57.

1274 The final report of the *ad hoc* Working Group on EIA titled '*Ad hoc* arbetsgruppen for miljökonsekvensbeskrivningar (MKB) under Nordiska Ministerrådet 1989-1998' is published in the EIA section of Nordregio's home page at http://www.nordregio.se [12.3.2001]. See also a report in Swedish on how to continue the work developed by the *Ad Hoc* Working Group on EIA, 'Ett nordiskt MKB- nätverk. Rapport om nätverksuppbyggnad', 1999.

1275 The home page of Nordregio is at http://www.nordregio.se [22.2.2001].

1276 Nordregio's home page, in the EIA section (Background and purpose) states: 'The Nordic EIA Network with focus on Environmental Impact Assessment (EIA) and Strategic Environmental Assessment (SEA) was launched in January 1999. The network is hosted by Nordregio - the Nordic Centre for Spatial Development in Stockholm and financed by the Nordic Council of Ministers. This network initiative has grown out of the Nordic EIA and SEA networking experiences of the last 8 years by the ad hoc group of EIA under the Nordic Council of Ministers. Aims of the network: The aims of the network are to support research and development of EIA and SEA with Nordic relevance and strengthen the link between environmental considerations and regional development planning and programmes. The network shall create an arena for its participants to exchange ideas and experience through seminars, and conferences and by supporting research studies in the field of EIA and SEA'.

1277 The name of the publication is 'A Comparative Study of Nordic EIA Systems; Similarities and Differences in National Implementation' 2001.

1278 Ibid. It is stated (p. 73): 'Iceland has not ratified the Espoo Convention. However, there is a requirement in the Icelandic EIA Act No. 106/2000 (Article 18), that if the development is likely to have considerable environmental impacts in another state within the European Economic Area, the Planning Agency shall send that state a description of the development, with other available information of the possible impacts across borders and provide an opportunity to act on the case'.

1279 Ibid. It is stated (p. 72): 'In Denmark transboundary effects are included in Art. 9 of the Ministerial Order No. 428: "If the regional planning authority determines that a project covered by Art. 3 [of the Planning Act] is likely to have significant effects on the environment in another state

perspective of how the states' EIA systems are capable of managing the transboundary EIA procedure. Another important development is that the Nordic points of contact of the Espoo Convention meet regularly, usually in connection with the other meetings related to the Convention, to discuss pending or forthcoming proposed activities within the Nordic states as well as whether these activities have given rise to any special issues that should be considered.[1283]

---

or if another state is likely to be significantly effected requests information concerning the project, the regional planning authority shall inform the Minister for the Environment and Energy of this as soon as possible." In these cases the Minister for the Environment and energy has the power to establish ad-hoc a procedure to ensure that Article 7 in the EU directive will be fulfilled as well as the content of the Espoo Convention'.

[1280] Ibid. It is stated (p. 73): 'According to the Norwegian Planning and Building Act and the EIA Regulation, if a project is likely to have a significant adverse environmental impact in another state, the competent authority must send a notification to the appropriate authority in the said state (Art. 33-9 of the Planning and Building Act and Art. 10 of the regulation). The notification shall contain information on the possible transboundary impact, the studies proposed to assess this further and the types of decision that shall be based on an approved EIS. The notification shall be sent at the same time as it is available for comments in Norway and a reasonable time limit shall be set for the state to indicate whether it wishes to participate in the environmental assessment procedure. If the state wishes to participate in the environmental assessment procedure, the competent Norwegian authority shall accept comments from its authorities and interest parties that shall be treated in the same way as any other comments on the notification'.

[1281] Ibid. It is stated (p. 73): 'In Chapter 6 of the Swedish Environmental code it is stated that if an activity or measure is likely to have a significant environmental impact in another country, the responsible authority designated by the Government (the Swedish Environment Protection Agency) shall inform the competent authority in that country about the planned activity or measure. Furthermore, the country concerned and the citizens who are affected shall be given the opportunity to take part in a consultation procedure concerning the application and the EIS. There is reference to the Espoo Convention regarding the procedures (sections 4-6 of the Ordinance [1998:905] on EIS)'.

[1282] Ibid. It is stated (p. 72): 'In Finland the treatment of projects that have environmental impacts in the territory of another state is covered in Section 14 and 15 of the EIA Act 468/1994. It is the responsibility of the Ministry of the Environment to see to that the notification and consultation functions connected with the EIA are in accordance with the agreement as referred to in paragraph 1: "If a project is likely to have significant environmental impact in territory under the jurisdiction of another state, the co-ordinating authority shall supply the Ministry of the Environment with an assessment programme without delay for notification to the other state in accordance with an agreement as referred to above". In such cases, the authorities, persons or associations in a state party to the agreement referred to in Section 14, paragraph 1 of the EIA Act, shall be provided an opportunity to participate in an assessment procedure in accordance with the act'. For an examination of the capability of the Finnish EIA system to manage a transboundary EIA procedure, see Koivurova 1997 II, pp. 184-194.

[1283] Information obtained from the Finnish Ministry of the Environment via an e-mail communication (6.3.2001).

None of these processes have led to binding norms in international law; the 1997 EIA Guidelines and the 1997 Offshore Guidelines establish legal obligations but only obligations of attempt, as was noted above.[1284] This is in direct contrast to the Antarctic Cooperation process in which detailed rules covering EIA were adopted as part of a legally binding 1991 Protocol to the Antarctic Treaty of 1959.[1285] Chapter 11 of the EIA Guidelines nevertheless provides useful obligations of attempt for the Arctic states as to how to organize their transboundary EIA procedures. Above all, it is required that all activities assessed according to the national EIA legislation should be screened also from the viewpoint of whether transboundary impacts are likely.[1286] Thus, all activities to which a national EIA procedure is applied should be screened in view of likely transboundary impacts in the Arctic context. In addition, lower thresholds may be needed for those activities listed in the Espoo Convention if practiced in the Arctic conditions.[1287]

As was shown above, one of the problems of the Espoo Convention is that it does not set a clear deadline for the initiation of the transboundary EIA procedure. It is even possible that if the national EIA procedure is initiated only after the EIS, the

[1284] See *supra* chapter III (3.3.).

[1285] The Protocol, firstly, prohibits all mining activity in the Antarctic (Article 7 of the Protocol) and sets out a requirement of EIA (Article 8) detailed in Annex I. Hence, all activities must go through an EIA procedure. The system of the Annex is built on three levels of magnitude of potential impact: activities determined as causing less than a minor or transitory impact, activities causing minor or transitory impact and activities causing more than a minor or transitory impact. If an activity is considered to have less than minor or transitory impact, it can proceed without any EIA procedure (Article 1 (2)). Unless it can be determined that an activity will have less than a minor or transitory impact, or it can be determined without any preliminary impact assessment that an activity will have more than a minor or transitory impact, an initial environmental evaluation must take place (Article 2). If an initial environmental evaluation indicates that it is likely that only a minor impact will take place, the project can proceed, but if it is determined that it is likely that more than a minor or transitory impact will occur, a comprehensive environmental evaluation in accordance with Article 3 must take place with detailed requirements from the evaluation (Article 3 (2)).

[1286] Paragraph 8 of Section 11 of the EIA Guidelines.

transboundary EIA procedure must also be initiated at this time, making it impossible to include foreign actors and foreign impact in the preparation of the EIS document. The EIA Guidelines provide a stricter approach here. The origin state should initiate the transboundary EIA procedure in a very early phase of its national EIA procedure and include foreign impact and actors already at this phase.[1288] Another failure of the Espoo Convention is that it does not require that the EIS contain transboundary impact. The EIA Guidelines require that already in the scoping phase of the national EIA procedure, potential transboundary impact should be identified and methods to be used for assessing them should be agreed upon between the concerned states. The EIA Guidelines also urge cooperation in the implementation of the transboundary EIA procedures taking place in the Arctic.[1289]

The Espoo Convention provides for a basic right for all those private legal subjects of the affected state located in the area likely to be affected to participate in the transboundary EIA procedure, just as the private legal subjects of the origin state may also participate. The tenth paragraph of the EIA Guidelines contains an important adaptation norm for the Espoo Convention: 'Communities in the area of anticipated impacts should be given an opportunity to participate, irrespective of their location relative to the border'. In the Arctic context, these communities normally are indigenous peoples, as referred to in the following paragraph.[1290] This differs from the participation rights of the Espoo Convention, which are granted to 'natural or

---

[1287] Ibid.

[1288] Ibid., paragraph 4.

[1289] Ibid., paragraphs 7 and 8.

[1290] Ibid. Paragraph 10 reads: 'The Inuit Circumpolar Conference, the Sami Council and the Indigenous Peoples Secretariat are accredited non-governmental organizations on the Arctic Council, and which are active in several arctic countries. They may thus provide useful links to the public on both sides of the border.

legal persons', not to communities. Hence, it is required that the Arctic states include the indigenous peoples' communities in transboundary EIA procedures.[1291]

3. Conclusion

One can hardly avoid the feeling that the development of the Espoo Convention in the Arctic has not been as speedy as might have been inferred from the results of the meetings of both the signatories to the Espoo Convention and parties to the Arctic cooperation process. When the Espoo Convention was concluded in 1991, it contained great promise for managing situations of potential transboundary pollution caused by large-scale stationary activities in the Arctic; this optimism derived from the fact that all eight Arctic states signed the Convention. Moreover, Arctic cooperation promoted from the very beginning the ratification of the Convention by the Arctic states as well as the adaptation of the Espoo Convention to Arctic circumstances.[1292]

Although the United States and Canada signed the Convention, they started to prepare their own convention on transboundary EIA under the auspices of the NACEC. The draft TEIA was concluded in 1997, but as yet it has not been possible to make it a legally binding international agreement. The United States has not shown any sign that it intends to ratify the Convention although it has not withdrawn its signature yet. Canada ratified the Convention, but made an extensive reservation which severely limits its obligations in some areas. Yet, as was argued above,[1293] this is not the case in the Arctic since the reservation does not concern the Arctic areas of Canada; this would justify a recommendation that the Espoo Convention be fully applied by Canada and the Nordic states in the Arctic.

[1291] Ibid.

[1292] See *supra* chapter V (2.2.).

Two visions can be outlined for the Espoo Convention in the Arctic: the immediate future and the near future. First of all, the Espoo Convention applies in its entirety between the Nordic Arctic states except Iceland. When Iceland becomes a party to the Convention, the Convention will govern the transboundary EIA procedure between the five Nordic Arctic states. In addition, all five Nordic Arctic states are obligated to implement the EIA Directive as amended in 1997, either as member states of the EC or as part of the EEA Agreement. The Directive harmonizes the national EIA procedures of the Nordic Arctic states to a considerable degree.

It is therefore no wonder that the Nordic Council of Ministers and the *ad hoc* group for EIA established a project to study the 'Coordinated Application of the Espoo Convention'[1294]. Husby and Tesli state: 'we recognize that a group such as the Nordic ad hoc group for EIA should be well suited to play a central role relative to the exchange and continuous evaluation of the framework and guidelines for a common Nordic application of the Convention'[1295]. Although the Nordic *ad hoc* group has been terminated, the points of contact of the Nordic states to the Espoo Convention meet regularly to discuss future proposed activities which may have an impact on other Nordic states.[1296] In addition, an institution operating under the Nordic Council of Ministers, Nordregio, coordinates the research-work related to Nordic EIA's.

[1293] See *supra* chapter V (2.3.).

[1294] Husby and Tesli 1999, p. 57.

[1295] Ibid., p. 79. Husby and Tesli (1999, note 1, p. 58) also state: 'The project is carried out as a network cooperation involving partners appointed by the environmental authorities in each of the countries'.

[1296] See *supra* chapter V (2.7.).

It would seem possible to have similar kind of cooperation in transboundary EIA procedures for the Barents region as well,[1297] for although the Russian Federation is not yet a party to the Espoo Convention, it is a party to the 1992 Industrial Accidents Convention, which covers much the same activities as the Espoo Convention and contains a very similar transboundary EIA procedure. When Iceland becomes a party to the Espoo Convention, transboundary EIA procedure will apply between all the states that take part in the cooperation in the Barents Euro-Arctic Council. It may be possible to have a similar kind of coordination within the inter-state meetings of the ministerial meetings and/or in the environmental protection working group operating under the cooperation at the provincial level, i.e., informal discussions on which activities are anticipated in this region or have already been subjected to prior procedures. The environmental protection working groups of the Arctic Council are also well suited to this task, namely the Emergency Prevention, Preparedness and Response (EPPR), the Conservation of Arctic Flora and Fauna (CAFF) and the Protection of the Arctic Marine Environment (PAME). In fact, for instance in the EPPR meeting last year, the parties presented certain offshore hydrocarbon activities which will operate in their offshore areas.[1298]

As mentioned above, the Espoo Convention does not apply as yet between neighboring states in the North American region. Before these states become parties to the Convention, it can be recommended that Canada and the United States apply the draft TEIA in their northern regions. This recommendation is viable because it

---

[1297] For a comprehensive study in Norwegian on the applicability of the Espoo Convention in the relations between the Nordic countries, as well as the Russian Federation, see Måreng 1996.
[1298] The 2000 Report of the meeting of the Emergency Prevention, Preparedness and Response (EPPR) working group. The report can be found on the EPPR home page at http://www.arctic-council.org/eppr.asp [30.3.2001].

seems that Canada and the United States were willing to conclude the draft TEIA as a binding international agreement, but Mexico objected to it. Some signs exist that the sub-units of these (con)federal states are willing to apply the draft TEIA voluntarily.[1299] Hopefully, the Yukon and Alaska can also reach such an understanding, taking into account that it is highly likely that onshore or offshore hydrocarbon development will take place in these areas in the near future.[1300] It should also not be forgotten that there are some signs of hope that negotiations on a draft TEIA will start anew.[1301]

From the viewpoint of the second vision - the foreseeable future - it can be expected that all eight Arctic states will become parties to the Espoo Convention. At the moment, the Russian Federation and the United States have not become parties to the treaty, but have not withdrawn their signatures either. Iceland will become a party to the Convention soon.[1302] If this scenario is realized, the Arctic application of the Convention will become a reality. As explained above, the Arctic cooperation process

---

1299 According to an officer of the United States Fish and Wildlife Service, Alaska and Yukon could decide that they will cooperate voluntarily on the basis of the draft TEIA. There have been occasions where this has occurred on the border of the United States and Mexico (e-mail communication 12.3.2001).

1300 The northeastern part of Alaska, near the border of Canada, has been designated as a nature protection area called National Arctic Wildlife Refuge (NAWR) of which one part, part 1002, is still open for oil and gas development. The Canadian government has designated its northwestern areas as nature protection areas, and resists any plans of the United States to initiate hydrocarbon activities in the area of 1002. As stated by Chance (1995): 'In mid-summer of 1995, Canada's ambassador to the U.S., Raymond Chretien, led a lobbying campaign with U.S. Senators arguing that the region should be left pristine to "ensure the future of the shared wildlife populations of the Arctic coastal plain." Other contacts between Canada and the U.S. right up to the the presidency have stressed the former's desire to see the Arctic Refuge Coastal Plain at least retain its present status, or more preferably, be designated as part of the larger wilderness area'. In the letter to the Committee hearing in the Senate, Chretien also resorted to the due diligence principle as obligating the United States (author's personal copy). The history of the NAWR can be found on the home page of the United States Fish and Wildlife Service at http://www.r7.fws.gov/nwr/arctic/issues1.html [4.4.2001]. Recently, the new president of the United States, George Bush junior, announced the energy program of his administration, which included the drilling of oil in the NAWR. Information obtained from the Finnish newspaper Lapin Kansa (18.5.2001, p. 18).

1301 See *supra* chapter IV (4.1.2.).

has already tried to adapt the execution of EIA procedures to Arctic conditions through the EIA Guidelines in general as well as the Offshore Guidelines in particular.[1303] There also exists a website distributing reliable and up-to-date information on the Arctic EIAs.[1304]

Especially important here is Chapter 11 of the EIA Guidelines, which provides some obligations of attempt for adapting the Espoo Convention to Arctic conditions. It will be recalled that Article 31 (3) of the Vienna Convention, which sets out the norms of treaty interpretation, provides that:

> There shall be taken into account, together with the context:
> (b) any subsequent practice in the application of the treaty which establishes the agreement of the parties regarding interpretation.

If the Arctic states start resorting to these obligations of attempt when applying the Espoo Convention in the Arctic, they can be seen as providing an 'agreement of the parties regarding interpretation' of the obligations stipulated by the Espoo Convention.

The key issue in the transboundary screening phase of the Espoo Convention in the Arctic is whether the activities are conceived as restricted to a certain individual activity only or viewed more comprehensively, with a large-scale activity seen as one part of the development of a certain area.[1305] Since most of the Arctic is pristine and lacks human presence, a planned large-scale natural resource exploitation there is bound to engender a number of other activities. For instance, extracting minerals

---

[1302] According to an officer involved in the implementation of the Espoo Convention in the Arctic, Iceland will become a party perhaps already before 2002. E-mail communication, 26.3.2001.
[1303] See *supra* chapter IV (3.3.1.).
[1304] The home page of the Arctic Environmental Impact Assessment (ARIA) is at http://finnbarents.urova.fi/aria/ [25.4.2001].

requires the construction of access roads, the miners need to live near the mine, and the development of an extensive community is likely.[1306] In the transboundary screening phase of the Espoo Convention in the Arctic, this means that it is useful to conceive the planned activity as comprehensively as possible. As provided in Chapter 5 (2) of the EIA Guidelines:

> Cumulative impact assessment at the project level, along with an understanding of environmental impacts at the resource and land use planning level, helps set that project and its impacts in a broader ecological and development context.[1307]

In the application of the Espoo Convention, this means that the transboundary EIA procedure would be initiated even if the main proposed activity did not qualify within the list of Appendix I, since the whole complex of connected developments within the region would definitely cause significant adverse transboundary impacts.

Another important element in the transboundary screening procedure of the Espoo Convention in the Arctic is the extension of the Convention to military activities. As provided in the EIA Guidelines:

1305 As provided by Section 5 (1) of the EIA Guidelines: 'accumulation of changes to the environment caused by human activities (e.g. past, existing and proposed activities, including activities associated with the project under assessment)'.

1306 A good example of an EIA procedure that takes into account such considerations is the United States' federal EIA procedure, the National Environmental Policy Act (NEPA). As stated in the NACEC home page (at http://www.cec.org/pubs_info_resources/law_treat_agree/summary_enviro_law/publication/index.cfm?varlan=english [16.3.2001]), the lead agency 'must consider three types of actions: connected actions, cumulative actions, and similar actions. Connected actions are those where: (1) one action automatically triggers another action; (2) an action cannot proceed unless other actions are taken previously or simultaneously; or (3) the actions are interdependent parts of a larger action and depend on the larger action for their justification. Cumulative actions are actions which, when viewed with other proposed actions, have cumulatively significant impacts. Similar actions are actions which, when viewed with other reasonably foreseeable or proposed agency actions, have similarities that provide a basis for evaluating their environmental consequences together, such as common timing or geography'.

1307 Paragraph 1 of Section 5 (2) of the EIA Guidelines.

> In the Arctic, EIA should be applied to activities associated with the exploitation of both renewable and non-renewable natural resources, public use, **military activities** and the development of infrastructure for different purposes that may cause significant environmental impacts.[1308] (emphasis added)

As was examined above, the Espoo Convention provides the only prior transboundary procedure that gives the possibility to include military activities within its scope. Although military activities have not been included in the list of Appendix I, and thus no presumption exists that the transboundary EIA procedure of the Convention should be applied, it is still possible that in certain circumstances the Convention requires the states parties to initiate the procedure. As mentioned, if the criteria provided by Appendix III clearly point in the direction that significant adverse transboundary impact is likely, the origin state is obligated to commence the transboundary EIA procedure of the Espoo Convention with regard to military activities as well. The potentially affected state is entitled to initiate discussions with the origin state on how to proceed with the matter by virtue of Article 2 (5).

One of the problems of the Espoo Convention is that it does not harmonize the national EIA procedures of the contracting states much; it only requires that the procedure be able to produce an EIS with a certain minimum content and that the public be given some kind right to participate in the procedure. In contractual relations, this means that the parties' obligations may differ depending on the quality of the national EIA procedure. This lack of reciprocity of treaty obligations was accepted by the parties to the Espoo Convention through their not requiring stricter harmonization of national EIA procedures. As a consequence, the deadline for the initiation of the transboundary EIA procedure of the Espoo Convention is when the national EIA procedure requires the public to be informed. Importantly, the EIA Guidelines stipulate that: 'The Country of origin should thus provide information at

[1308] Ibid., Section 2 (2).

an early stage in the assessment process'[1309]. If this obligation of attempt is observed, the transboundary EIA procedure is initiated in an early stage of the assessment process. This would evidently increase the possibilities for the potentially affected Arctic state and its public to participate efficiently in the transboundary EIA procedure and contribute significantly to the study on potential transboundary impact. One important aspect of the transboundary EIA procedure of the Espoo Convention in the Arctic is public participation. As the living conditions, economy and presence of indigenous peoples entail special requirements for public participation, one important element of impact assessment, including transboundary impact assessment, should be the assessment of social impact, i.e., systematic studies covering the potential economic effects of the proposed activity, changes in the social structure etc.[1310] Especially important in the Arctic context is the participation of indigenous peoples. Most of the indigenous peoples of the Arctic live in the area of many Arctic states and have also established international organizations to further their cross-border interests. As was shown above, these international organizations of Arctic indigenous peoples have been accorded the status of permanent participants in the Arctic Council. The international nature of the indigenous peoples of the Arctic is also acknowledged in the EIA Guidelines:

> The Inuit Circumpolar Conference, the Sami Council and the Indigenous Peoples Secretariat are accredited non-governmental organizations on the Arctic Council, and which are active in several arctic countries. They may thus provide useful links to the public on both sides of the border.[1311]

1309 Ibid., Section 11 (4).

1310 Ibid. In Section 3 (1) of the EIA Guidelines it is provided: 'all potential environmental, socio-cultural and economic impacts, especially impacts on the traditional uses of resources and livelihoods of indigenous people and also the consideration of alternatives' should be included in the scope of the impact studies in the Arctic. See also Harrop and Nixon 1999, pp. 66-71. They distinguish between the following sectors that can be covered by the assessment of social impact: health and safety, the economy, cultural and urban resources, aesthetic characteristics, regional growth and infrastructure, and population characteristics.

1311 Section 11 (10) of the EIA Guidelines. It should be noted that at the moment there are also other international organizations of indigenous peoples accepted as permanent participants in the Arctic Council. These are the Arctic Athabascan Council (AAC), the Aleut International Association

Indigenous communities in the area of anticipated impact should be given an opportunity to participate, irrespective of their location relative to the border. When applying the Espoo Convention in the Arctic, the Arctic states should consult the international organizations of indigenous peoples, an approach which could provide, especially via their traditional knowledge, a more comprehensive view of what the impact might be on the whole area of concern to the indigenous peoples living on both sides of the border.

(AIA), the Gwich'in Council International, the Saami Council, the Inuit Circumpolar Conference (ICC), the Russian Association of Indigenous Peoples of the North (RAIPON). Important is also the Indigenous Peoples Secretariat (IPS) established in the Nuuk meeting. See *supra* chapter III (1.1.).

VI CONCLUDING REMARKS

Three major events can be identified which have been important for the development of EIA norms in the Arctic. First, the Cold War was evidently a major factor inhibiting the development of EIA norms in the region. Before the end of the Cold War, there was only one inter-governmental cooperation process which produced any significant impact on EIA norms in the Arctic: the cooperation between the Nordic states. Many international environmental protection treaties were concluded as part of this cooperation, in particular the Nordic Environment Protection Convention in 1974. This Convention, evidently ahead of its time, provided a good framework for managing potential transboundary pollution between the four Nordic states. Although the EC Directive of 1985 also provided a sound basis for managing potential transboundary pollution between states as well as for strict harmonization of the national EIA procedures of the member states, Denmark was the only Arctic state that was a member of the EC. At the time, for the most part the EIA norms that existed before the end of the Cold War were bilateral initiatives concentrating on potential transboundary pollution from a certain set of human activities or from certain harmful substances, or specifying certain environmental media, such as the agreements covering Arctic transboundary watercourses.

A sweeping change occurred during the period when the Cold War ended. Since two Arctic states, the United States and the Soviet Union, had been the two superpowers of the Cold War, fighting each other on every front, the end of this 'war' provided new opportunities for international cooperation in general, and international environmental protection and Arctic cooperation in particular. On the initiative of Finland, the eight Arctic states were able to conclude an informal arrangement, the AEPS document, that paved the way for international efforts for the protection of the Arctic environment in 1991, a development that culminated in the establishment of the Arctic Council in 1996.

From an international legal point of view, there has been uncertainty as to the legal status of this cooperation. In general terms, international cooperation has witnessed a considerable number of new forms of cooperation between international legal subjects. On the one hand, there are formal treaties which contain vague and general provisions, so-called framework treaties. On the other, there are inter-governmental institutions, such as a conference of parties to a treaty, for which it is hard to determine whether the applicable law is international institutional law or the law of treaties. In addition, there are organizations whose legal status is hard to determine such as the Organization on Security and Cooperation in Europe (OSCE). Another of the new forms of cooperation is the class of instruments referred to here as informal arrangements, which are characterized by the absence of norms that regulate the instrument itself and thus cannot be considered typical international treaties. In the main, informal arrangements contain obligations of attempt and political statements.

Although there exist differing views in the scholarly world on the legal status of the AEPS document, the prevailing opinion is that the instrument is in some manner binding. The majority perceive it as softly binding, some as politically binding but only very few as non-binding altogether. Most scholars consider that the norms adopted in this cooperation process require only modest environmental protection measures. On the other hand, the available comments on the legal status of the Arctic Council indicate that it is perceived as operating outside international law altogether.

The purpose of this thesis is to provide an analytic perspective on the legal status of Arctic cooperation. The idea held by the majority of commentators that the AEPS document is somehow binding and somehow soft is best treated within the basic principle of international law, *pacta sunt servanda*. If a commitment with a view to adhere to it in the future has been reached, it is a commitment binding in international law. Significantly, however, international law provides flexibility within this binary conception: binding commitments can take the form of direct orders, obligations of attempt and permissions.

In legal terms, a difference exists between the environmental protection provisions and instruments adopted within this Arctic cooperation process, on the one hand, and the other norms, e.g., those regulating the forms of this cooperation, on the other. The environmental protection provisions and instruments can be seen as providing for obligations of attempt only, whereas the other norms can be qualified in main to be direct orders. The norms regulating how the cooperation is to continue, rules on decision-making and participation and organizational mandates are direct orders. Indeed, although some might disagree, it is rather hard to show that the Arctic states did not intend to continue the cooperation in the forms decided by the Declaration establishing the Arctic Council, for the text of the Declaration gives every indication to the contrary.

Another trend which has been important for the development of the EIA norms applicable in the Arctic is that the goal of sustainable development has become a principal goal of international society in the situations studied in this thesis. The goal started to develop in the beginning of the 1980s, providing a framework principle for reconciling two important social values of international society, environmental protection and economic development. This change was reflected in the Brundtland Report of 1987 and culminated in the Rio Conference on Environment and Development in 1992. One milestone in this regard from the viewpoint of international law was the ICJ decision endorsing the goal of sustainable development in the *Gabcíkovo Nagymaros* case in 1997. Although sustainable development cannot be seen as part of *lex lata*, it can most appropriately be seen as a goal of international society having normative relevance in differing ways as has been examined above.[1312]

The third significant trend was the development of EIA procedure and in particular transboundary EIA. The EIA procedure as a national EIA procedure was already adopted in the beginning of the 1970s. The spread of these EIA procedures made it possible for states to conclude treaties and other legally relevant documents on transboundary EIA in which foreign actors and impacts could be linked to the national EIA procedure of the origin state. Nevertheless, it was the goal of sustainable development and its entry onto the scene in international relations that had a major impact on the success of EIA procedures and transboundary EIA procedures as well, for the EIA procedure was seen as one of the main tools for implementing the goal. As a consequence, it became feasible to conclude the 1991 Espoo Convention, the most advanced treaty on transboundary EIA. Importantly from the viewpoint of the Arctic cooperation process, the Espoo Convention was signed by all eight states participating in that cooperation.

There were thus three main developments that coincided in the year 1991: the general international political setting made it possible to initiate Arctic-wide cooperation; the goal of sustainable development provided new direction for international normative activity, challenging the ideas that international environmental protection was exclusively transboundary in character or separate from the other normative fields of international law; and the Espoo Convention was signed, providing Arctic-wide possibilities for extending national EIA procedures to other Arctic states and their private legal subjects. As was shown in the foregoing chapters, the opportunities provided by these developments are still waiting to be realized, since the Espoo

[1312] See *supra* chapter IV (3.2.)

Convention has not become the pan-Arctic regime it could be, but governs only certain relations between the eight Arctic states. The Arctic international cooperation process, now operating under the Arctic Council, has produced some relevant obligations of attempt for these situations, but no direct orders.

On the other hand, as was shown in chapter V, the regime of EIA norms now applicable in the Arctic has progressed in both quantity and in quality. The regional advances made especially in the Nordic region and between the United States and Canada were complemented by the multilateral treaties of the 1990s which applied to the whole Arctic region. Especially important are the multilateral treaties concluded within the UN ECE. Significantly, the older treaties are still in force, and the body of EIA norms has thus grown. This gradual accumulation of applicable EIA norms in the Arctic seems to be the most promising sign for the management of the planning phase of stationary activities in general, and situations of potential transboundary pollution between the Arctic states in particular, inasmuch as the Arctic Council and the Espoo Convention have not yet realized their full potential.

Many situations studied in the present thesis are not governed by treaty procedures. This applies in particular to situations in which potential pollution from activity in the Arctic states may cause damage to international areas or potential pollution is caused by planned stationary military activities, but treaty gaps exist in all areas of the situations studied in this thesis. Herein lies the function of generally applicable standards, the goal of sustainable development and especially the valid principles of customary international law, the principles of due diligence and equitable utilization. They serve to fill the gaps in treaty law. On the other hand, the problem with the due diligence principle is that it cannot be invoked in the planning stage of the stationary activity but only when actual damage is caused. It is precisely treaty procedures that operationalize the guidelines of the due diligence principle in the planning stage; the practical influence of the due diligence principle in the Arctic may thus be uncertain in situations which are not governed by treaty procedures.

Could there be prospects for a formal international treaty in the field of EIA norms which fills the existing gaps and takes into account the special Arctic conditions? Indeed, some scholars have proposed that treaties specializing in different aspects of protection of the Arctic environment should be agreed upon by the Arctic states.[1313] It would seem, though, that Arctic cooperation has been possible only via the informal arrangements. Since the international political setting of this cooperation process has been very difficult, consisting as it does of states with varying interests and with a dominant position in world affairs in general, it does not seem very likely that the Arctic states could adopt a formal international treaty as a basis for their cooperation process.[1314]

In fact, it can be regarded as significant that the Arctic cooperation process has succeeded in becoming as institutionalized as it now is. To the present author, it seems that no major changes in the Arctic cooperation process are likely in the foreseeable future. The way forward in the field of international environmental law, EIA norms in particular, is to make use of the large number of international treaties already available and to try to adapt these to the Arctic conditions through national measures of the Arctic states and/or informal arrangements, in other words, to continue in the forms already adopted by the eight Arctic states.

[1313] See, e.g., Brubaker 1993, pp. 276-447 and Rothwell 1995 II, pp. 301-312.
[1314] As pointed out by Young (1997, p. 264): 'The Arctic is dominated by major powers, including Canada, Russia, and especially the United States, which tend to resist the application of regulatory arrangements to themselves, even when they act to impose such arrangements on others. Unlike Antarctica, most of the Arctic is fully integrated into the jurisdictional domains of nation states, a fact that makes international regulation seem particularly intrusive'.

SPECIAL TERMINOLOGY USED IN THE THESIS

Terminology Related to Chapter III:

Direct orders: 'Direct orders' obligate states to accomplish what is required in the norm-sentence, for instance, to establish an EIA procedure.

Obligations of attempt: Norms in this category only require that an attempt be made to conform to the requirements prescribed in the norm-sentence. Obligations of attempt can be paraphrased as 'to try to' and are expressed through modals such as 'should' and 'ought to'.

Informal Arrangements: Informal instruments concluded between international legal subjects that are binding in international law but provide only the minimum of requirements, such as obligations of attempt.

Legally Relevant Documents: Written normative instruments that are non-binding in international law but are still legally relevant; these documents are produced by international legal subjects and other international actors alike.

Terminology Related to Chapter IV:

EIA Norms: Norms of international law that apply to the planning stage of a proposed environmentally harmful stationary activity.

Prior procedure: a procedure which requires states to establish governmentally controlled procedures in their own territorial jurisdiction for the planning phase of environmentally harmful stationary activities.

EIA procedure: a procedure which requires states to establish governmentally controlled procedure that evaluates the likely effects of a proposed project on the environment, alternatives to the proposal and measures to be adopted to protect the environment before a licence is given to undertake the activity.

Screening procedure: A procedure forming the phase of an EIA procedure in which it is decided whether an EIA procedure is to be applied to a certain individual activity.

Scoping procedure: A procedure forming the phase of an EIA procedure in which the public and relevant governmental authorities are entitled to make comments on which issues should be taken up in the assessment studies.

Environmental Impact Statement (EIS): the document that records the results of the impact studies.

Prior transboundary procedures: procedures that require the states concerned, i.e., the origin state and the affected state, to communicate on the potential transboundary impacts of a proposed activity.

Transboundary EIA procedures: prior transboundary procedures that aim to integrate the foreign impacts and foreign actors into the operation of the national EIA procedures of the origin state.

TABLE OF CASES (Cases decided by the International Court of Justice (ICJ) can also be found on its home page at http://www.icj-cij.org/ [4.5.2001])

Legal Status of Eastern Greenland, Judgment of April 5th, 1933 (PCIJ Series C, No. 62; Series A./B., Fascicule No. 53).

Trail Smelter case of 1938 (and 1941) (Canada/United States of America). All the relevant documents related to this case (Summary, report, agreements and awards) can be found from International Environmental Law Reports, Vol. 1, Early Decisions (edited by Robb, Cairo A.), pp. 231-331.

Corfu Channel case, Judgment of April 9th, 1949: I.C.J. Reports 1949, p.4.

Reservations to the Convention on Genocide, Advisory Opinion: I.C.J. Reports 1951, p. 15.

Fisheries case, Judgment of December 18th, 1951: I.C.J. Reports 1951, p. 116.

Lake Lanoux case. Award of Arbitral Tribunal, 16 November 1957. The award can be found from International Environmental Law Reports, Vol. 1, Early Decisions (edited by Robb, Cairo A.), pp. 332-385.

North Sea Continental Shelf, Judgment, I.C.J. Reports 1969, p. 3.

Aegean Sea Continental Shelf, Judgment, I.C.J. Reports 1978, p.3.

Military and Paramilitary Activities in and against Nicaragua (Nicaragua v. United States of America), Merits, Judgment, I.C.J. Reports 1986, p. 14.

Passage Through the Great Belt (Finland v. Denmark), Order of 10 September 1992, I.C.J. Reports 1992, p. 348.

Maritime Delimitation in the Area between Greenland and Jan Mayen, Judgment, I.C.J. Reports 1993, p. 308.

Maritime Delimitation and Territorial Questions between Qatar and Bahrain, Jurisdiction and Admissibility, Judgment, I.C.J. Reports 1994, p. 112.

Request for an Examination of the Situation in Accordance with Paragraph 63 of the Court's Judgment of 20 December 1974 in the Nuclear Tests (New Zealand v. France) Case, I.C.J. Reports 1995, p. 288.

Legality of the Threat or Use of Nuclear Weapons, Advisory Opinion, I.C.J. Reports 1996, p. 226.

Oil Platforms (Islamic Republic of Iran v. United States of America), Preliminary Objection, Judgment, I.C.J. Reports 1996, p. 803.

Gabcikovo-Nagymaros Project (Hungary/Slovakia), Judgment, I.C.J. Reports 1997, p. 7.

Southern Bluefin Tuna Cases (New Zealand v. Japan; Australia v. Japan) International Tribunal for the Law of the Sea, Year 1999, (27 August 1999), list of Cases No:s 3 and 4, requests for provisional measures. The case can be found on the home page of the International Tribunal for the Law of the Sea at http://www.un.org/Depts/los/ITLOS/Order-tuna34.htm [14.2.2001].

INDEX OF TREATIES AND INFORMAL ARRANGEMENTS

The index contains not only treaties that are in force but also informal arrangements as defined above, treaties that are not yet in force and directives of the EC. The treaties in Finnish can also be found on the home page of Finlex at http://finlex.om.fi/sops/ [13.3.2001] and treaties in English on the home page of United Nations Treaty Series at http://untreaty.un.org/ [13.3.2001].

The 1867 Treaty Concerning the Cession of the Russian Possessions in North America by His Majesty the Emperor of All the Russian to the United States of America. The Treaty can be found on the Yale Law school's Avalon Project at http://www.yale.edu/lawweb/avalon/diplomacy/treatywi.htm [13.3.2001].

The 1909 Treaty Between the U.S. and Great Britain Relating to Boundary Waters, and Questions Arising Between the United States and Canada (the 1909 Boundary Waters Treaty) The Boundary Waters Treaty can be found via the homepage of the International Joint Commission, a joint organ of the parties established by Article VII of the Boundary Waters Treaty at http://www.ijc.org/ijcweb-e.html [27.4.2001].

The 1920 Treaty Relating to Svalbard. The Treaty is reproduced in Vol. 18 American Journal of International Law, Suppl., 1924, pp. 199-208. The Finnish version can be found at http://finlex.om.fi/sops/ [13.3.2001], in this case SopS 15/1925.

The 1929 Agreement Between Sweden and Norway Concerning Certain Issues Over Water Rights. The Agreement can be obtained from the Swedish treaty series (Överenskommelser med främmande makter, No. 10 (1931)). The name of the Agreement is author's translation from Swedish.

The 1951 Treaty Establishing the European Coal and Steel Community (ECSC). The Treaty can be found on the home page of the EUROPA at http://europa.eu.int/abc/obj/treaties/en/entoc29.htm [18.2.2001].

The 1957 Treaty Establishing the European Atomic Energy Authority (EURATOM). The Treaty can be found on the home page of the EUROPA at http://europa.eu.int/abc/obj/treaties/en/entoc38.htm [18.2.2001].

The 1957 Treaty Establishing the European Economic Community. The Treaty can be found on the home page of EUROPA at http://europa.eu.int/abc/obj/treaties/en/entoc05.htm [18.2.2001]. Amended as the 1992 Treaty Establishing the European Community through the 1992 Treaty on European Union. (Consolidated version incorporating the changes made by the Treaty of Amsterdam, signed on 2 October 1997 is reproduced in Official Journal C 340, 10.11.1997, pp. 173-308)). The amended Treaty can be found on the Europa home page at http://europa.eu.int/eur-lex/en/treaties/index.html [16.5.2001].

The 1957 ILO Convention No. 107 Concerning the Protection and Integration of Indigenous and Other Tribal and Semi-Tribal Populations in Independent Countries. The Convention can be found on the home page of the ILO at http://ilolex.ilo.ch:1567/public/english/docs/convdisp.htm [28.4.2001].

The 1957 Agreement between Norway and the Soviet Union Concerning the Sea Frontier in the Varangerfjord of 15 February 1957 and Protocol of 29. Reproduced in Charney and Alexander 1993, pp. 1786-1789.

The 1957 Agreement (with annexed charts) on the Utilization of Water-Power on the Pasvik (Paatso) River (between Norway and the Union of Soviet Socialist Republics). The Agreement can be found via the United Nations Treaty Series (UNTS) on-line service at http://untreaty.un.org/ [13.3.2001), registration No. 4522.

The 1958 Convention on the Territorial Sea and the Contiguous Zone. The Convention can be found on the multilaterals project –home page at http://www.fletcher.tufts.edu/multi/chrono.html#51 [13.3.2001]. SopS 6-7/1965.

The 1958 Continental Shelf Convention. The Convention can be found on the multilaterals project –home page at http://www.fletcher.tufts.edu/multi/chrono.html#51 [13.3.2001]. SopS 6-7/1965.

The 1959 Antarctic Treaty. The Treaty can be found on the multilaterals project – home page at http://www.fletcher.tufts.edu/multi/chrono.html#51 [13.3.2001]. SopS 31/1984.

The 1961 Agreement between the United States and Canada regarding the Columbia River Basin: Cooperative Development of Water Resources. Obtained via the Oceana on-line service at http://www.oceanalaw.com/ [17.2.2001].

The 1962 Agreement Concerning Co-operation Between Denmark, Finland, Iceland, Norway and Sweden (UNTS No. 6262). SopS 28/1962.

The 1963 Treaty Banning Nuclear Weapon Tests in the Atmosphere, in Outer Space and Underwater. UNTS No. 6964. SopS 1/1964.

The 1964 Agreement Between the Finnish Republic and the Soviet Union Concerning Frontier Watercourses (translation in UNTS, No. 7804). SopS 26/1965.

The 1966 International Covenant on Civil and Political Rights. The Covenant can be found from 6 I.L.M. 368 (1967). SopS 7-8/1976.

The 1968 Exchange of Letters Constituting an Agreement Concerning the Construction and Maintenance of a Bridge Over the River Torne at Pello (Overtornea). The Agreement can be found via the UNTS, No. 9376. SopS 71/1968.

The 1969 Vienna Convention on the Law of Treaties. The Convention can be found from 8 I.L.M. 679 (1969). SopS 32-33/1980.

The 1971 Agreement Concerning Frontier Rivers (with annexes). The Agreement can be found via the UNTS, No. 11827. SopS 53/1971.

The 1971 Agreement Concerning Co-Operation in Measures to Deal with Pollution of the Sea by Oil (Denmark, Finland, Norway and Sweden). UNTS, No. 11793. SopS 41/1971.

The 1971 Agreement Between the Government of Canada and the Government of Norway on Sealing and the Conservation of the Seal Stocks in the North-west Atlantic. The Agreement can be found on the home page of the Office of the Auditor General of Canada at http://www.oag-bvg.gc.ca/domino/env_commitments.nsf/all/B4EAA1B372110C87852566480061DA8D?OpenDocument [14.3.2001].

The 1972 Agreement on Cooperation in the Field of Environmental Protection between the United States of America and the Union of Soviet Socialist Republics. The text is reproduced in 11 I.L.M. 761 (1972).

The 1972 Convention for the Prevention of Marine Pollution by Dumping from Ships and Aircraft. The Convention can be found from International Environmental Law: Primary Materials 1991, pp. 172-178 and 11 I.L.M. 262 (1972). SopS 35-36/1979.

The 1973 Agreement on the Conservation of Polar Bears. The Agreement is reproduced in 13 I.L.M. 13 (1974).

The 1974 Agreement between the Government of the Kingdom of Denmark and the Government of Canada Relating to the Delimitation of the Continental Shelf between Canada and Greenland. The text is reproduced in 13 I.L.M. 506 (1974).

The 1974 Agreement Relating to the Establishment of Joint Pollution Contingency Plans for Spills of Oil and Other Noxious Substances. The Agreement can be found on the NACEC home page at http://www.cec.org/pubs_info_resources/law_treat_agree/transbound_agree/SourceFiles/F15-AGREEMENT.HTML [13.3.2001].

The 1974 Nordic Environmental Protection Convention (the 1974 NEPC). The english text of the NEPC is reproduced in 3 I.L.M. 591 (1974). SopS 74-75/1976.

The 1974 Paris Convention for the Prevention of Marine Pollution from Land-Based Sources (The 1974 Paris I Convention). The text is reproduced in 13 I.L.M. 352 (1974).

The 1975 Agreement Between the United States of America and Canada Relating to the Exchange of Information on Weather Modification Activities (the 1975 Weather Agreement). The text is reproduced in 14 I.L.M. 589 (1975).

The 1976 Guidelines for Communication Between Finland, Norway, Sweden and Denmark on Security Issues Related To the Nuclear Installations Constructed Near the Border (the 1976 Nordic Guidelines Treaty). SopS 19/1977.

The 1977 United Nations Convention on the Prohibition of Military or Any Other Hostile Use of Environmental Modification Techniques. The Convention can be found from Documents on the Laws of War, pp. 377-386 and 16 I.L.M. 88 (1977). SopS 60/1978.

The 1978 Agreement between the United States and Canada on Great Lakes Water Quality. The Agreement can be found on the home page of the IJC at http://www.ijc.org/ijcweb-e.html [13.3.2001].

The 1978 Agreement between Norway and the Soviet Union on a Temporary Practical Arrangement for Fishing in an Adjacent Area in the Barents Sea with Attached Protocol of 11 January 1978. The Agreement can be found from Counter-Memorial Submitted by the Government of the Kingdom of Norway, 11 May 1990, Vol. II, Annexes, Annex 62, Jan Mayen Case.

The 1979 Convention on Long-Range Transboundary Air Pollution. UNTS No. 21623. The text can also be found on the home page of the UN ECE at http://www.unece.org/env/lrtap/ [13.3.2001]. SopS 15/1983.

The 1979 Directive on the Conservation of Wild Birds. The Directive is reproduced in OJ No. L 103/27, 79/409/EEC.

The 1981 Agreement on the Continental Shelf between Iceland and Jan Mayen. The Agreement is reproduced in 21 I.L.M. 1222 (1982).

The 1981 Borderwater Commission Agreement (Agreement Between Finland and Norway on Finnish Norwegian Borderwater Commission). SopS 32/1981.

The 1981 Memorandum of Intent Between the Government of the United States of America and the Government of Canada Concerning Transboundary Air Pollution. The text is reproduced in 20 I.L.M. 690 (1981).

The 1982 United Nations Convention on the Law of the Sea (the 1982 LOSC). The text is reproduced in 21 I.L.M. 1261 (1982). SopS 49-50/1996.

The 1982 Treaty amending, with regard to Greenland, the Treaties establishing the European Communities, Brussels, 13 Mar. 1984 EC 19 (1985); [1985] OJ L29/1. The Treaty with Greenland was amended and extended in December 1994; [1994] OJ L351/1.

The 1983 Agreement Between the Government of the Kingdom of Denmark and the Government of Canada for Cooperation Relating to the Marine Environment for co-operation relating to the marine environment (the 1983 Marine Agreement). The text is reproduced in 23 I.L.M. 269 (1984).

The 1984 Treaty relating to the Skagit River and Ross Lake, and the Seven Mile Reservoir on the Pend d'Oreille River, with Annex. Obtained via the oceana on-line service.

The 1985 Exchange of Notes Constituting an Agreement on Building a Bridge Across River Muonio (Finland and Sweden). The Agreement can be found via the UNTS, No. 24074. SopS 34/1985.

The 1985 Council Directive of 27 June 1985 on the Assessment of the Effects of Certain Public and Private Projects on the Environment (the 1985 EIA Directive). The Directive is reproduced in OJ No. L 175/40, 85/337/EEC.

The 1986 Convention on Early Notification of a Nuclear Accident of 1986. The Convention is reproduced in 6 I.L.M. (1986) 1370-1376. SopS 98/1986.

The 1986 Convention on Assistance in the Case of a Nuclear Accident or Radiological Emergency. The Convention is reproduced in 6 I.L.M. (1986) 1377-1386. SopS 82-83/1990.

The 1987 Agreement Between the Government of Canada and the Government of the United States of America on the Conservation of the Porcupine Caribou Herd (the 1987 Caribou Agreement). The text can be found at http://arcticcircle.uconn.edu/ANWR/anwrint-agreement.html [7.12.2000].

The 1988 Agreement on Arctic Cooperation and Exchange of Notes Concerning Transit of Northwest Passage. The Agreement can be found from 28 I.L.M. 141 (1989).

The 1988 Agreement between the United States and Canada Concerning Cooperation in Flood Control. Obtained via the oceana on-line service.

The 1989 ILO Convention No. 169 Concerning Indigenous and Tribal Peoples in Independent Countries (the 1989 ILO Convention). The text is reproduced in 28 I.L.M. 1382 (1989). The Convention can also be found on the home page of the ILO at http://ilolex.ilo.ch:1567/public/english/docs/convdisp.htm [28.4.2001].

The 1989 Agreement Between the United States and the Soviet Union Concerning Cooperation in Combatting Pollution in the Bering and Chukchi Seas in Emergency Situations. The Agreement obtained via the Oceana on-line Service at http://www.oceanalaw.com/ [17.2.2001].

The 1989 Agreement Between Denmark, Finland, Norway and Sweden on Cross-Border Cooperation in Preventing or Limiting the Damage in Emergency Situations Having Effect on Humans, Property or the Environment (authors own translation from a Finnish text). SopS 21/1992.

The 1989 Agreement on Joint Fishing Regulations for the Tana River Fishing District (with annexes). The Agreement can be found via the UNTS, No. 27809.

The 1990 Maritime Boundary Agreement Between the United States of America and the Union of the Soviet Socialist Republics. The text is reproduced in 29 I.L.M. 941 (1990).

The 1990 International Convention on Oil Pollution Preparedness, Response and Co-operation. UNTS No. 32194.

The 1991 Declaration on the Protection of the Arctic Environment and the 1991 Arctic Environmental Protection Strategy (Referred to as the AEPS document). The AEPS is reproduced in 30 I.L.M. 1624 (1991). The Declaration and the AEPS can be found on the home page of the Arctic Council at http://www.arctic-council.org/aeps_docs.asp [16.2.2001].

The 1991 Agreement Between the Government of the United States of America and the Government of Canada on Air Quality (the 1991 Air Quality Agreement). The text is reproduced in 30 I.L.M. 676 (1991). The Air Quality Agreement can also be found on via the International Joint Commission's home page at http://www.ijc.org/ijcweb-e.html [15.12.2000].

The 1991 Protocol to the Antarctic Treaty. The text is reproduced in 30 I.L.M. 1455 (1991).

The 1991 Convention on Environmental Impact Assessment in a Transboundary Context (the 1991 Espoo Convention) The Convention is reproduced in 30 I.L.M. 800 (1991) and can be found on the home page of the UN ECE at http://www.unece.org/env/eia/ [30.4.2001]. SopS 67/1997.

The 1992 United Nations Framework Convention on Climate Change. The text is reproduced in 31 I.L.M. 849 (1992). SopS 61/1994.

The 1992 Action Program Between Finland and Russian Federation with a view to Reduce Pollution and Implement Water Protection in the Baltic Sea Area as well as Other Areas Near the Border of Finland and Russian Federation (the 1992 Action Treaty) (Author's own translation). SopS 77/1992.

The 1992 Agreement on Cooperation in Research, Conservation and Management of Marine Mammals in the North Atlantic (Norway, Iceland, Greenland and Faroe Islands). The Agreement can be found on the home page of the the North Atlantic Marine Mammals Commission at http://www.nammco.no/Agreement.htm [13.3.2001].

The 1992 Convention on the Protection of the Marine Environment of the Baltic Sea Area. The Convention can be found on the home page of the Helsinki Commission (Baltic Marine Environment Protection Commission) at http://www.helcom.fi/ [21.2.2001]. SopS 2/2000.

The 1992 Maastricht Treaty on European Union. Consolidated version incorporating the changes made by the Treaty of Amsterdam, signed on 2 October 1997, is reproduced in Official Journal C 340, 10.11.1997, pp. 145-172. The Treaty can also be found on the Europa home page at http://europa.eu.int/eur-lex/en/treaties/index.html [16.5.2001].

The 1992 Treaty Establishing the European Community. The 1957 Treaty Establishing the European Economic Community was amended through the Treaty on European Union. The amended Treaty can be found on the Europa home page at http://europa.eu.int/eur-lex/en/treaties/index.html [16.5.2001] (Consolidated version incorporating the changes made by the Treaty of Amsterdam, signed on 2 October 1997 is reproduced in Official Journal C 340, 10.11.1997, pp. 173-308)).

The 1992 Agreement on Cooperation in the Field of Environmental Protection Between the Finnish Republic and the Russian Federation (author's own translation). SopS 2/1992.

The 1992 Agreement Between Denmark, Finland, Norway and Sweden on Cross-Border Cooperation in Preventing or Limiting the Damage in Emergency Situations Having Effect on Humans, Property or the Environment (authors own translation). SopS 21/1992.

The 1992 Agreement between the Government of Canada and the Government of the Russian Federation on Cooperation in the Arctic and the North. The Agreement can be found on the home page of the Office of the Auditor General of Canada at http://www.oag-bvg.gc.ca/domino/env co...bdadb88b3882565d90080dab3?OpenDocument [9.2.2001].

The 1992 Convention on the Transboundary Effects of Industrial Accidents (the 1992 Industrial Accidents Convention). The Convention can be found on the home page of the UN ECE at http://www.unece.org/env/teia/english/homepage.html [28.4.2001] or 31 I.L.M. 1330 (1992).

The 1992 Convention on the Protection and Use of Transboundary Watercourses and International Lakes (the 1992 Transboundary Watercourses Convention). The Convention can be found from 31 I.L.M. 1312 (1992) as well as on the home page of the UN ECE at http://www.unece.org/env/water/ [28.4.2001]. SopS 71/1996. The 1999 Protocol on Water and Health can be found from the same site.

The 1992 Convention on the Prohibition of the Development, Stockpiling and Use of Chemical Weapons. UNTS No. 33757. The text is reproduced in 32 I.L.M. 800 (1993). SopS 17-18/1997.

The 1992 Convention on Biological Diversity. The Convention can be found from 31 I.L.M. 818 (1992). The Convention can also be found at http://www.biodiv.org/convention/articles.asp [28.4.2001]. SopS 78/1994.

The 1992 Convention for the Protection of the Marine Environment of the North-East Atlantic (the 1992 Paris II Convention). The Convention can be found from 32 I.L.M. 1069 (1993) and on the home page of the Commission established by the Convention at http://www.ospar.org/eng/html/welcome.html [28.4.2001].

The 1992 Council Directive 92/43/EEC of 21 May 1992 on the Conservation of Natural Habitats of Wild Fauna and Flora. Official Journal L 206, 21/5/1992.

The 1992 Agreement on European Economic Area. Official Journal L 001, 03/01/1994, p. 494.

The 1993 Convention on Civil Liability for Damage Resulting from Activities Dangerous to the Environment. The text is reproduced in 32 I.L.M. 1228 (1993) and can also be found on the home page of the Council of Europe at http://conventions.coe.int/treaty/EN/cadreprincipal.htm [18.2.2001].

The 1993 Nuuk Declaration on Environment and Development in the Arctic. The Declaration can be found on the home page of the Arctic Council at http://www.arctic-council.org/aeps_docs.asp [16.2.2001].

The 1993 ILO Convention No. C 174 of Major Industrial Accidents. The Convention can be found on the ILO home page at http://ilolex.ilo.ch:1567/scripts/convde.pl?query=C174&query0=C174&submit=Display [18.2.2001].

The 1993 Agreement Between the Government of Canada and the Government of the Russian Federation Concerning Environmental Cooperation. The Agreement can be found on the home page of the Office of the Auditor General of Canada at http://www.oag-bvg.gc.ca/domino/env_commitments.nsf/80cacb4d2ed8b3a3052565da0067c393/d766264ab25de0f48525667b006fe58d?OpenDocument [13.3.2001].

The 1994 North American Agreement on Environmental Cooperation (the 1994 Environmental Cooperation Agreement). The text is reproduced in 32 I.L.M. 1480 (1993) and can also be found on the home page of the North American Commission for Environmental Cooperation (NACEC) established through this agreement at http://www.cec.org/home/index.cfm?varlan=english [28.4.2001].

The 1994 Energy Charter Protocol on Energy Efficiency and Related Environmental Aspects. The Energy Charter Treaty and the Protocol are reproduced in 34 I.L.M. 445 (1995).

The 1994 Agreement on Cooperation in the Prevention of Pollution of the Environment in the Arctic Between the United States and the Soviet Union. Obtained via the Oceana on-line service.

The 1994 Agreement on Cooperation in the Field of Protection of the Environment and Natural Resources, with Annex between the United States and the Russian Federation. Obtained via the Oceana on-line service.

The 1994 Agreement Relating to the Implementation of part XI of the United Nations Convention on the Law of the Sea of 10 December 1982. The text is reproduced in 33 I.L.M. 1309 (1994) and can also be found on the home page of the ISA at http://www.un.org/Depts/los/unclos/closindx.htm [13.3.2001]. SopS 49-50/1996.

The Agreement on Cooperation in the field of protection of the Environment and Natural Resources, with Annex, of 1994. Obtained via the oceana on-line service.

The 1994 Agreement Between the Government of the Kingdom of Norway and the Government of the Russian Federation on Cooperation and Response to Oil Pollution in the Barents Sea. Obtained from the Norwegian Ministry of Foreign Affairs.

The 1994 Convention on Nuclear Safety. The text is reproduced in 33 I.L.M. 1514 (1994). SopS 74/1996.

The 1995 Agreement between the Kingdom of Denmark and the Kingdom of Norway relating to the Delimitation of the Continental Shelf in the Area between Jan Mayen and Greenland and of the Boundary between Fisheries Zones in the Area. Reproduced in Charney and Alexander 1998, pp. 2524-2525.

The 1996 Administrative Arrangement for Cooperation and the Exchange of Information in Nuclear Regulatory Matters, with Appendices) (the 1996 Nuclear Agreement; Canada and the United States). Obtained via the oceana on-line service.

The 1996 United Nations Comprehensive Test Ban Treaty. The text is reproduced in 35 I.L.M. 1439 (1996).

The 1996 Inuvik Declaration on Environmental Protection and Sustainable Development in the Arctic. The Declaration can be found on the home page of the Arctic Council at http://www.arctic-council.org/aeps_docs.asp [16.2.2001].

The 1996 Declaration on the Establishment of the Arctic Council. The Declaration is reproduced in 35 I.L.M. 1385-1390 (1996). It can also be found on the home page of the Arctic Council at http://www.arctic-council.org/f_documents.asp [16.2.2001].

The 1996 Council Directive 96/61/EC of 24 September 1996 Concerning Integrated Pollution Prevention and Control. OJ No L 257, 10.10.1996, p. 26.

The 1996 Council Directive 96/82/EC of 9 December 1996 on the Control of Major-Accident Hazards Involving Dangerous Substances. OJ L 10, 14.1.1997, p. 13.

The 1997 Alta Declaration on the Protection of the Arctic Environmental Protection Strategy. The Declaration can be found on the home page of the Arctic Council at http://www.arctic-council.org/aeps_docs.asp [16.2.2001].

The 1997 Guidelines for Environmental Impact Assessment (EIA) in the Arctic (the 1997 EIA Guidelines). The Guidelines can be found on the home page of the Arctic Environmental Impact Assessment (ARIA) at http://finnbarents.urova.fi/aria/ [1.6.2001].

The 1997 Arctic Offshore Oil and Gas Guidelines (the 1997 Offshore Guidelines). The Guidelines can be found on the home page of the Protection of the Arctic Marine Environment (PAME) –home page at http://www.grida.no/pame/FRreports.htm [1.6.2001].

The 1997 Transboundary Environmental Impact Assessment Treaty (the 1997 Draft TEIA Treaty). The draft can be found on the NACEC home page at http://www.cec.org/pubs_info_resources/law_treat_agree/pbl.cfm?varlan=english [19.2.2001].

The 1997 Joint Convention on the Safety of Spent Fuel Management and on the Safety of Radioactive Waste Management (the 1997 Joint Convention). The text is reproduced in 36 I.L.M. 1431 (1997).

The 1997 Council Directive 97/11/EC of 3 March 1997 Amending Directive 85/337/EEC on the Assessment of the Effects of Certain Public and Private Projects on the Environment (the 1997 Amendment Directive). OJ No. L 073, 14/03/1997.

The 1997 Convention on Non-Navigational Uses of International Watercourses (the 1997 Watercourses Convention). The text is reproduced in 36 I.L.M. 710 (1997).

The 1998 Iqaluit Declaration. The Declaration can be found on the home page of the Arctic Council at http://www.arctic-council.org/f_documents.asp [16.2.2001].

The 1998 Convention on Access to Information, Public Participation in Decision-Making and Access to Justice in Environmental Matters (the 1998 Aarhus Convention). The text is reproduced in 38 I.L.M. 517 (1999) and can also be found on the home page of the UN ECE at http://www.unece.org/env/pp/ [28.4.2001].

The 1998 Agreement Between the Government of the Kingdom of Denmark Together with the Greenland Home Rule Government, on the One Hand, and the Government of the Republic of Iceland, on the Other Hand, Concerning the Delimitation of the Continental Shelf and the Fishery Zones in the Area Between Greenland and Iceland. An unofficial translation of the text of the agreement was obtained from the Royal Danish Ministry of Foreign Affairs. A copy obtained by the author from the foreign ministries of both Denmark and Iceland.

The 2000 Memorandum of Understanding Between the Government of Norway and the Government of the Russian Federation on Cooperation on search and rescue as well as Warning of Serious Incidents The Memorandum can be found on the home page of ODIN at http://odin.dep.no/ud/engelsk/p10002187/032001-990361/index-dok000-b-n-a.html [6.2.2001].

The 2000 Barrow Declaration on the Occasion of the Second Ministerial Meeting of the Arctic Council. The Declaration can be found on the home page of the Arctic Council at http://www.arctic-council.org/whatsnew.asp [16.2.2001].

The 2000 Agreement on between the United States of America and the Government of the Russian Federation on the Conservation and Management of the Alaska-Chukotka Polar Bear Population. Agreement was obtained by the author from the United States Department of the Interior, Fish and Wildlife Service.

The 2001 Stockholm Convention on Persistent Organic Pollutants. Final draft can be obtained from the address at http://www.chem.unep.ch/pops/POPs_Inc/dipcon/convtext/disclaimer.htm [25.5.2001].

MONOGRAPHS, ARTICLES ETC. ATTRIBUTABLE TO AUTHORS

Ahlström, Christer

The Status of Multilateral Export Control Regimes – An Examination of Legal and Non-Legal Agreements in International Co-Operation. Iustus Förlag AB, Uppsala 1999.

Akehurst, Michael

A modern Introduction to International Law. Sixth edition 1987, third impression (1991). Harper Collins Academic. Mackays of Chatham PLC, Chatham, Kent.

Alexander, Lewis M.

'Canada-Denmark (Greenland), Report No. 1-1, (Summary)' in Vol I, International Maritime Boundaries (edited by Charney, J.I. & Alexander L.M.), pp. 371-378. The American Society of International Law 1993. Martinus Nijhoff Publishers.

Alfredsson, Gudmundur

'Minimum Requirements for a New Nordic Sami Convention' in 68 Nordic Journal of International Law (1999), pp. 397-411.

Anaya, S. James

Indigenous Peoples in International Law. Oxford University Press 1996.

Anderson, D. H.

Iceland – Norway (Jan Mayen), Report No. 9-4 (Summary)' in Vol. II International Maritime Boundaries (edited by Charney, J.I. & Alexander L.M.), pp. 1755-1760. The American Society of International Law 1993. Martinus Nijhoff Publishers.

'Norway – Soviet Union, Report No. 9-6 (Summary)' in Vol. II International Maritime Boundaries (edited by Charney, J.I. & Alexander L.M.), pp. 1781-1784. The American Society of International Law 1993. Martinus Nijhoff Publishers.

'Denmark (Greenland) – Norway (Jan Mayen), Report No. 9-19 (Summary and Background to the Case)' in Vol. III International Maritime Boundaries (edited by Charney, J.I. & Alexander L.M.), pp. 2507-2520. The American Society of International Law 1998. Kluwer Law International.

Archer, Clive & Scrivener, David

'International Cooperation in the Arctic Environment' in The Arctic: Environment, People, Policy (edited by Nuttall M. & Callaghan T.V.), pp. 601-619. Overseas Publishers Association 2000.

Armstrong, Terence & Roberts, Brian & Swithinbank, Charles

Illustrated Glossary of Snow and Ice (second edition). Scott Polar Research Institute 1973.

Aust, Anthony

Modern Treaty Law and Practice. Cambridge University Press 2000.

Barsh, Russel Lawrence

'Indigenous Peoples in the 1990s: From Object to Subject of International Law' in Vol. 7 Harvard Human Rights Journal (1994), pp. 33-86.

Baxter, Richard R.

'International Law in 'Her Infinite Variety'' in 29 International & Comparative Law Quarterly (1980), pp. 549-566.

Benvenisti, Eyal

'Collective Action in the Utilization of Shared Freshwater: The Challenges of International Water Resources Law' in Vol. 90, No. 3 American Journal of International Law (1996), pp. 384-415.

Berg, Axel

'Nordic Council and Nordic Council of Ministers' in Encyclopedia of Public International Law, pp. 261-263. Elsevier Science Publishers B.V. 1983.

Bernhardt, Rudolf

'Treaties' in Encyclopedia of Public International Law (editor Bernhardt, R.), pp. 459-464. Elsevier Science Publishers B.V. 1984.

Birnie, Patricia W. & Boyle, Alan E.

International Law & the Environment. Clarendon Press Oxford 1992.

Bloom, Evan

'The Establishment of the Arctic Council' in 93 American Journal of International Law (1999), pp. 712-722.

Boscek, Boleslaw Adam

'The Arctic Ocean and the New Law of the Sea' in Vol. 29 German Yearbook of International Law (1986), pp. 154-189.

Bourne, Charles B.

'The Case Concerning the Gabcikovo-Nagymaros Project: An Important Milestone in International Water Law' in Vol. 8 Yearbook of International Environmental Law (1997), pp. 6-12.

Bowett, D.W.

'Reservations to Non-Restricted Multilateral Treaties' in 48 British Yearbook of International Law (1976-1977), pp. 67-92.

Boyd, Susan B.

'The Legal Status of the Arctic Sea Ice: A Comparative Study and a Proposal' in Vol. XXII The Canadian Yearbook of International Law (1984), pp. 98-152.

Boyle, Alan

'Nuclear Energy and International Law: An Environmental Perspective' in 60 the British Yearbook of International Law (1989), pp. 257-313.

'The Gabcíkovo-Nagymaros Case: New Law in Old Bottles' in Vol. 8 Yearbook of International Environmental Law (1997), pp. 13-20.

'Codification of International Environmental Law and the International Law Commission: Injurious Consequences Revisited' in International Law and Sustainable Development: Past Achievements and Future Challenges (edited by Boyle and Freestone), pp. 61-85. Oxford University Press 1999.

Boyle, Alan & Freestone, David

'Introduction' in International Law and Sustainable Development: Past Achievements and Future Challenges (edited by Boyle and Freestone), pp. 1-19. Oxford University Press 1999.

Broms, Bengt

'The Nordic Convention on the Protection of the Environment' in Transboundary Air Pollution; International Legal Aspects of the Co-operation of States (edited by Flinterman, Kwiatkowska, Lammers), pp. 141-152. Martinus Nijhoff Publishers 1986.

Brownlie, Ian

System of the Law of Nations: State Responsibility (Part I). Clarendon Press, Oxford, 1983.

'General Course on Public International Law by Ian Brownlie' in Recueil des Cours (Collected Courses of the Hague Academy of International Law) 1995, pp. 9-228. Kluwer Law International.

Principles of Public International Law (Fifth edition, 1998). Clarendon Press, Oxford 1998. Oxford University Press.

Brubaker, Douglas R.

Marine Pollution and International Law: Principles and Practice. Belhaven Press 1993.

The Legal Status of the Russian Baselines in the Arctic. INSROP Working Paper No. 37 – 1996, IV.3.1, 1996.

Brunnee and Toope

Environmental Security and Freshwater Resources: A Case Study for International Ecosystem Law' in 5 Yearbook of International Environmental Law (1994), pp. 41-76.

Brölmann, C.M. & Zieck M.Y.A

'Indigenous Peoples' in Peoples and Minorities in International Law (edited by Brölmann, C.M. & Lefeber R. & Zieck M.Y.A.), pp. 187-220. Kluwer Academic Publishers 1993.

Byers, Michael

'Conceptualizing the Relationship Between Jus Cogens and Erga Omnes Rules' in Vol. 66, Nos. 2-3 Nordic Journal of International Law (1997), pp. 211-239.

Caron, David D.

'Toward an Arctic Environmental Regime' in Ocean Development and International Law, Volume 24 (1993), pp. 377-392.

Castro, Paulo Canelas de
'The Judgment in the Case Concerning the Gabcikovo-Nagymaros Project: Positive Signs for the Evolution of International Water Law' in Vol. 8 Yearbook of International Environmental Law (1997), pp. 21-31.

Chance, Norman
Special Report 1995 at http://borealis.lib.uconn.edu/ArcticCircle/ANWR/anwrpreface.html [4.4.2001].

Charney, Jonathan I.
'Third State Remedies for Environmental Damage to the World's Common Spaces' in International Responsibility of Environmental Harm (editors Francioni, F. & Scovazzi, T.), pp. 149-177. Graham & Trotman 1991.

Charney, Jonathan I. & Alexander, Lewis M.
International Maritime Boundaries. Vol I and Vol II, 1993. Martinus Nijhoff Publishers. International Maritime Boundaries. Vol III, 1998. Kluwer Law International.

Chinkin, Christine
'The Challenge of Soft Law: Development and Change in International Law' in 38 International & Comparative Law Quarterly (1989), pp. 850-867.

'A Mirage in the Sand? Distinguishing Binding and Non-Binding Relations Between States' in 10 Leiden Journal of International Law (1997), pp. 223-247.

Churchill, R.R. & Lowe A.V.

The Law of the Sea. Third edition 1999. Juris Publishing. Manchester University Press.

Churchill, Robin R. & Ulfstein, Geir

'Autonomous Institutional Arrangements in Multilateral Environmental Agreements: A Little-Noticed Phenomenon in International Law' in Vol. 94, No. 4 American Journal of International Law (2000), pp. 623-659.

Connelly, Robert G.

'The UN Convention on EIA in a Transboundary Context: A Historical Perspective' in 19 Environmental Impact Assessment Review 1999, pp. 37-46.

Corke M.J. & Holmes C.H.

Final Report: Evaluation of New Gas Supply Options for Finland. Prepared for Ministry of Trade and Industry 2000. The report can be found on the home page of the Ministry at http://www.vn.fi/ktm/eng/3/maakaasu.htm [6.3.2001].

Corten Olivier & Schaus Annemie

'La Responsabilité Internationale des États-Unis pour les Dommages Causés par les Précipitations Acides sur le Territoire Canadien' in XXVII Canadian Yearbook of International Law 227 (1989), pp. 239-261.

Danilenko, G.M.

Law-making in the International Community. 1993 Kluwer Academic Publishers.

Decaux, Emmanuel

'L'Affaire de la Délimitation Maritime dans la Région Située entre le Groenland et Jan Mayen (Danemark c. Norvegè) Arrêt de la C.I.J. du 14 Juin 1993' in Annuaire Francais De Droit International XXXIX (1993), pp. 495-513.

Dedichen, L., Eliasson, D., Hökborg, K., Johansson, M. & Norberg, S.

The European Economic Area: EEA Law; A Commentary on the EEA Agreement (1st edition). CE Fritzes AB 1993.

Derry, T.K.

A history of Scandinavia: Norway, Sweden, Denmark, Finland & Iceland. Published by the University of Minnesota Press 1979.

Doeker Günther & Gehring, Thomas

'Private or International Liability for Transnational Environmental Damage – The Precedent of Conventional Liability Regimes' in No. 1, Vol. 2, Journal of Environmental Law (1990), pp 1-16.

Duffy, Quinn R.

The Road to Nunavut; the Progress of the Eastern Arctic Inuit Since the Second World War. Mcgill-Queen's University Press 1988.

Dworkin, Ronald

Law's Empire. The Belknapp Press of Harvard University Press 1986.

Dzidzornu, David M.

'Environmental Impact Assessment Procedure Through the Conventions' in Vol. 10, No. 1 European Environmental Law Review (2001), pp. 15-26.

Eaton, M.R.

'Common Foreign and Security Policy' in Legal Issues of the Maastricht Treaty (editors O'Keeffe, D. & Twomey P.M.), pp. 215-225. Chancery Law Publications, London 1994.

Ebbesson, Jonas

Svensk tillämpning av den nordiska miljöskyddskonventionen. Om inverkan på nationell rätt av den nordiska miljöskyddskonventionen och principen on icke-diskriminering. Naturvårdsverket Rapport 4063, 1992.

Compatibility of International and National Environmental Law. Iustus Förlag AB, Uppsala 1996. Printed and Bound in Sweden by Graphic Systems AB, Gothenburg, 1996.

'The Notion of Public Participation in International Environmental Law' in Vol. 8 Yearbook of International Environmental Law (1997), pp. 51-97.

'Innovative Elements and Expected Effectiveness of the 1991 EIA Convention' in 19 Environmental Impact Assessment Review (1999), pp. 47-55.

Elferink, Alex G. Oude

The Law of Maritime Boundary Delimitation: A Case Study of the Russian Federation. 1994 Kluwer Academic Publishers.

'The Law and Politics of the Maritime Boundary Delimitations of the Russian Federation: Part 2' in the International Journal of Marine and Coastal Law, Vol. 12, No. 1, March (1997), pp. 5-37.

Elias, T.O.

The Modern Law of Treaties. A.V. Sijthoff International Publishing Company B.V. 1974. Printed in the Netherlands.

Fabo, Edward de & Björn, Lars Olof

'Impact of Ultraviolet-B Radiation in the Arctic' in The Arctic: Environment, People, Policy (edited by Nuttall M. & Callaghan T.V.), pp. 555-573. Overseas Publishers Association 2000.

Falk, Richard A.

'On the Quasi-Legislative Competence of the General Assembly' in 4 American Journal of International Law (1966), pp. 782-792.

Fawcett, James

'The Legal Character of International Agreements' in 30 British Yearbook of International Law (1953), pp. 381-400.

Fitzmaurice, John

Security and Politics in the Nordic Area. Published by Avebury, Gower Publishing Company 1987.

Fitzmaurice, Malgosia

'The Finnish-Swedish Frontier Rivers Commission' in Hague Yearbook of International Law (1992), pp. 33-67.

'International Environmental Law as a Special Field' in Vol. XXV Netherlands Yearbook of International Law (1994), pp. 181-226.

'The Variety of Theoretical bases of Responsibility for Environmental Damage: Introduction' in Global Biodiversity, pp. 49-98. 3rd Finnish Conference on International Environmental Law. Ius Gentium ry. 1996. Hakapaino, Helsinki.

'The Sámi People: Current Issues Facing an Indigenous People in the Nordic Region' in Vol. VIII Finnish Yearbook of International Law (1997), pp. 200-243.

Francioni, Fransesco

'International Cooperation for the Protection of the Environment: The Procedural Dimension' in Environmental Protection and International Law (editors Lang, W, Neuhold, H and Zemanek, K), pp. 203-221. Graham & Trotman 1991.

'International 'Soft Law': a Contemporary Assessment' in Fifty Years of the International Court of Justice (editors Malgosia Fitzmaurice and Vaughan Lowe), pp. 167-178. Cambridge University Press 1996.

Franckx, Erik

Maritime Claims in the Arctic; Canadian and Russian Perspectives. Martinus Nijhoff Publishers 1993 (Kluwer Academic Publishers).

Fuentes, Ximena

'The Criteria for the Equitable Utilization of International Rivers' in 67 British Yearbook of International Law (1996), pp. 337-412.

'Sustainable Development and the Equitable Utilization of International Watercourses' in 69 British Year Book of International Law (1998), pp. 119-200.

Furman, Eeva & Hildén, Mikael

Environmental Impact Assessment – A Tool for Sustainable Development in the Arctic. Speech given in the Third International Conference of Parliamentarians of the Arctic Region Salekhard, April 22-24, 1998. The text can be found on the home page of the Standing Committee of Arctic Parliamentarians at http://www.grida.no/parl/salek/index.htm [19.2.2001].

Furman, E., Hilden, M., Soveri U.-R. & Brinkman, I.

'Assessment Across Borders – Experiences of Countries and International Institutions on Implementation of the Espoo Convention' in Proceedings from the 3rd Nordic EIA/SEA Conference, Karlskrona, Sweden 22 – 23 November 1999. The article is published in the Nordregio's home page at http://www.nordregio.se [23.2.2001] in the EIA section.

Garner, Bryan A.

A Dictionary of Modern Legal Usage (second edition).Oxford University Press 1995.

Gates, David M.

Energy and Ecology. 1985 by Sinauer Associates Inc. Publishers.

Gayim, Eyassu

The UN Draft Declaration on Indigenous Peoples. Assessment of the Draft Prepared by the Working Group on Indigenous Populations. Publication of the Northern Institute for Environmental and Minority Law at the University of Lapland 1994. Juridica Lapponica, No. 13.

Gilpin, Alan

Environmental Impact Assessment (EIA): Cutting Edge for the Twenty-First Century. Cambridge University Press 1995.

Gizewski, Peter

'Military Activity and Environmental Security: The Case of Radioactivity in the Arctic' in Vol. 21, No. 4, Northern Perspectives (1993-1994). The article can be found on the home page of the Canadian Arctic Resources Committee at http://www.carc.org/pubs/v21no4/military.htm [15.2.2001].

Goldblat, Jozef

'The Nuclear Non-Proliferation Régime: Assessment and Prospects' in 256 Collected Courses of the Hague Academy of International Law (1995), pp. 9-192.

Hakapää, Kari

Marine Pollution in International Law; Material Obligations and Jurisdiction with Special Reference to the Third United Nations Conference on the Law of the Sea (1981). Academia Scientiarum Fennica. Karprint Ky, Huhmari 1981.

'Some Observations on the Arctic Waters and the New Law of the Sea' in Legal Problems in the Arctic Regions (1990), pp. 67-77. Publication of the Institute for Nordic Law at the University of Lapland. Lakimiesliiton kustannus.

'Arktiset merialueet kansainvälisessä oikeudessa' in Oikeuskirja (editor Martikainen, P.), pp. 87-105. Lapin yliopiston oikeustieteellisiä julkaisuja sarja B 25, Kitee 1995, Pandecta Oy.

'Vessel-Source Pollution in the UN Law of the Sea Convention: Some Assessment as of today' in Liber Amicorum Bengt Broms; Celebrating His 70th Birthday 16 October 1999, pp. 97-117. PMS Print Oy Helsinki 1999. Publications of the Finnish Branch of the International Law Association, No. 9.

Handl, Gunther

'Territorial Sovereignty and the Problem of Transnational Pollution' in No. 1, Vol. 69 American Journal of International Law (1975), pp. 50-76.

'State Liability for Accidental Transnational Environmental Damage by Private Persons' in 74 American Journal of International Law (1980), pp. 525-566.

'Environmental Security and Global Change: The Challenge to International Law' in Environmental Protection and International Law (editors Lang, W & Neuhold, H & Zemanek, K), pp. 59-87. Graham & Trotman 1991.

Hannikainen, Lauri

Peremptory Norms (Jus Cogens) in International Law; Historical Development, Criteria, Present Status. Finnish Lawyers' Publishing Company, Helsinki 1988.

'Position of Indigenous People, Especially the Scandinavian Sámi, in Decision Making Processes' in Proceedings of the Arctic Opportunities Conference (Editor Manfred Lange), pp. 315-326. 1996 Arctic Centre. Printed by Lapin Yliopistopaino.

Harders, Enno Johannes

Regionaler Umweltschutz in der Arktis. Die Völkerrechtliche Pflicht zur Zusammenarbeit und die Drittbindungswirkung eines Regionalmeerabkommens. Nomos Verlagsgesellschaft, Baden-Baden 1997.

Harrop, Owen D. & Nixon, Ashley J.

Environmental Assessment in Practice. Routledge Environmental Management Series 1999.

Hart, H.L.A.

The Concept of Law. Oxford University Press 1961. Reprinted in 1993.

Heininen, Lassi

'Environmental Threats of Military Presence in Northern Waters and the Arctic' in Arctic Environmental Problems (edited by Heininen, L.), pp. 42-69. Tampere Peace Research Institute Occasional Papers No. 41, 1990.

Sotilaallisen läsnäolon ympäristöriskit arktiksessa. Kohti Arktiksen säätelyjärjestelmää. Rauhan- ja konfliktintutkimuslaitos. Tutkimuksia No. 43, 1991, Tampere.

'The Conflict of Interests Between the Environment and Military Strategy in Northern Waters and the Arctic' in Perspectives on Environmental Conflict and International Relations (edited by Käkönen J.), pp. 55-71. Tampere Peace Research Institute 1992.

Euroopan pohjoinen 1990-luvulla; moniulotteisten ja ristiriitaisten intressien alue. Lapin yliopisto, yhteiskuntatieteiden tiedekunta ja Lapin yliopisto, Arktinen keskus. Lapin yliopistopaino, Rovaniemi 1999.

Heininen, L., Jalonen, O.-P. & Käkönen, J.

Expanding the Northern Dimension. Tampere Peace Research Institute, research report No. 61, 1995.

Hellsten, Kai

'Hydro-Power Plants in the North' in The Legal Status of the Individual in Nordic Environmental Law, pp. 179-187. Publication of the Northern Institute for Environmental and Minority Law 1994. Juridica Lapponica, No. 10.

Hepola, Matti

'Mietteitä Suomen ja Ruotsin välisen rajajokisopimuksen uudistamisesta' in Artikla ry 20 vuotta (toim. Kopperoinen, S. & Koskinen, S. & Lönnfors, M. & Pulkamo M.), pp. 145-174. Lapin yliopistopaino, Rovaniemi 1999.

Hicks, Jack

'The Nunavut Land Claim and the Nunavut Government: Political Structures of Self-Government in Canada's Eastern Arctic' in Dependency, Autonomy, Sustainability in the Arctic (edited by Petersen, H. & Poppel, B.), pp. 21-53. Ashgate Publishing Ltd 1999.

Hildengard, Hartmut

'A Fresh Look at Soft Law' in 10 European Journal of International Law (1999), pp. 499-515.

Hilpinen, Risto

'Introduction to the Second Impression' in Deontic Logic: Introductory and Systematic Readings (edited by Hilpinen, R.), pp. xi-xvii. Published by Reidel Publishing Company. (First published in 1971), reprinted in 1981 with new introduction.

Hohmann, Harald

Precautionary Legal Duties and Principles of Modern International Environmental Law. The Precautionary Principle: International Environmental Law Between Exploitation and Protection. Graham & Trotman 1994.

Horn, Frank

Reservations and Interpretative Declarations to Multilateral Treaties. Elsevier Science Publishers B.V., 1988.

Husby, Stig Roar & Tesli, Arne

'Eia in a Transboundary Context: Principles and Challenges for a Coordinated Nordic Application of the Espoo Convention' in 19 Environmental Impact Assessment Review (1999), pp. 57-84.

Jennings, Robert & Watts, Arthur

Oppenheim's International Law. Ninth Edition 1992, Vol. 1 Peace. Longman Group UK Limited.

Johnson, Stanley P.

The Earth Summit; the United Nations Conference on Environment and Development (UNCED). Graham & Trotman/Martinus Nijhoff 1993. Printed and bound in Great Britain by Hartnolls Ltd, Bodmin, Cornwall.

Jumppanen, Pauli

'Environmental Aspects of the Exploitation of Arctic Oil and Gas Reserves' in Arctic Environmental Problems (edited by Heininen, L.), pp. 70-84. Tampere Peace Research Institute Occasional Papers No. 41, 1990.

Karvo, Ulla

'Kestävän kehityksen merkityksestä' in Juhlajulkaisu; oikeustieteen ylioppilaiden yhdistys Artikla ry. 15 vuotta, pp. 279-300. Artikla ry:n kustannus Pandecta oy. Kiteen paino, ky, Kitee 1994.

Kent, Penelope

Law of the European Union (second edition). Pearson Professional Limited 1996.

Kish, John

The Law of International Spaces. A.W. Sijthoff International Publishing Company, N.V. 1973.

Kiss, A.

'La Convention Nordique sur l'Environment' in XX Annuaire Francais de Droit International (1974), pp. 808-814.

'Les Traités-Cadres: une Technique Juridique Caractéristique du Droit International de L'Environment' in XXXIX Annuaire Francais de Droit International (1993), pp. 792-797.

Kiss, A. & Shelton D.

Manual of European Environmental Law. Grotius Publications Limited 1993.

Klabbers, Jan

'Informal Agreements in International Law: Towards a Theoretical Framework' in V Finnish Yearbook of International Law (1994), pp. 267-387.

The Concept of Treaty in International Law. Kluwer Law International 1996. Referred to as Klabbers 1996 I.

'The Redundancy of Soft Law' in 65 Nordic Journal of International Law (1996), pp. 167-182. Referred to as Klabbers 1996 II.

'Some Problems Regarding the Object and Purpose of Treaties' in Vol. VIII the Finnish Yearbook of International Law (1997), pp. 138-160.

'Presumptive Personality: The European Union in International Law' in International Law Aspects of the European Union (editor Koskenniemi M.), pp. 231-253. Kluwer Law International 1998.

'Accepting the Unacceptable? A New Nordic Approach to Reservations to Multilateral Treaties' in Vol. 69, No. 2 Nordic Journal of International Law (2000), pp. 179-193.

Klein, Eckart

'Treaties, Effect of Territorial Changes' in Encyclopedia of Public International Law, pp. 473-476. Elsevier Science Publishers B.V. 1984.

Klein-Chesivoir Carole

'Avoiding Environmental Injury: The Case for Widespread Use of Environmental Impact Assessments in International Development Projects' in Vol. 30 Virginia Journal of International Law (1990), pp. 517-551.

Knapp, Gunnar

'The Population of the Circumpolar North' in The Arctic: Environment, People, Policy (edited by Nuttall M. & Callaghan T.V.), pp. 459-483. Overseas Publishers Association 2000.

Koivurova, Timo

'The Future of the Nordic Environment Protection Convention' in 66 Nordic Journal of International Law 1997, pp. 505-526. Referred to as Koivurova 1997 I.

'The Transnational EIA procedure of the Espoo Convention' in Vol. VIII Finnish Yearbook of International Law (1997), pp. 161-199. Referred to as Koivurova 1997 II.

'The Legal Status of International Arctic Cooperation' in Security in the European North; from 'Hard' to 'Soft' (editors Lassi Heininen & Gunnan Lassinantti), pp. 143-160. Publishers: Arctic Centre, University of Lapland and the Olof Palme International Center 1999. Printed by Lapin yliopistopaino, Rovaniemi 1999.

'Yleinen kansainvälinen ympäristöoikeus ja arktinen ympäristönsuojelu' in Oikeus 4/2000, pp. 463-472.

Kolodkin A., Mirovitskaya N. & Roginko Alexei

'The Arctic: Arena for International Environmental Security' in Ocean Yearbook 14 (1993), pp. 277-289.

Korhonen, Seija

Experience Gained from the Transboundary EIA: Tornio Ferrochrome and Stainless Steel Works (28.4.1998). Author's personal copy obtained from the Finnish Ministry of the Environment.

Koskenniemi, Martti

'International Pollution in the System of International Law' in XVII Oikeustiede Jurisprudentia 1984, pp. 93-181.

'Peaceful Settlement of Environmental Disputes' in Vol. 60 Nordic Journal of International Law (1991), pp. 73-92.

'Breach of Treaty or Non-Compliance? Reflections on the Enforcement of the Montreal Protocol' in Vol. 3 Yearbook of International Environmental Law (1992), pp. 123-162.

Kuokkanen, Tuomas

'Itämeren ja Pohjois-Amerikan suurtenjärvien suojelusopimukset' in Kansainvälinen ympäristöoikeus, pp. 77-136. Ius Gentium ry. Suomen ympäristöoikeustieteen seuran julkaisuja, 10. Helsinki 1988.

'Ympäristövaikutusten arviointimenettely kansainvälisissä sopimuksissa' in Global Biodiversity; 3rd Finnish Conference on International Environmental Law, pp. 165-172. Ius Gentium ry. 1996. Hakapaino, Helsinki.

International Law and the Environment: Variations on a Theme. Printed in Tummavuoren Kirjapaino Oy, Vantaa 2000.

Kuusiniemi, Kari

'Sic tuo utere –periaate ja immissiot' in Kansainvälinen ympäristöoikeus (toim. Hollo, E.J. & Parkkari, J.K. & Vilander, M), pp. 118-136. Suomen ympäristöoikeustieteen Seura ry:n julkaisuja 23. Gummerus kirjapaino 1994.

Kuusisto, Paula

'Arktisen alueen rajaaminen' in TERRA 112:3 (2000), pp. 163-167.

Kwiatkowska, Barbara

The 200 Mile Exclusive Economic Zone in the New Law of the Sea. 1989 by Kluwer Academic Publishers, Martinus Nijhoff Publishers.

Lakhtine, W.

'Rights Over the Arctic' in 24 American Journal of International Law (1930), pp. 703-717.

Lange, Manfred A.

'Integrated Global Change Impact Assessments' in The Arctic: Environment, People, Policy (edited by Nuttall M. & Callaghan T.V.), pp. 517-553. Overseas Publishers Association 2000.

Langlais, Richard

Status Report on their work June 2000. A Report for the Standing Committee of Parliamentarians of the Arctic Region. The article can be found on the home page of the Standing Committee of Arctic Parliamentarians at http://www.grida.no/parl/ [8.2.2001].

Lavalle, Roberto

'About the Alleged Customary Law Nature of the Rule of Pacta Sunt Servanda' in No. 1-2, Vol. 33 Austrian Journal of Public and International Law (Österreichische zeitschrift für öffentliches recht und völkerrecht) (1982), pp. 9-28.

Lefeber, Rene

Transboundary Environmental Interference and the Origin of State Liability. 1996 Kluwer Law International.

Leigh, Kathy

'Liability for Damage to the Global Commons' in Vol. 14 Australian Yearbook of International Law (1992), pp. 129-156.

Lowe, Vaughan

'Do General Rules of International Law Exist?' in 9 Review of International Studies (1983), pp. 207-213.

'Sustainable Development and Unsustainable Arguments' in International Law and Sustainable Development: Past Achievements and Future Challenges (edited by Boyle and Freestone), pp. 19-37. Oxford University Press 1999.

Långström, Tarja

'The Dissolution of the Soviet Union in the Light of the 1978 Vienna Convention on Succession of States in Respect of Treaties' in State Succession: Codification Tested Against the Facts (editors Eisemann, P.M. & Koskenniemi M.), pp. 723-826. Hague Academy of International Law. Martinus Nijhoff Publishers 2000.

Magraw, Daniel Barstow

'Transboundary Harm: The International Law Commission's Study of "International Liability"' in No. 2, Vol. 80 American Journal of International Law (1986), pp. 305-330.

Malanczuk, Peter

'Some Critical Thoughts in the Light of the Rio Conference' in Sustainable Development and Good Governance (edited by Ginther K. & Denters E. & Waart P.), pp. 23-52. Kluwer Academic Publishers 1995.

Akehurst's Modern Introduction to International Law (seventh revised edition). 1997 Routledge.

Malvia, R.A.

'State Responsibility for Environmental Damage Beyond Territorial Waters' in Vol. 27, No. 1 The Indian Journal of International Law (January-March 1987), pp. 30-49.

Manner E.J. & Metsälampi, Veli-Martti

The Work of the International Law Association on the Law of International Water Resources (editors Manner E.J. & Metsälampi, Veli-Martti). Publications of the Finnish Branch of the International Law Association, No. 6. Vammala 1988. Vammalan kirjapaino.

Marauhn, Thilo

'Towards a Procedural Law of Compliance Control in International Environmental Relations' in No. 3 Zeitshchrift für ausländisches öffentliches Recht und Völkerrecht (1996), pp. 696-731.

McRae, Donald M.

'Arctic Sovereignty: Loss by Dereliction' in Vol. 22, No. 4, Northern Perspectives (1994-1995). Article can be found on the home page of the Canadian Arctic Resources Committee at http://www.carc.org/pubs/v22no4/loss.htm [15.2.2001].

Mustavuori, Jaana-Mirjam

'Vuotoksen tekoaltaan pitkä varjo: Onko laki niin kuin sitä luetaan?' in Oikeus 3/1995, pp. 295-303.

Myntti, Kristian

Minoriteters och urfolks politiska rättigheter. En studie av rätten for små minoriteter och urfolk till politiskt deltagande och självbestämmande. Utgiven av Nordiska Institutet for Miljö- och Minoritesträtt. Juridica Lapponica No. 18. Rovaniemi 1998.

Måreng, Ine

Når Miljovirkninger krysser landegrenser; Forebyggning av grensoverskridande miljovirkninger ved mellomstatlig samarbeid om konsekvensutredninger. Hovedfagsoppgrave i Ressursgeografi og landskapsokologi, Geografisk Institutt, Universitetet i Oslo, 1996.

Mähönen, Outi

Projects with Major Environmental Impacts in (Sub-)Arctic Areas of Finland: Planning and Decision-Making. Case Study: Environmental Impact Assessment of the Sokli Mining Project. September 7-9.1994, Inari, Finland. Author's personal copy obtained from the Finnish Ministry of the Environment.

Nardin, Terry

Law, Morality and the Relations of States. 1983 by Princeton University Press.

Nollkaemper, André.

'EEC Directive 85/337 in an International Context' in the Implementation of the EEC Directive 85/337/EEC on Environmental Impact Assessments 1992' in Beiträge zum Umweltrecht, Heft 3 (edited by Steiger, Heinhard), pp. 149-175.

'The Contribution of the International Law Commission to International Water Law: Does It Reverse the Flight from Substance?' in Vol. XXVII Netherlands Yearbook of International Law (1996), pp. 39-73.

Nordström, Nina

The State Duty to Inform on Transboundary Environmental Accident. Research project 'Legal Responses to Environmental Harm', Project Publication No. 25. Åbo Akademi 1996. Rättsvetenskapsliga Institutionen Ser. A:467.

Nuttall, Mark

'Indigenous Peoples' Organisations and Arctic Environmental Co-operation' in The Arctic: Environment, People, Policy (edited by Nuttall M. & Callaghan T.V.), pp. 621-637. Overseas Publishers Association 2000.

O'Connell, D. P.

The International Law of the Sea, Volumes I (1982) and II (1984). Clarendon Press, Oxford. Published by Oxford University Press, New York.

Ojala, Olli

'Environmental Actions in the Barents Region' in Europe's Northern Dimension: the Bear Meets the South (edited by Heininen, L & Langlais, R.), pp. 153-158. University of Lapland Press 1997.

Okowa, Phoebe N.

'Procedural Obligations in International Environmental Agreements' in 67 British Yearbook of International Law (1996), pp. 275-336.

State Responsibility for Transboundary Air Pollution in International Law. Oxford University Press 2000.

Osherenko, Gail & Young, Oran

The Age of the Arctic: Hot Conflicts and Cold Realities. Cambridge University Press 1989.

Ostreng, Willy

The Northern Sea Route – Economic Prospects and Challenges. Speech given in the Third International Conference of Parliamentarians of the Arctic Region Salekhard, April 22-24, 1998. The text can be found on the home page of the Standing Committee of Parliamentarians of the Arctic Region at http://www.grida.no/parl/salek/index.htm [19.2.2001].

Peters, H-J

'The Significance of Environmental Precaution in the Environmental Impact Assessment Directive' 7 European Environmental Law Review (1996), pp. 210-213.

Pharand, D.

'The Arctic Regions in International Law' in New Directions in International Law pp. 257-277; Essays in Honour of Wolfgang Abendroth. 1982 bei Campus Verlag, Frankfurt/Main.

Canada's Arctic Waters in International Law, 1988. Cambridge University Press, 1988. Published by the Press Syndicate of the University of Cambridge.

The 1991 Draft Treaty proposal. The proposal can be found on the home page of the Canadian Arctic Resources Committee at http://www.carc.org/pubs/v19no2/5.htm [9.2.2001]. Excerpted from the

Canadian Institute of International Affairs, National Capital Branch, The Arctic Environment and Canada's International Relations (Ottawa: Canadian Arctic Resources Committee, 1991), pp.AI-A10.

Phillips, Charles

'Nordic Co-Operation for the Protection of the Environment Against Air Pollution and the Possibility of Transboundary Private Litigation' in Transboundary Air Pollution; International Legal Aspects of the Co-operation of States (edited by Flinterman, Kwiatkowska, Lammers), pp. 153-180. Martinus Nijhoff Publishers 1986.

Pisillo-Mazzeschi, Riccardo

'Forms of International Responsibility for Environmental Harm' in International Responsibility of Environmental Harm (editors Francioni, F. & Scovazzi, T.), pp. 15-35. Graham & Trotman 1991.

Preisler, Bent

A Handbook of English Grammar on Functional Principles. Aarhus University Press, 1992.

Prescott J. R. V.

The Maritime Political Boundaries of the World, 1985. Methuen & Co. Ltd. University Press Cambridge.

Rehof , Lars Adam

'Human Rights and Self-government for Indigenous Peoples' in Vol. 61 Nordic Journal of International Law (1992) pp. 19-42.

Reuter, Paul

Introduction to the Law of Treaties (2 Rev. Ed.). The Graduate Institute of International Studies, Geneva 1995.

Richardson, Elliot L.

'Jan Mayen in Perspective' in 82 The American Journal of International Law (1988), pp. 443-459.

Roginko, Alexei Yu

The NSR in the Context of Arctic Military and Ecological (Environmental) Security. Summary of Working Paper No. 13, 1995. The article can be found on the home page of the INSROP at http://www.fni.no/insrop/INSROPSummary_of_Working_Paper_No_131 9.html [29.1.2001].

Rosas, Allan

'Issues of State Liability for Transboundary Environmental Damage' in Vol. 60 Nordic Journal of International Law (1991), pp. 29-47.

Rosenne, Shabtai

Developments in the Law of Treaties 1945-1986. Cambridge University Press 1989.

Ross, Alf

On Law and Justice. Stevens & Sons Limited (1. edition 1958) California Library Reprint Series Edition 1974. University of California Press.

Rothwell, Donald R.

'The Arctic Environmental Protection Strategy and International Environmental Cooperation in the Far North' in Vol. 6 Yearbook of International Environmental Law (1995), pp. 65-106. Referred to as Rothwell 1995 I.

'International Law and the Protection of the Arctic Environment' in Vol. 44, Part 2, International and Comparative Law Quarterly (1995), pp. 280-312. Referred to as Rothwell 1995 II.

The Polar Regions and the Development of International Law (1996). Printed in Great Britain at the University Press, Cambridge. Published by the Press Syndicate of the University of Cambridge.

Saastamoinen, Salla

'YVA-sääntelyn kansainväliset kehitysvaiheet kohti kansallista sääntelyä' in Global Biodiversity; 3rd Finnish Conference on International Environmental Law, pp. 211-223. Ius Gentium ry. 1996. Hakapaino, Helsinki.

Samson, Paul

Thin Ice; International Environmental Cooperation in the Arctic. First published 1997 by Pacific Press. Printed and bound by Antony Rowe LTD.

Sands, Philippe

'Introduction' in Chernobyl: Law and Communication. Transboundary Nuclear Air Pollution – the Legal Materials, pp. 1-49. Grotius Publications Limited 1988.

'International Law in the Field of Sustainable Development' in 65 British Yearbook of International Law (1994), pp. 303-381.

Sands, P. & Werksman, J.

'Procedural Aspects of International Law in the Field of Sustainable Development: Citizens' Rights' in Sustainable Development and Good Governance (edited by Ginther K. & Denters E. & Waart P.), pp. 178-204. Kluwer Academic Publishers 1995.

Schachter, Oscar

'The Twilight Existence of Nonbinding International Agreements' in 2 American Journal of International Law (1977), pp. 296-305.

International Law in Theory and Practice, 1991. Martinus Nijhoff Publishers (Kluwer Academic Publishers). The Chapter on 'Protecting the Environment' is in pp. 362-388. This Chapter has also been published in 'The Emergence of International Environmental Law' in 44 Journal of International Affairs (1991), pp. 457-470.

Schermers, Henry G. & Blokker Niels M.

International Institutional Law; Unity within Diversity. 1995 Kluwer Law International (Third Revised Edition).

Schrijver, Nico

Sovereignty over Natural Resources; Balancing Rights and Duties. Cambridge University Press 1997.

Scovazzi, Tullio

'La Ligne de Base de la Mer Territoriale dans L'arctique Canadien' in XXXIII Annuaire Francais de Droit International (1987), pp. 663-680.

Scrivener, David

'Arctic Environmental Cooperation in Transition' in Polar Record 35 (192), pp. 51-58 (1999).

Shaw, Jo

Law of the European Union (Third Edition, 2000). Palgrawe Law Masters.

Sheate, William

'Amending the EC Directive (85/337/EEC) on Environmental Impact Assessment' in March, European Environmental Law Review (1995), pp. 77-81.

Environmental Impact Assessment: Law and Policy; Making an Impact II. Cameron May 1996. Printed by Regent Communications.

'The Environmental Impact Assessment Amendment Directive 97/11/EC – a Small Step Forward' in August/September, European Environmental Law Review (1997), pp. 235-243.

Shelton, Dinah

'Introduction: Law, Non-Law and the Problem of 'Soft Law'' in Commitment and Compliance; the Role of Non-binding Norms in the International Legal System, pp. 1-18 (edited by Dinah Shelton). American Society of International Law 2000. Published in the United States by Oxford University Press Inc., New York.

Sifakis, Antonios

'Precaution, Prevention and the Environmental Impact Assessment Directive' in Vol. 7, No. 12 European Environmental Law Review (1998), pp. 349-352.

Sinclair, Ian

The Vienna Convention on the Law of Treaties (second, revised and enlarged edition). Melland Schill Fund 1984. Manchester University Press.

Spiry, Emmanuel

'From 'Self-Determination' to a Right to 'Self-Development' for Indigenous Groups' in 38 German Yearbook of International Law (1995), pp. 129-152.

Springer, Allen L.

The International Law of Pollution : Protecting the Global Environment in a World of Sovereign States. Quorum Books Westport, Connecticut 1983.

Stokke, Olav Schram

'Environmental Threats in the Arctic' in Arctic Environmental Problems (edited by Heininen, L.), pp. 22-41. Tampere Peace Research Institute Occasional Papers No. 41, 1990.

'The Making of Norwegian Antarctic Policy' in Governing the Antarctic: The Effectiveness and Legitimacy of the Antarctic Treaty System (edited by Schram O.S. & Vidas D.), pp. 384-408. Cambridge University Press 1996.

Stokke, Olav Schram & Tunander Ola

The Barents Region: Cooperation in Arctic Europe (edited by Schram & Tunander). International Peace Research Institute and the Fridtjof Nansen Institute 1994. Sage Publications Ltd.

Stokke, Olav Schram & Vidas D.

Governing the Antarctic: The Effectiveness and Legitimacy of the Antarctic Treaty System (edited by Schram O.S. & Vidas D.). Cambridge University Press 1996.

Suopajärvi, Leena

Vuotos- ja Ounasjoki –kamppailujen kentät ja merkitykset Lapissa. Acta Universitatis Lapponiensis 37. Lapin yliopisto Rovaniemi 2001.

Tennberg, Monica

'The Rovaniemi Process and New International Actors' in Finnish Essays on Arctic Issues (edited by Olli-Pekka Jalonen), pp. 133-153. Tampere Peace Research Institute, Occasional Papers No. 54, 1994.

The Arctic Council. A Study in Governmentality. University of Lapland, Rovaniemi 1998. Lapin yliopisto, Rovaniemi 1998, yliopistopaino.

Theutenberg, Bo Johnson

The Evolution of the Law of the Sea; A Study of Resources and Strategy with Special Regard to the Polar Areas (1984). Published by Tycooly International Publishing Ltd. Printed in Ireland by Irish Elsevier Printers Ltd., Shannon.

Thirlway, Hugh

'The Law and Procedure of the International Court of Justice 1960-1989 (Part Three)' in 62 The British Yearbook of International Law (1991), pp. 1-75.

Thornberry, Patrick

'Some Implications of the UN Declaration on Minorities for Indigenous Peoples' in Indigenous and Tribal Peoples' Rights – 1993 and After (edited by Gayim, E. & Myntti, K.), pp. 46-91. Publication of the Northern Institute for Environmental and Minority Law 1995. Juridica Lapponica No. 11.

Timtchenko, Leonid

Quo Vadis Arcticum?; The International Law Regime of the Arctic and Trends in its Development. Kharkiv, State University Press 'Osnova' 1996.

Tirronen, Tanja

Finland has Started to Examine its Possibilities to Ratify the ILO Convention No. 169 on Indigenous Peoples – The question of Land Rights. Summary Memorandum contributing to the seminar organized by the Northern Institute for Environmental and Minority Affairs on May 4-5 2001 at the University of Lapland. The Seminar was titled 'ILO-CONVENTION NO.169 AND ITS RATIFICATION BY NORWAY, FINLAND AND SWEDEN – The Question of Sami Rights over Traditional Lands, Waters and Natural Resources'. The article can be found from the home page of the Northern Institute for Environmental- and Minority Law at http://www.urova.fi/home/pyvi/ILO-full.htm [6.6.2001].

Tupamäki, Matti

Valtion rikosoikeudellisen toimivallan ulottuvuus kansainvälisessä oikeudessa. Publications of the Finnish Branch of the International Law Association, No. 8. Helsinki 1999.

Ulfstein, Geir

The Svalbard Treaty; From Terra Nullius to Norwegian Sovereignty, 1995. Scandinavian University Press, Oslo 1995.

Utton, Albert E.

'The Arctic Waters Pollution Prevention Act, and the Right of Self-Protection' in International Environmental Law (edited by Teclaff & Utton), pp. 140-153. 1974 by Praeger Publishers, Inc.

Verville, Elizabeth G.

'United States – Soviet Union, Report No. 1-6 (Summary)' in Vol I, International Maritime Boundaries (edited by Charney, J.I. & Alexander L.M.), pp. 447-452. The American Society of International Law 1993. Martinus Nijhoff Publishers.

Vidas, Davor

'The Polar Marine Environment in Regional Cooperation' in Protecting the Polar Marine Environment; Law and Policy for Pollution Prevention (ed. by Vidas, Davor), pp. 78-103. The Fridtjof Nansen Institute 2000. Published by the Press Syndicate of the University of Cambridge.

Vierdag, E.W.

'The Time of the 'Conclusion' of a Multilateral Treaty: Article 30 of the Vienna Convention on the Law of Treaties and Related Provisions' in 1988 British Yearbook of International Law, pp. 75-111.

'The International Court of Justice and the Law of Treaties' in Fifty Years of the International Court of Justice (editors Malgosia Fitzmaurice and Vaughan Lowe), pp. 145-166. Cambridge University Press 1996.

Villiger, Mark E.

Customary International Law and Treaties. Dordrecht Martinus Nijhoff 1985.

Virally, Michel

'Sur la Notion D'accord' in Le Droit International en Devenir; Essais Écrits au Fil des Ans (1990), pp. 135-145. Presses Universitaires de France. First published in Festschrift für Rudolf Bindschedler, Berne, Staempfli & Cie S.A., 1980, pp. 159-172.

Vukas, Budislav

'United Nations Convention on the Law of the Sea and the Polar Marine Environment' in Protecting the Polar Marine Environment; Law and Policy for Pollution Prevention (ed. by Vidas, Davor), pp. 34-56. The Fridtjof Nansen Institute 2000. Published by the Press Syndicate of the University of Cambridge.

Wendt, Frantz

Cooperation in the Nordic Countries: Achievements and Obstacles. Published for the Nordic Council by Almqvist & Wiksell International 1981.

Wetterstein, Peter

'Recent Trends in the Development of International Civil Liability' in Vol. 60 Nordic Journal of International Law (1991), pp. 49-71.

Wyatt, Derrick & Dashwood, Alan

European Community Law (Third edition). Published in 1993 by Sweet & Maxwell Limited.

Young, Oran

'Sustainable Development in the Arctic: Operationalizing the Arctic Council' in Europe's Northern Dimension: the Bear Meets the South (edited by Heininen, L & Langlais, R.), pp. 259-277. University of Lapland Press 1997.

The Structure of Arctic Cooperation: Solving Problems/Seizing Opportunities. A paper prepared at the request of Finland in preparation for the Fourth Conference of Parliamentarians of the Arctic Region, Rovaniemi, 27-29 August 2000. Can also be found on the home page of the Standing Committee of Arctic Parliamentarians at http://www.grida.no/parl/ [7.2.2001].

Zemanek, Karl

'State Responsibility and Liability' in Environmental Protection and International Law (editors Lang, W, Neuhold, H and Zemanek, K), pp. 203-221. Graham & Trotman 1991.

SOURCES OF INTERNATIONAL LEGAL MATERIALS

Documents on the Laws of War. Edited by Roberts, Adam & Guelff, Richard. Clarendon Press Oxford (Second Edition (with revisions and editions)1989).

International Environmental Law: Primary Materials. Edited by Molitor, Michael R. Kluwer Law and Taxation Publishers 1991.

Basic Documents of International Environmental Law, Vol. 1: the Important Declarations. Edited by Hohmann, Harald. Graham & Trotman Ltd, 1992.

The Raoul Wallenberg Compilation of Human Rights Instruments. Edited by Melander G. & Alfredsson, G. 1997 Kluwer Law International.

International Environmental Law Reports. Vol. 1, Early Decisions. Edited By Robb, Cairo A.R. Lauterpacht Research Centre for International Law 1999. Published by the Press Syndicate of the University of Cambridge.

REPORTS AND PAPERS OF ORGANIZATIONS, STUDY GROUPS ETC. (the documents related to the United Nations are excluded from the section because of their large number and they can thus be found from the respective footnotes)

First Preliminary Report of the Committee on Legal Aspects of Long-Distance Air Pollution of International Law Association (Rapporteur Dr. Johannes G. Lammers). Report of the 61st Conference (1984), pp. 377-413.

Report of the Committee on the Legal Aspects of the Conservation of the Environment (Rapporteur Prof. Dietrich Rauschning). Report of the 60th Conference (1982), pp. 157-182.

Protecting the Arctic Environment; Report on the Yellowknife Preparatory Meeting. Ottawa 1990.

Environmental Impact Assessment: For Better Planning in Finland (editors Salonen, L. & Paukkunen, M.). Ministry of the Environment 1996. Edita Helsinki 1996.

AMAP Assessment Report: Arctic Pollution Issues. Published by Arctic Monitoring and Assessment Programme (AMAP), Oslo 1998.

Ad hoc arbetsgruppen for miljökonsekvensbeskrivningar (MKB) under Nordiska Ministerrådet 1989-1998 -report can be found on the EIA section of the Nordregio's home page at http://www.nordregio.se [12.3.2001].

Ett nordiskt MKB-nätverk. Rapport om nätverksuppbyggnad. TemaNord 1999: 526. Nordisk Ministerråd, Kobenhavn 1999.

Practical Application of the UN/ECE Convention on Environmental Impact Assessment in a Transboundary Context, final report of a workshop. The Finnish Environment 406, Ministry of the Environment, Land Use Department. Edita Ltd. Helsinki 2000.

Oikeusministeriön säädöshankkeet (1.2.2001). Oikeusministeriö.

A Comparative Study of Nordic EIA Systems - Similarities and Differences in National Implementation (edited by Bjarnadóttir, H.). Stockholm, Nordregio 2001. (Nordregio Report 2001:1). Printed by Elanders Gotab, Stockholm 2001.

INTERNET ADDRESSES OF ORGANIZATIONS RELEVANT IN THE ARCTIC

INTERNATIONAL ARCTIC ORGANIZATIONS:

Inter-governmental Cooperation:

Arctic Council (AC)
http://www.arctic-council.org/ [14.3.2001].

Barents Euro-Arctic Council
http://www.barents.no/ [14.3.2001].

The Finnish Norwegian Borderwater Commission
http://www.vyh.fi/kvasiat/rajavesi/snrajave.htm [29.4.2001].

Nordic Council of Ministers
http://www.norden.org/web/3-1-raad/3-1-5-nmr/uk/index.asp [14.3.2001].

Northern Forum
http://www.northernforum.org/ [14.3.2001].

Northwest Atlantic Fisheries Organization (NAFO)
http://www.ices.dk/ [14.3.2001].

The North Atlantic Marine Mammals Commission (NAMMCO)
http://www.nammco.no/Default.htm [14.3.2001].

International inter-parliamentary Cooperation:

Westnorthern Council (Vestnordisk Råd)
http://www.vestnordisk.is/ [15.3.2001].

Standing Committee of Parliamentarians of the Arctic Region
http://www.grida.no/parl/ [15.3.2001].

Nordic Council
http://www.norden.org/nr/sk/3-1-4-0-om_nr.asp [15.3.2001].

Indigenous peoples organizations:

Arctic Athabascan Council (AAC):
http://www.arcticathabaskancouncil.com/ [23.4.2001].

The Aleut International Association (AIA)
http://finnbarents.urova.fi/aria/in_links.asp [28.4.2001].

Gwich'in Council International
http://finnbarents.urova.fi/aria/in_links.asp [28.4.2001].

Saami Council
http://www.samicouncil.org/ [12.3.2001].

The Inuit Circumpolar Conference (ICC)

http://www.inusiaat.com/icc.htm [24.4.2001].

The Russian Association of Indigenous Peoples of the North (RAIPON)
http://www.raipon.org/ [23.4.2001].

Indigenous Peoples Secretariat (IPS)
http://www.arcticpeoples.org/index.htm [20.3.2001].

Non-Governmental Organizations:

The International Arctic Science Committee (IASC)
http://www.iasc.no/ [15.3.2001].

GENERAL INTERNATIONAL ORGANIZATIONS HAVING ARCTIC INTERESTS:

The European Union's Programme of Northern Dimension
http://europa.eu.int/comm/external_relations/north_dim/index.htm [14.3.2001].

The United Nations Environment Program (UNEP)
http://www.grida.no/arctic.htm [14.3.2001].

The Arctic Programme of World Wildlife Fund (WWF)
http://www.ngo.grida.no/wwfap/ [12.3.2001].

International Atomic Enercy Agency (IAEA)
InternationalArctic Seas Assessment Project: http://www.iaea.org/worldatom/Periodicals/Bulletin/Bull391/specialreport.html [13.3.2001].

The International Council for the Exploration of the Sea (ICES)
http://www.ices.dk/ [13.3.2001].

The International Whaling Commission (IWC)
http://ourworld.compuserve.com/homepages/iwcoffice/ [14.3.2001].

Index

For Product Safety Concerns and Information please contact our EU representative GPSR@taylorandfrancis.com Taylor & Francis Verlag GmbH, Kaufingerstraße 24, 80331 München, Germany

Batch number: 10725991

Printed by Printforce, the Netherlands